Spring Meeting Poster Session and Networking Reception 2020

Held at the 2020 AIChE Spring Meeting and 16th Global Congress on Process Safety

Online
17 - 21 August 2020

ISBN: 978-1-7138-2350-6

Printed from e-media with permission by:

Curran Associates, Inc.
57 Morehouse Lane
Red Hook, NY 12571

CURRAN ASSOCIATES INC. proceedings.com

Some format issues inherent in the e-media version may also appear in this print version.

Printed with permission by Curran Associates, Inc. (2021)

For permission requests, please contact AIChE
at the address below.

AIChE
120 Wall Street, FL 23
New York, NY 10005-4020

Phone: (800) 242-4363
Fax: (203) 775-5177

www.aiche.org

Additional copies of this publication are available from:

Curran Associates, Inc.
57 Morehouse Lane
Red Hook, NY 12571 USA
Phone: 845-758-0400
Fax: 845-758-2633
Email: curran@proceedings.com
Web: www.proceedings.com

TABLE OF CONTENTS

590872 Horsehead Corporation Zinc Explosion

Abstract:

A fatal explosion occurred on July 22nd, 2010 at the Horsehead Corporation Zinc Refinery in Monaca, Pennsylvania. The accident resulted in the deaths of two workers and serious injury to a third. The United States Chemical Safety Board conducted a thorough investigation to determine the cause of the disaster and provide recommendations on how to prevent future accidents. From the investigation, it was determined that a cadmium distillation column, a process unit used to separate cadmium and other relatively low boiling impurities, exploded. It was determined that the sump, the structure through which the bottom product drained, became clogged; this resulted in liquid zinc flooding the distillation unit. The large accumulation of liquid led to the development of pressures and temperatures above the unit's design specifications and ultimately the failure of the system. Though this was the physical initiating event of the actual accident process, many other factors played a role in causing the accident.

The design itself used for the distillation column, specifically the sump, had specifications that were determined to be hazardous in prior investigations across the globe. The company failed to address issues with the design even though multiple nonfatal explosions had occurred at the same facility in the years prior to the disaster. The facility considered such accidents as normal and never addressed the problem. If proper safety protocol had been followed and these near misses investigated, the 2010 explosion may have never occurred. The Horsehead facility also did not follow proper management of change (MOC) protocol in the construction of the cadmium distillation column, and it did not educate employees on how to identify and evaluate hazards while operating the process. All these factors led to the eventual explosion on July 22nd, 2010.

585879 Development of a Comprehensive Procedure Writers' Guide Framework

Abstract:

Procedures are considered to be an integral part of safe operations within the process industry, as they serve as safeguards to inform workers of proper operating methods and prevent workers from engaging in unsafe and/or undesired actions. A number of industrial incidents stemming from poorly written procedures have highlighted pitfalls within one of the most critical elements of procedure development, the writers' guides. Previous studies have shown that there are many inconsistencies between writers' guides of different companies, with only a few elements, such as general procedure goals and proper writing styles, appearing in a majority of them. Many other critical elements, particularly those related to human factors and safety/environmental responsibilities, appear in few writers' guides or are poorly addressed. This work seeks to provide a comprehensive framework for the development of writers' guides across process industries, so as to have consistent content within procedures that addresses the issues that can lead to unsafe actions and incidents. As part of this framework, a procedure life-cycle has been developed and proposed based on a literature review of system development life-cycles from a number of industries, including software development and architecture. This life-cycle illustrates the stages of the procedure development process that are influenced by the writer's guide as well as the elements that are used to develop the writer's guide. Using the procedure development life-cycle as a guide, this framework has been developed based on regulatory elements and industry best practices as well as human factors guidelines.

586526 Justifying Improved Management of Loading Hoses

Abstract:

Justifying Improved Management of Loading Hoses

Abstract

There are several sources of loading hose failure rate data. However in general these data sources provide failure rates without regard to the situation in which the hose is used. One of the most vexing aspects of this subject is that there are several variables associated with loading hose specification, usage, maintenance, replacement and more that are known or suspected to have a dramatic effect on the failure rate, but there is no method for quantifying these variables. For this reason it would be useful to develop failure rates that are situation-specific, which in turn plants could use to justify improvements in their hose management policies. In the absence of actual data, this paper intends to solicit expert opinion from industry prior to or during the conference, and publish the results of the survey and suggested failure rate modifiers at the 2020 conference or in an issue of Process Safety Progress.

The conference presentation will present the current published data, as well as the proposed modification factors. To the extent that it is possible to do so without prejudicing 'real time' responses to the survey, the presentation will discuss general features of any survey results received in advance of the conference.

NOTE: This is being submitted for poster session only. BakerRisk has already submitted its allotted number of abstracts for the LPS presentation slots, and this abstract is not intended to be taken in preference to either of those two slots. In its current form, it is probably more suited as a poster.

587021 Machine Learning Based Quantitative Structure-Property Relationship Prediction of Lower Flammability Limit

Abstract:

This study used lower flammability limit (LFL) data and ten calculated molecular descriptors data of 78 pure chemical compounds to construct Quantitative structure-property relationship (QSPR) models. Four machine learning methods, k-nearest neighbors (*k*-NN), support vector machine (SVM), random forest (RF) and boosting, are applied to QSPR models to improve prediction accuracy. Prediction errors and accuracy are compared with traditional multiple linear regression (MLR) models. A novel cross validation method, 10-fold cross validation method, is also used to increase the data usage and prediction reliability. Result shows that models generated by machine learning methods have a significantly lower root mean square error (RMSE) than traditional MLR method in the test dataset. Machine learning based models can be used as substitution methods to improve UFLs predictability of chemical compounds

Using API RP571 in PHAs

Michael H. Oliver
EN Engineering
8308 Lake Bonita Road, Catlettsburg, KY 41129
moliver@enengineering.com

Prepared for Presentation at
American Institute of Chemical Engineers
2020 Spring Meeting and 16th Global Congress on Process Safety
Houston, TX
March 30 – April 1, 2020

Using API RP571 in PHAs

Michael H. Oliver
EN Engineering
8308 Lake Bonita Road, Catlettsburg, KY 41129
moliver@enengineering.com

Keywords: PHA, API RP571, Event-based failure

Abstract

API-RP571, Degradation Mechanisms Affecting Fixed Equipment in the Refining Industry, addresses approximately 60 degradation mechanisms that can affect equipment in petroleum refineries. As the title suggests, the mechanisms are well known in refineries and several of them can also occur in other industries. The overwhelming majority of these mechanisms are "time-based" but there are at least 3 "event-based" mechanisms that should be taken into consideration when conducting a PHA. Those three are low temperature brittle fracture, chloride stress corrosion cracking (Cl-SCC) and polythionic acid stress corrosion (PASCC).

Since all three of these mechanisms are likely to occur during startup, shutdown, upset conditions or maintenance activities they need to be taken into consideration when conducting a PHA with special attention to the actual materials of construction.

With the enormous increase in hydroprocessing capacity in the refining industry in recent decades, there has been a corresponding increase in the quantity of piping and equipment fabricated from austenitic stainless steels which are susceptible to both Cl-SCC and PASCC.

1 Introduction

API Recommended Practice 571 addresses Damage Mechanisms Affecting Fixed Equipment in the Refining Industry and covers 60+ damage mechanisms, the clear majority of which are time based but there are a handful of event-based mechanisms and as such may arise during a PHA depending on the nature of the process being reviewed RP571 is a very well organized document. Damage mechanisms common to a variety of industries are addressed in section 4.0 while those specific to refining and petrochemical industries are in section 5.0. Additionally, section 5.2 includes PFDs showing where the various damage mechanisms may be found in the process.

It is not the intention of this paper to delve into the technical details of the three event-based mechanism but rather the need for heightened awareness of these relatively common failures so they can be addressed during a PHA.

2 Low Temperature Brittle Fracture

A brittle fracture is the fracture of a metallic object without prior plastic deformation.

Brittle fracture typically occurs in carbon steels (CS) below the ductile-brittle transition temperature. Probably the most common CS used in refinery and petrochemical plant construction is A106 for piping which along with the flange equivalent (A105) and the plate equivalent for vessels (A285) have virtually the same composition.

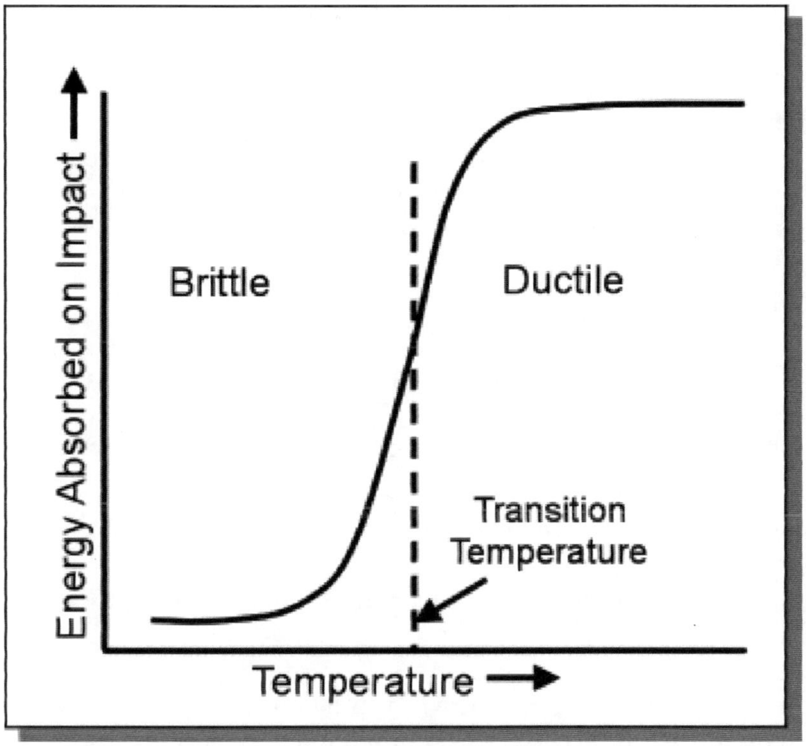

Figure 1. Temperature vs. Energy Absorbed on Impact

The actual ductile to brittle transition temperature for A106 starts at about 50 deg F (10 deg C). Compare that to the following atmospheric boiling points for some light hydrocarbons:

Table 1. Atmospheric Boiling Point of Some Light Hydrocarbons

Hydrocarbon	BP deg F	BP deg C
Propane	-44	-42.1
n-Butane	31	-0.5
i-Butane	11	-11.7

It is not unreasonable to conclude that in abnormal operating conditions, especially in the case of auto-refrigeration (when light hydrocarbons might boil off at near atmospheric pressure) that piping components constructed of A106 material will easily reach temperatures below ductile-brittle transition thus being susceptible to brittle fracture.

A notable low temperature brittle fracture incident in recent history is the Esso Longford (Australia) gas plant incident in September, 1998 which resulted in 2 fatalities plus 8 injured. In that incident, a heat exchanger was cooled to sub transition temperatures and then quickly exposed to high temperature resulting in a major loss of containment.

A review of the piping specification for 3 major oil companies reveals the following:

- Company A allows A106 and A105 flanges down to -20 deg F (-29 deg C) which is consistent with ASME non-shock ratings
- Company B allows A106/A105 only down to 32 deg F (0 deg C), more conservative than ASME
- Company C allows A106/A105 only down to 32 deg F (0deg C), also more conservative than ASME
- Companies B and C also require additional impact test specs for components with wall thickness greater than 0.5 inch

Companies B and C have factored their experience into the ASME piping code and their internal piping spec reflects that experience with more conservative low temperature constraints on A106/A105 components. Those 2 companies (B and C) specify "fine grained steels" for operation below 32 deg F (0 deg C).

What are "fine grained steels"? Those are steels with a finer grain structure and are much less susceptible to low temperature brittle fracture down to -58 deg F (-50 deg C). Typically for piping that would be A333, A350 for flanges and A516 plate for pressure vessels.

Learning point: if during a PHA, it is determined that equipment components could be cooled below the ductile-brittle transition temperature due to startup, shutdown or upset conditions, a review should be conducted of materials of construction to determine if the materials are susceptible to low temperature brittle fracture and if so, recommendations should be made to avoid failure such as heat up before pressure up or material replacement. Age of the equipment may also be a factor. Equipment manufactured to ASME B&PV Code, Section VIII, Div 1 prior to December, 1987 may be more susceptible.

Historical note: during World War 2, the US mass produced 2700 “Liberty Ships” to transport goods to the UK and Europe. The construction time for these ships was typically 42 days. About a third suffered (sometimes catastrophic) hull fractures and more than a few actually cracked in half. This was because the steel used had a relatively high ductile to brittle transition temperature (worse than A106) and in the cold water of the North Atlantic, the hulls became brittle and cracked usually starting at a stress point like a port hole.
The Titanic suffered a similar fate.

3 Chloride Stress Corrosion Cracking

Chloride stress corrosion cracking (Cl-SCC) like Polythionic Acid Stress Corrosion Cracking (PASCC) are included in the broader category of Environmental Cracking. Both these degradation mechanisms are event-based and usually result in sudden failure often with catastrophic results.

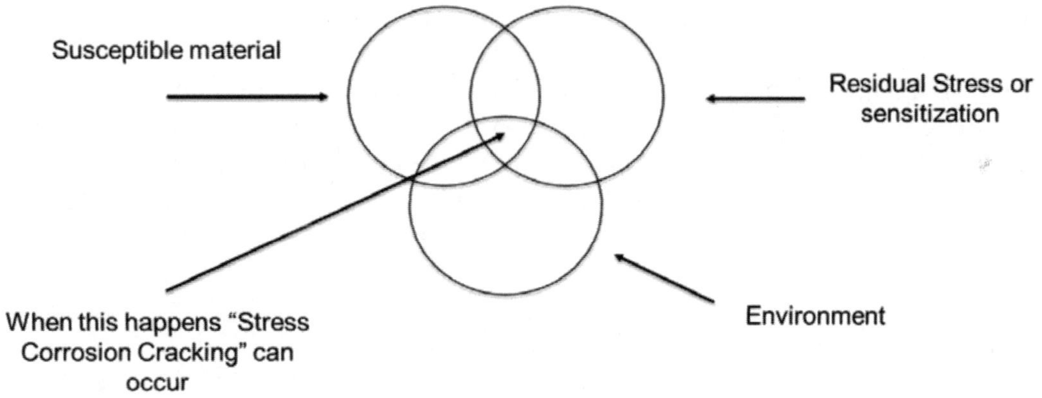

Figure 2. Stress Corrosion Cracking Model

In the case of Chloride Stress Corrosion Cracking and Polythionic Acid Stress Corrosion Cracking, the susceptible materials are the same: austenitic stainless steels including the common 304 family (304, 321 and 347) and 316.

Critical factors for Cl-SCC are chloride content, pH, temperature, stress, O2 content and alloy composition. Drains are particularly susceptible during start up and shutdown if not purged.

The “stress or sensitization” component of the mechanism can be induced stress for example in piping around supports or in heat exchangers where tubing passes through baffles or tube sheets or more commonly, it is sensitization of the metal due to exposure to high temperatures especially in the heat affected zone (HAZ) of welds.

Technically, sensitization is due to chromium carbide precipitates which form resulting in depletion of chromium around grain boundaries. In conditions which may allow formation of water in the presence of chlorides, cracking tends to occur along the grain boundaries and it can happen rapidly.

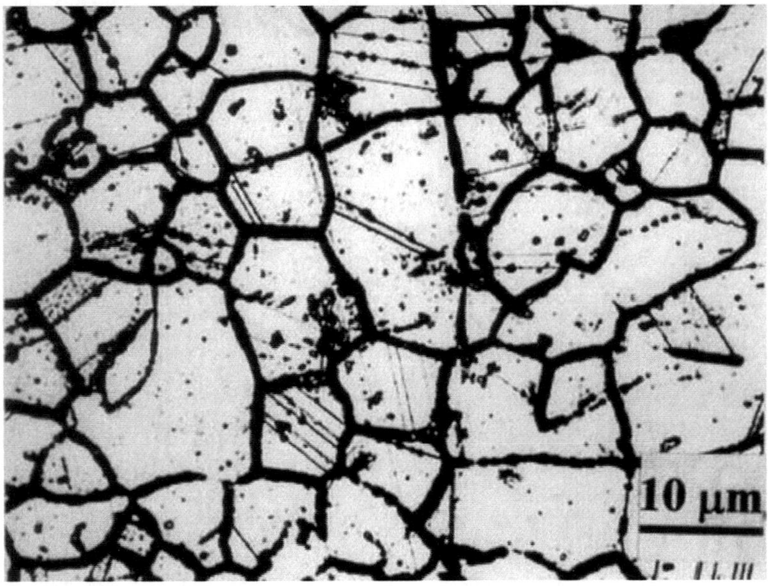

Figure 3. Chromium Carbide Precipitates Around Grain Boundaries

Referring to the SCC model in Figure 2, two of the three components, susceptible material and stress/sensitization are almost certainly not variables. That leaves the "environment" component of the mechanism. That's where particular attention must be devoted during the PHA. Recommendations during the PHA must exclude the possibility of that particular environment from forming internally and externally, especially during startup, shutdown and upsets.

4 Polythionic Acid Stress Corrosion Cracking

There are similarities between Cl-SCC and PASCC in that the susceptible material and stress/sensitization are common to the two, but the environment is very different. In the presence of sulfur containing compounds and/or H2S, a sulfide film forms inside piping and equipment. These are conditions typically found in virtually all hydroprocessing units that remove sulfur. If that film is exposed to air, the combination of O2 and moisture react with that sulfide film to form polythionic acid which can cause stress corrosion cracking in austenitic stainless steels. As with Cl-SCC, these failures can occur hours after exposure to an ambient air environment. This environment is unlikely to occur during any phase of operation, but it is certainly a concern when maintenance activities are underway that may expose sensitized piping and equipment to ambient air. Fortunately, the solution to this problem is potentially simpler than for the other two damage mechanisms discussed in this paper. NACE has produced a recommended practice RP0170 that describes a soda ash wash procedure that should be used before exposing austenitic SS piping and equipment to air during maintenance activities. This procedure has been successfully applied for many years and should be a recommendation from any PHA on a hydroprocess unit containing susceptible materials.

5 Summary and Conclusion

API RP571 is a valuable reference when conducting PHAs especially when concerned about event-based failures. Three common event-based failures include low temperature brittle fracture, chloride stress corrosion cracking and polythionic acid stress corrosion cracking. Recent experiences with EN Engineering clients highlight the diversity regarding company piping specifications and while the 3 examples cited may not be representative of the industry they highlight the need to always consider materials of construction when reviewing abnormal operations such as startup, shutdown and upsets.

Process Safety Capability Building as a Strategic Priority within Organizations

Kristle M. Thompson, PMP, Trinity Consultants, Inc., Houston, TX, USA

Abstract

Capability building provides organizations with the opportunity to develop skill-building approaches, that, if executed strategically, provide a powerful framework for organizational sustainability. This paper is designed for individuals and organizations interested in learning how to build and maintain expertise in process safety. Trinity Consultants will share best practices on how Process Safety Management (PSM) and Risk Management Program (RMP) capability is achieved via targeted recruitment, intensive onboarding and training, well-defined career development tracks, program element standardization, industry best practices, knowledge management, and employee recognition.

Introduction

Nearly 30 years ago the Occupational Safety and Health Administration (OSHA) published the Process Safety Management (PSM) Standard (1910.119). The goal of the PSM standard is to provide guidance on the management of hazards associated with processes using highly hazardous chemicals that have the potential to cause catastrophic incidents in the workplace and surrounding environment.

As we fast-forward to the present, you will find that many organizations still struggle with effectively identifying, evaluating, and controlling process hazards and risks to ensure safe and reliable operations. To address these process hazards, organizations need to develop the necessary expertise, judgement, and initiative to properly implement and maintain an effective PSM program.[1] One of the primary contributing factors that relates to this industry challenge is the level at which process safety is viewed as a strategic priority within organizations by senior leadership. Ensuring senior leadership actively supports process safety through its investment strategy is imperative to the success of a PSM/RMP program. [2]

Trinity's leadership actively supports PSM within the organization and has gained, through direct facility experience nationwide, the depth and breadth of PSM/RMP knowledge. Trinity will share best practices acquired as a result of working with client PSM/RMP programs that are staffed by process safety teams. The teams vary in size, technical expertise, and organizational structures across a broad cross-section of industries.

Specifically, Trinity will discuss how to develop a solid process safety team through:

- Targeted recruitment of process safety professionals;
- Intensive onboarding and training programs;
- Well-defined tailored career development tracks;
- Development of internal standards;

- Development and use of adopted templates from Recognized and Generally Accepted Good Engineering Practices (RAGAGEP);
- Extensive coaching through high-frequency project peer reviews; and
- Employee recognition programs.

1 Targeted Recruitment

Targeted recruitment of safety professionals with global and domestic industry-specific PSM expertise is the first step to building PSM/RMP capability in an organization. Over the past five years, there has been a shift in how organizations prioritize PSM and recruit for PSM candidates, particularly with catastrophic incidents capturing more attention of the public and media outlets. Within organizational structures, what may have once been lumped in with the occupational safety group is now distinguished as its own standalone PSM group within the Environmental, Health, and Safety (EHS) organization. Additionally, more stringent PSM competency requirements have been established by organizations, given the weight of responsibility in a PSM role now requiring science or engineering degrees, and some level of experience in process operations, process hazard analysis (PHA), and PSM/RMP program implementation. Lastly, you will find that there is a competitive market in Process Safety, where salaries are typically higher for Process Safety professionals in comparison to general EHS professionals.

Now that we understand how the landscape of PSM recruiting has shifted in organizations, let's focus on some of the burning questions around targeted recruitment.

How does my organization's HR department target solid process safety candidates for new-hire consideration?

Trinity's reputation for excellence has been shaped by their consultants' intimate knowledge and expertise of industry and EHS regulatory requirements [3], which makes targeted recruitment critical for its organization. Candidates are typically found through the LinkedIn recruiter job boards. Additionally, referrals play an important role in how the organization secures top talent, as referrals tend to be more reliable and improve the quality of candidates. Regardless of how the candidates are initially targeted, each must move through a rigorous recruitment process to include a series of probing technical phone interviews, face-to-face screenings with senior leadership, and full day on-site interviews with technical PSM/RMP staff and leadership.

The ideal candidate must be able to demonstrate competence, preferably across diverse industries domestically and globally, as indicated in the PSM Hiring Competency Checklist below (Figure 1). The candidate must also convey an energy and drive to succeed in the role; the ability to meticulously execute PSM/RMP projects; and have the capability to quickly learn and master any identified PSM/RMP gaps that may have been noted during the recruitment process.

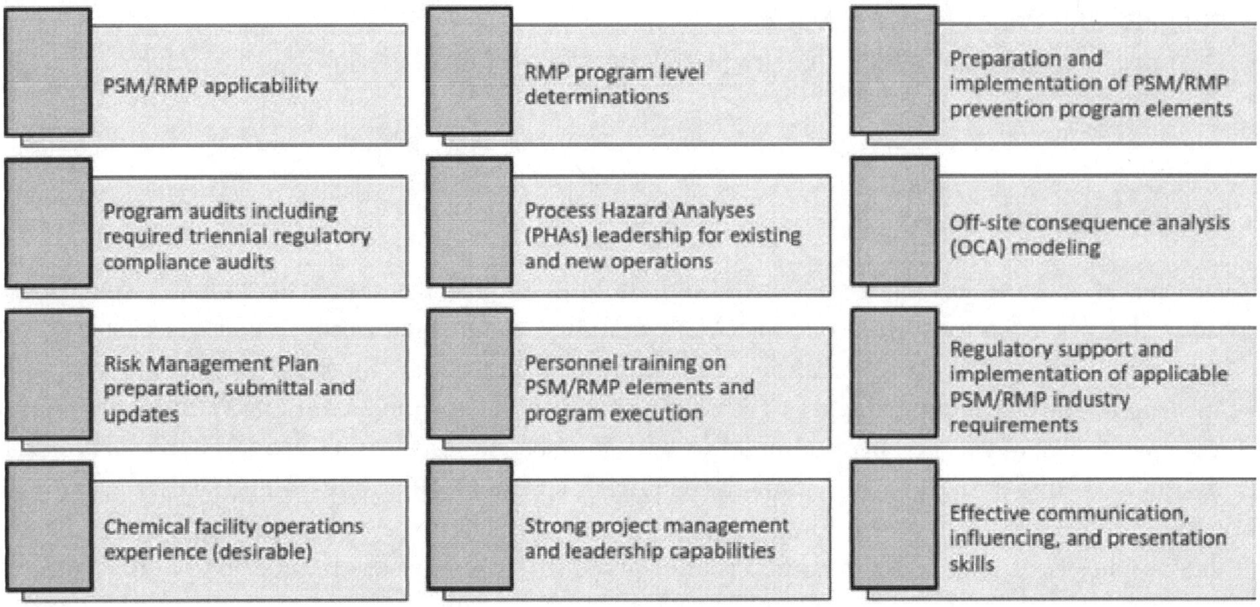

Figure 1. PSM Hiring Competency Checklist

One of the recruiting challenges for this role is identifying candidates who not only know how to lead and delegate work, but also know how to technically execute PSM/RMP work. An effective way of evaluating a candidate's depth and breadth of technical and leadership capabilities, in addition to the conventional interview questioning model, is through demonstration. This can be in the form of a case study or career presentation.

Trinity is an organization of Doers, Leaders, and Sellers. Each consultant regardless of seniority must demonstrate flexibility and be able to perform technical work, effectively lead projects, and create business development opportunities. The world of consulting is fast-paced and often requires varied skills and deliverables.

The concept of Doers, Leaders, and Sellers in the consulting industry is analogous to PSM/RMP-related industry roles as described in Figure 2 below.

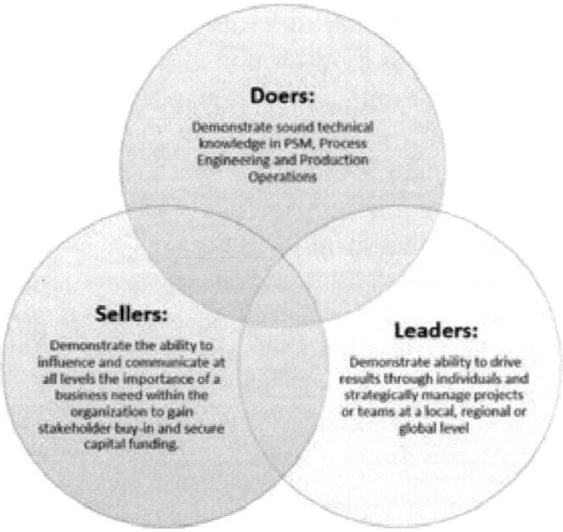

Figure 2. Description of Industry Doers, Leaders, and Sellers

The consulting recruitment model may not directly align with your organization's recruitment strategy but serves as a good starting point to landing solid candidates. A key step for internal or external recruitment is to identify which of the above roles is **required** for the current position, and which skills can be **developed** for career growth from that role. This ensures both a good initial fit for the company and ongoing career satisfaction for the candidate.

What if my organization relies on external EHS/Engineering placement firms for recruitment of PSM Candidates?

Trispyrian is a technical search firm that focuses on placement of engineering, operations, and EHS professionals, and has developed the technical knowledge and network to efficiently find the right candidates for their client organization's need. [4]

One obstacle that often appears in the PSM/RMP recruitment process is the recruiter's (internal or external) failure to understand the differences between roles supporting *Process Safety* versus *Occupational Safety*. Technical search firms ensure that they include the following aspects in their recruiting process.

- Technical screening – The unique ability to screen a candidate's technical knowledge to narrow your interview group to a handful of highly qualified professionals that meet the exact criteria of the role. Ensure the recruiters have been trained to understand the requirements associated with OSHA PSM and EPA RMP regulations, their project work types, technical requirements, and your internal standards to increase candidate fit.

- Simplified hiring process – Understanding facility operations, both simple and complex, provides insight into the skills and qualities desired by hiring managers, thereby simplifying the hiring process.

- Candidate network – Obtains a broad and deep network of potential candidates with proven skill sets, and ongoing partnerships with industry clients for more than a decade.

- Search option flexibility – Offers multiple options to meet your organization's specific staffing needs, ranging from direct hire to long-term contract assignments. For direct hire placements, the placement firm screens and presents a list of the most qualified candidates, and coordinates phone and face-to-face interviews to explore the fit between the candidate and the position. When direct hire is not feasible, the placement firm associates can be placed in long-term contract assignments, and at any point, should an organization decide to extend an offer for permanent employment, the placement firm will facilitate the hiring process.[4]

Selecting an experienced PSM/RMP savvy search firm that values client relationships, experienced in applicable regulations, and understands your operations will allow your organization to find and recruit top, qualified candidates.

What if my organization is small and doesn't have the PSM resources or finds it difficult to execute PSM?

Implementation of the PSM/RMP elements can be quite complex and resource-intensive, particularly for smaller organizations. By evaluating the critical technical skills (similar process experience, PSM/RMP program knowledge) and roles (Doer, Seller, Leader) based on your specific organization and its applicable processes, you can determine the technical areas that may benefit from third-party expertise.

Let's use the Process Hazard Analysis (PHA) element as an example. OSHA states that "an effective and comprehensive process hazard analysis shall be performed by a team with expertise in engineering and process operations, and the team shall include at least one employee who has experience and knowledge specific to the process being evaluated. Also, one member of the team must be knowledgeable in the specific process hazard analysis methodology being used". Given this requirement, the agency acknowledges that some organizations, particularly smaller ones, may not have the staff expertise to perform a PHA, and therefore, the organization may need to hire an engineering or consulting company to perform the analysis. [5]

If your organization does not have the staff to address all PSM/RMP compliance areas, establish a good working relationship with a consulting firm who specializes in PSM/RMP compliance, remains abreast of key regulatory developments, and is highly skilled in PSM/RMP program implementation for companies varying in size, skillset, and organizational structure. Site resources can focus on maintaining compliance with the PSM/RMP program once established for ongoing site changes, and can utilize third-party resources for larger task such as PHA review, compliance audits, and program updates based on regulatory changes.

2 Intensive Onboarding and Training Programs

An intensive onboarding and training program inclusive of on-the-job exposure, as well as internal and external competency training requirements is the second step in developing PSM/RMP capability. Once the qualified candidate has been selected and hired via the targeted recruitment process, it is imperative that the employee is immersed in a well-structured and rigorous onboarding and training program, and that training continues to assist the employee in growing through the life of their career. A key in determining your critical training needs is to identify the tasks that internal staff will perform and start there with your training efforts. Tasks that will be performed by external resources may only require a high level of knowledge by internal staff to review completed work products.

Employee Orientation

All new hires, not just the PSM/RMP team, should participate in an initial orientation. The orientation process should include a general overview of the facility, review PSM job-specific roles and responsibilities, and provide an overview of the organization's mission and safety culture. For example, employees may be informed of emergency exits, muster points, how to access process safety information and PSM-related procedures, and the proper safety-reporting protocols, such as reporting an injury or unsafe condition. [6]

On-the-Job PSM Competency Training

Competency training will be used to address gaps between critical skills, the candidate's experience, and the organization's specific procedures. The process by which internal and external training is provided for a PSM employee may vary by organization, but let's take the training of a PSM Engineer as an example.

Critical on-the-job training for a PSM Engineer would involve both skill-based training and awareness-based training. Skill-based training demonstrates the actual hands-on procedures necessary to perform a specific task [6], such as conducting an incident investigation or facilitating a PHA utilizing an industry recognized methodology. A great way for PSM Engineers to get up to speed with their role is to shadow senior staff. For example, a new Process Safety Engineer may support a lead auditor in the documentation review for the facility compliance audit, or she/he may take advantage of any PHA scribing opportunities to gain competency exposure to the PHA process. Facilitating a Management of Change (MOC) team will familiarize staff with how the PSM program elements are interrelated, and how small, simple process changes can impact many program elements.

Awareness-based training includes general policies, hazard recognition and expectation for maintaining a safe work environment [6]. For example, a Process Safety Engineer should have a high-level familiarity of the intent of the OSHA PSM and EPA RMP regulations, particularly at the start of a role.

The ongoing combination of skill and awareness-based training will ultimately build a PSM Engineer who is well-versed in PSM/RMP regulations, and capable of delivering a solid PSM/RMP program.

Internal PSM Competency Training

Some organizations may have specific internal competencies that must be mastered before carrying out certain PSM element tasks. In keeping with the same PSM Engineer example used above, a PSM Engineer may have to complete a rigorous PHA certification process to facilitate a PHA for their assigned process unit/operation. A PHA certification process may include a rigorous multi-day internal

PHA course and exam covering company-specific hazards, hazard methodologies, risk matrices, and risk acceptance criteria. Additionally, the PSM Engineer may be required to participate in at least two (2) PHAs as a scribe/observer and/or have shared facilitation responsibility with guidance from a more experienced PHA facilitator for a set number of complex operations before securing the PHA certification. As a precursor to or in lieu of an internal PHA course, some facilities utilize a formal multi-day PHA course from a reputable industry trade association, which leads into the next topic of External PSM Competency Training.

External PSM Competency Training

The Center for Chemical Process Safety's (CCPS) HAZOP Studies and Other PHA Techniques for Process Safety and Risk Management course at the American Institute of Chemical Engineers (AICHE) Academy is an industry-recognized course for PHA facilitators. After this rigorous three-day course, PSM engineers and technical professionals will: understand the obligations under the OSHA PSM and EPA RMP regulations; examine various hazard evaluation methods, including HAZOPs, how to select the hazard methodology appropriate for your facility; and lastly, how to determine the adequacy of safeguards and report findings and recommendations. [7]

Another great external course for PSM Engineers and technical professionals is CCPS' Process Safety Boot Camp. This is AICHE Academy's most popular course, and was developed for organizations who require quick ramp-up of staff knowledge in the fundamental concepts of Process Safety. The four-day course is intensive and covers key process areas, including risk-based process safety, process hazard analysis, regulatory compliance and process safety in plant and design operations. [7]

If there are state-specific PSM/RMP regulations for your locations, such as the New Jersey Toxic Catastrophe Prevention Act or the California Accidental Release Prevention and Cal/OSHA PSM programs, invest in external courses based on those regulations to ensure all federal and state requirements are also met.

3 Well-defined Tailored Career Development Tracks

The third step to building PSM/RMP capability is establishing well-defined, tailored career development tracks that reinforce subject matter expertise and support advancement within the organization. To this point, we have discussed what it takes to recruit, onboard, and train employees to be successful in their roles. Eventually, after achieving and maintaining success in their role, these employees will want to understand where the next opportunity lies within the organization, particularly those who are ambitious and expect to see career development opportunities to remain engaged and motivated in their role. It is imperative for organizations to play their part in supporting employees' career path by ensuring transparency and access to resources necessary for career development and progression, specifically:

- Job descriptions
- Well-defined competencies required for advancement
- Clear internal job application process

- On-the-job development opportunities
- Job shadowing and training
- Mentoring and supervisor coaching
- Transfer or lateral moves
- Formal succession planning [8]

A solid career program model should provide a roadmap for employees to strategically ascend in their career and capabilities. The program should promote growth vertically and laterally and ensures transparency in clearly defined career channels, grade levels, and job positions with associated responsibilities and competency requirements. To go a step further, organizations should have a promotion review board process dedicated to the advancement of its employees. Each employee should be provided with a promotion metrics summary matrix tailored to their selected career channel of interest, and this document should serve as a go-to reference for PSM professionals to identify and target any outstanding competency gaps, and potentially enable more expeditious advancement within the organization.

Career development program models and tracks will vary by organization based on company type and organizational structure. However, when employees have access to career development support resources, organizations will reap the benefits of employee engagement, success in their role and retention. Please reference Figure 3 below for a high-level example of two possible PSM career development tracks with potential cross-over, and cross training opportunities if PSM interests shift over the course of their career.

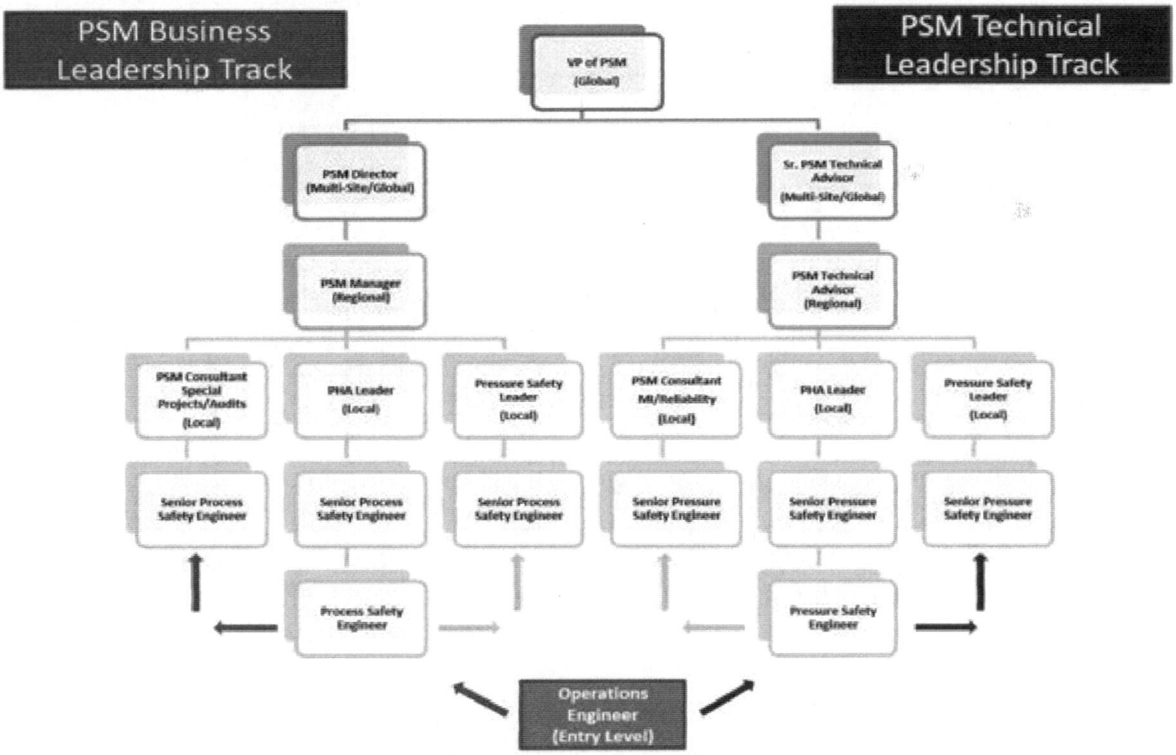

Figure 3. PSM Business Leadership Track vs. PSM Technical Leadership Track

4 Development of Internal Standards

Development of internal standards to produce a consistent quality work product across the PSM team is the fourth step in building capability. PSM/RMP personnel will find great value in the development of internal standardization within their team, primarily because it will ensure delivery of a quality work product, minimize project ambiguity and promote productivity within the organization. PSM elements recommended for internal standardization include the PHA, MOC, Pre-Startup Safety Review (PSSR), and Compliance Audit processes. Organizations that invest time to develop an internal standard on these processes will reap many benefits to include internal consistency in processes and risk evaluation.

For example, a robust PHA Standard Operating Procedure (SOP) should define requirements for:

- Personnel that facilitate, document, and approve PHAs
- Methodologies used to systematically identify, evaluate, and control hazards for covered processes subject to OSHA's PSM standard (codified in 29 CFR 1910.119) and/or EPA RMP standard (40 CFR Part 68)
- Methodologies for PHAs conducted under other scopes
- Quality Assurance / Quality Control (QA/QC) procedures for PHA reports

The SOP should also include a PHA Workflow to align with the steps of the PHA procedure as a best practice, a company-specific risk matrix, and risk acceptance criteria. Please reference Figure 4 for an example of a PHA Workflow.

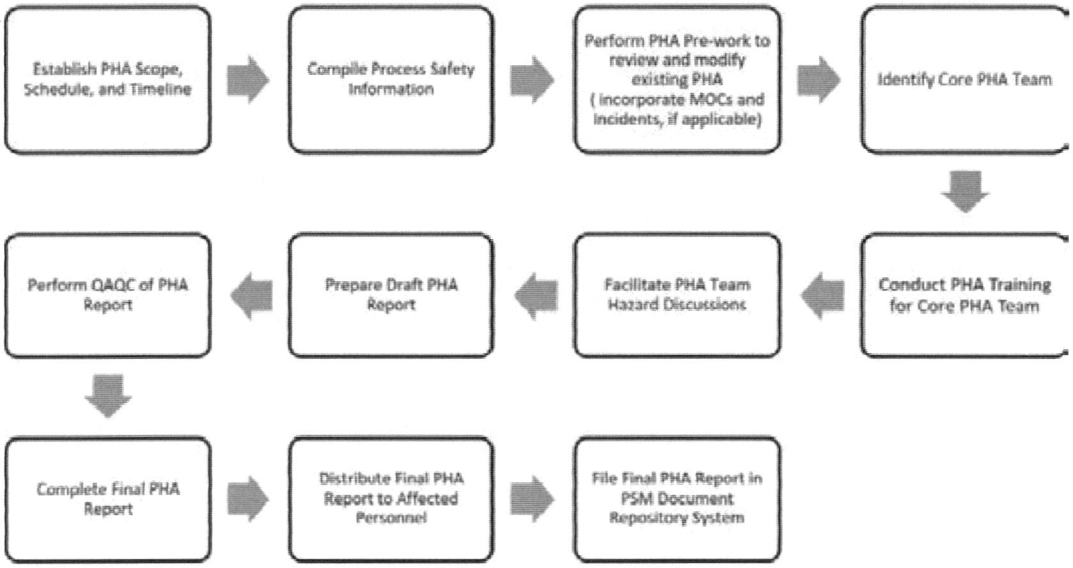

Figure 4. PHA Workflow

PSM Organizations that invest in internal standardization, where applicable, will find that it makes PSM information more accessible to employees, promotes transparency, and creates more confidence within the team [9] that a quality PSM/RMP work product will be achieved.

5 Development and use of Adopted Templates from Recognized and Generally Accepted Good Engineering Practices (RAGAGEP)

The fifth step to develop Process Safety capability is the use of adopted templates from Recognized and Generally Accepted Good Engineering Practices (RAGAGEP) to ensure all PSM projects are executed based on sound, safe, and reliable industry practices. According to OSHA, "the PSM standard allows employers to select the RAGAGEP that apply in their covered processes. Examples of RAGAGEP include widely adopted codes, consensus documents, non-consensus documents, and internal standards."[10]

Let's take an Ammonia Refrigeration facility for the first RAGAGEP example, where industry standard practices and templates are readily available for your organization's use. The International Institute for Ammonia Refrigeration (IIAR), the world's leading advocate for ammonia refrigerants, provides the most up-to date technical industry guidance to the ammonia refrigeration community [11], and serves as an acceptable source of RAGAGEP for non-refrigeration facilities that store or use anhydrous ammonia for process operations. IIAR provides technical expertise and comprehensive guidance in the area of PHAs. As a member of the IIAR community, your organization would have full access to the IIAR What-If/Checklist for Ammonia Refrigeration operations. This PHA template can be customized to accurately capture the site-specific hazards, and will serve to be an extremely valuable compliance tool as your organization implements OSHA's PSM PHA requirements.

For a second RAGAGEP example, look at CCPS's Guidelines for Auditing Process Safety Management Systems as a key industry guidance document to deliver a consistent, comprehensive, and quality PSM/RMP compliance audit for organizations that range in size, industry-type, or chemical process. Whether your organization performs internal audits or utilizes an external consultant to assess PSM/RMP compliance at the site, it is imperative that the compliance audit addresses the facility's level of compliance with the regulatory requirements of 29 CFR 1910.119 (PSM) and 40 CFR 68 (RMP), including evaluating compliance with incorporated standards such as 1910.252(a), 1910.38, and 1910.120 relating to fire prevention and emergency planning.

Additionally, the audit should include clear, unambiguous findings of compliance or non-compliance based on audited evidence, and document references to applicable regulations, RAGAGEP, or best practices for each finding and recommendation. For instance, if an audit was performed on the Mechanical Integrity (MI) program element, the auditor may want to confirm if the facility's pressure vessels are included in the MI program as cited in PSM 1910.119(j)(1)(i) and RMP 68.73(a)(1) regulations. Through discussion and evidence review, the auditor observes that the facility has not carried out pressure vessel inspections and is unaware of the appropriate RAGAGEP to reference for pressure vessel inspections. An appropriate recommendation based on this finding may be for the facility to develop an inspection schedule for all pressure vessels based on consensus RAGAGEP, or manufacturer recommendations, and ensure the pressure vessel inspection is completed by a certified inspector whose qualifications are in accordance with API-510 standard or another regulatory/consensus standard (i.e. certified National Board pressure vessel inspectors). Figure 5 below highlights the key components of a quality compliance audit.

Figure 5. Key Compliance Audit Components

A final consideration for delivering quality audits is to ensure and/or confirm, if externally performed, that the audit checklist template has been reviewed and updated on a defined frequency to ensure the checklist content remains in alignment with the ever-changing world of PSM/RMP compliance. A relevant example would be remaining informed of the many iterations of the EPA's RMP Rule and ensuring that the regulatory language and audit questions in an RMP Prevention Program Level 3 audit checklist template have been updated to reflect the requirements of the recently published December 2019 Final RMP Reconsideration Rule. Oversight of ongoing rule changes should be used to ensure your organizations PSM/RMP program procedures are current and require the facility to incorporate any new or updated regulations, such as the recent RMP emergency response coordination, notification, and exercise requirements.

Relying on RAGAGEP tools for your organization's covered processes is vital to the success of an PSM/RMP Program. Use of recognized industry practices provides a level of certainty in quality and processes and ensures that organizations are less susceptible to missteps [9] in implementing their PSM/RMP program.

6 Extensive Coaching via High-Frequency Peer Reviews

Extensive coaching through high-frequency peer reviews enabling effective transfer of knowledge and continuous improvement within the PSM team is step six. A peer review is a quality control measure, and is defined as a process in which professionals review each other's work to confirm quality, accuracy, and relevance.[12] All PSM organizations should encourage routine peer reviews of work products per internal Quality Management Systems (QMS) to promote continuous improvement. Peer feedback can also be focused on "how" tasks are performed to encourage staff growth in PHA, MOC and audit facilitation roles.

For example, Trinity has a robust QMS and its mission is "to develop and implement intelligent solutions and to provide high quality consulting services that efficiently achieve clients' business objectives. All employees are involved in our efforts to continuously improve our ability to meet and exceed customer expectations."[13]

The organization ensures high quality deliverables and guarantees continuous improvement in its staff's learning and development through peer reviews. Each project is subject to a detailed task review (Tier 1) at the peer level, and a high-level review (Tier 2) at the senior leadership level. The two tiers of quality checks, in the context of deliverables and services, are described below, and can be applied to your teams.

Tier 1: How "correct" is it?

- Are technical calculations correct?
- Are regulatory determinations correct?
- Have the correct procedures been applied or implemented?
- Is the work product free of spelling, grammar, and formatting errors?

Tier 2: How "good" is it?

- Does the work product present insightful and innovative solutions?
- Does the work product provide added value to the client, helping them meet business objectives?
- Is the work product well-written?
- Is the style, organization, and aesthetic presentation of the work product appropriate?
- What distinguishes this work product from competitors?
- Does the work product make a strong impression?[13]

Trinity's Knowledge Management System (KMS) is another critical piece that ties into ongoing staff coaching and peer reviews. The KMS provides consultants with the ability to collaborate, leverage expertise, and effectively capture, store, and access knowledge content on a software platform to enable more prompt, insightful solutions to clients.[14] This tool is similar to LinkedIn, Facebook and Twitter in that each consultant maintains their own online profile highlighting education, experience and specialized skills. Additionally, areas of expertise can be identified via hashtags, which are organized within a structured taxonomy. Other key benefits of the KMS system are listed below:

- Business process, technical, and regulatory questions can be directed to relevant peers through use of hash tags and/or targeting to specific communities
- Discussion threads occur in an open collaborative space, then are stored, and searchable
- Subject matter experts, who are validated and weighted by the software per their KMS participation, can be identified by hashtag-based searches
- Each consultant controls the content flowing into their newsfeed based on their connections to colleagues, membership in communities, or subscription to hashtags
- Consultants can also receive instant email notifications and/or daily digest emails of their KMS newsfeed content
- Consultants can earn recognition badges based on levels of KMS participation and contributions
- Community sites covering the organization's overhead departments, businesses, locations, and important subject-matter topics.

- Keyword based search capabilities to identify documents, people, and relevant past conversation threads within the KMS [14]

Implementation of a Quality Management System and Knowledge Management System are effective avenues for providing ongoing coaching and support. Without these systems, you run the risk of remaining stagnant in employee competency growth and miss out on continuous improvement opportunities to promote greater performance within the organization.

7 Employee Recognition Programs

The final step to developing PSM/RMP capability in your organization is the fostering of employee recognition programs that provide timely recognition through personal acknowledgement of positive behaviors. Employee recognition must be an inherent part of an organization's culture. Otherwise, you risk losing your best and most skilled assets, the employees, and this is critical when you think about the specialized skill-set of PSM professionals. Losing a top performer can create a major competency gap within in your organization and potentially impact its ability to effectively execute PSM/RMP work.

Employee acknowledgement serves as a key motivator for staff, which can lead to improved productivity and cultivate a culture of self-improvement. Employee recognition promotes employee retention, as staff may feel a greater sense of loyalty to an organization that consistently values employees. Lastly, employee recognition is a great morale booster, fostering a happier and more collaborative team environment. [15]

One of the most effective ways to recognize employees is by implementing a formal employee recognition program. An employee recognition program should provide both manager and employee initiated recognition. The program, as an example, can be a point-based recognition system, where each employee is allocated a set amount of points annually to distribute to peers. Employee-received points can be assigned a monetary value and can be redeemed in the calendar year or rolled over to the next calendar year. The recognition can be presented in the form of a personalized recognition card, highlighting the employee's stellar performance. As an extra incentive, your organization could allow top-point earning employees to enter into a drawing to win an additional bonus. [16]

While structured employee recognition programs are a great way to demonstrate appreciation, employee recognition doesn't always have to be as sophisticated or even in the form of a monetary reward. It can be as simple as a fun day away from the office, or a thank you email acknowledging an employee's outstanding work effort. The level of energy that your organization's leaders invest in employee recognition will make all the difference in the success and sustainability of your PSM/RMP team and program.

Conclusion

Capability building should be one of the highest strategic priorities within an organization, primarily because it is a process that is not always straight-forward or easy to tackle. This is especially evident for the PSM/RMP regulated industry, where poor judgement and lack of sound technical experience and leadership could be catastrophic not only for the business, but also for the employees and surrounding community. OSHA and EPA developed their respective PSM/RMP standards as guiding lights to aid organizations in the proper implementation and maintenance of an effective process safety program, but these standards alone are not enough. Senior leadership must demonstrate an unwavering commitment to building process safety capability and keep it at the forefront of their organization's strategic agenda to ensure PSM/RMP program success.

The following steps pave the path for capability building to successfully propel your PSM/RMP organization forward:

- Targeted recruitment of safety professional with global and domestic industry-specific PSM/RMP expertise
- Intensive onboarding and training programs inclusive of on-the-job exposure, internal and external competency training requirements
- Well-defined, tailored career development tracks designed to reinforce subject matter expertise and support advancement within the organization
- Development of internal standards to produce a consistent quality work product across the PSM/RMP team
- Development and use of adopted templates from RAGAGEP to ensure all work is executed based on sound, safe, and reliable industry practices
- Extensive coaching through high-frequency peer reviews allowing for effective transfer of knowledge and continuous improvement within the PSM team
- Employee Recognition programs that provide PSM teams with timely recognition through personal acknowledgement of positive behaviors

While your PSM/RMP team may vary in size, technical expertise, or organizational structures, these best practices will provide a well-documented roadmap and a high level of assurance that PSM/RMP capability is building within your organization.

References

1. Occupational Safety and Health Administration, Process Safety Management Guidelines for Compliance, Available at https://www.osha.gov/Publications/osha3133.html, Accessed on January 9, 2020.
2. Chemical Processing, Process Safety Begins in the Board Room, Available at https://www.chemicalprocessing.com/articles/2013/process-safety-begins-in-the-board-room/, Accessed on January 9, 2020.
3. Trinity Consultants, Statement of Qualifications for Process Safety Management and Risk Management Program, Available via Trinity Consultants Intranet System, Accessed on January 9, 2020.
4. Trinity Consultants, Trispyrian Services Sheet, Available via Trinity Consultants Intranet System, Accessed on January 9, 2020.
5. Occupational Safety and Health Administration, Process Safety Management for Small Business, Available at https://www.osha.gov/Publications/OSHA3908.pdf, Accessed on January 9, 2020.

6. Travelers, How to Onboard and Train Employees into a Safety Culture, Available at https://www.travelers.com/resources/workplace-safety/onboarding-and-training-employees-into-a-safety-culture, Accessed on January 9, 2020.
7. American Institute of Chemical Engineers, 2019 AICHE Academy Course Catalog, Available at www.aiche.org/academy, Accessed on January 9, 2020.
8. The Balance Careers, How to Make a Career Path Plan, Available at https://www.thebalancecareers.com/career-pathing-1918080, Accessed on January 10, 2020.
9. Planon Software, Five Reasons why Organizations can no longer avoid Standardization, Available at https://planonsoftware.com/us/whats-new/blog/five-reasons-why-organizations-can-no-longer-avoid-standardization/ , Accessed on January 10, 2020.
10. Occupational Safety and Health Administration, Process Safety Management for Petroleum Refineries: Lessons Learned from the Petroleum Refinery Process Safety Management National Emphasis Program, Available at https://www.osha.gov/Publications/OSHA3918.pdf , Accessed on January 10, 2020.
11. International Institute of Ammonia Refrigeration, Technology and Standards, Available at https://www.iiar.org/iiar/technology_and_standards/technology_and_standards.aspx, Accessed on January 10, 2020.
12. Medical News Today, What to know about Peer Review, Available at https://www.medicalnewstoday.com/articles/281528.php , Accessed on January 11, 2020.
13. Trinity Consultants, Quality Management System, Available via Trinity Consultants Intranet System, Accessed on January 11, 2020.
14. Trinity Consultants, Knowledge Management System, Available via Trinity Consultants Intranet System, Accessed on January 12, 2020.
15. Efront Learning, Employee Recognition in the Workplace: The Why and How, Available at https://www.efrontlearning.com/blog/2018/04/employee-recognition-workplace-benefits-ways.html, Accessed on January 12, 2020.
16. Trinity Consultants, Employee Recognition Program, Available via Trinity Consultants Intranet System, Accessed on January 12, 2020.

A human's perspective on human factors in bowtie modeling

Two visions

Both the bowtie method as well as human factors have become increasingly popular topics in risk management. The challenge remains, however, to discover the best way to include human factors in the bowtie method. The study of human factors occupies itself with researching the interaction between the user and his/her working environment. In this white paper two visions will be examined on how and why human factors can be included in the bowtie method. The two visions draw from Hollnagel's System-1 and System-2 thinking. Vision 1 (which is more closely related to Safety-I thinking) is in favor of the idea that human factors can be included as a threat as well as an escalation factor. Vision 2 (which is more closely related to Safety- II thinking) is in favor of the idea that human factors can only be included as an escalation factor. This article concerns itself first and foremost with qualitative bowties and not quantitative bowties, because bowtie thinking is, at its core, scenario-driven. This paper is intended for people who use bowtie modeling in their jobs, but who do not have any background in cognitive psychology. By the end of this white paper, we hope you have gained a better understanding of human factors and how to include these in the bowtie method.

The challenge remains

The challenge of including human factors in bowties can best be illustrated by means of an incident. Imagine the following scenario, a recurring operation of many industries is that their employees must travel on the road for work. In this outlined situation the driver and his passengers are driving from site to another location to drop off the work car after a long shift. The driver is tired and speeding to get home as he needed to catch his plane to go on vacation. However, he is driving on gravel roads that have gotten wet from rain. At one point, the driver is distracted and does not notice a large pool of water on the road. The right front wheel hits the pool of water causing the steering wheel to sharply pull to the right. The driver overcorrects, which results in the car going sideways. Control is lost, and the car impacts some bushes. No one is hurt, the car is damaged in the process, but can still be driven back to the drop off location.

How would the risks in this incident fit within the bowtie method? Which human factors are present in this incident setting and where do they fit in the bowtie model? How should these risks and human factors be described? What happens in case human factors are 'wrongly' placed in the bowtie model? To answer these questions and fully understand their answers, both the foundation of the method as well as the building blocks of the bowtie must be examined.

'A' bowtie method?

Barrier management is being used in multiple industries (McLeod et al., 2016), with the bowtie methodology on the rise in popularity. Bowtie methodology is applied mostly in high-hazard industries (nuclear, oil and gas, aviation etc.) but is relevant for nearly any industry that relies on barriers to prevent losses (healthcare, banking, public services, public utilities etc.). Differences in bowtie practices, however, persist and yet there is no agreed upon standard on "the" bowtie method. De Ruijter and Guldenmund (2016) suggest there are many different bowtie methods, and we should perhaps speak of "a" bowtie method. The method takes its name from the shape of the diagram that you create, which looks like a men's bowtie.

One of these methods is the Shell-bowtie risk evaluation method, which is the bowtie method that will be used in this paper. The Shell-bowtie is used to analyze and demonstrate multi-causal relationships in high risk scenarios. The main goal of using bowties is to communicate as simply and visually as possible where an organization's risks lay. A bowtie diagram does two things. First, a bowtie provides a visual summary of all plausible accident scenarios that could exist around a certain hazard. The knot in the middle is called a top event, which is the first 'loss of control moment' of the hazard.

Secondly, and most important, by identifying control measures or barriers, the bowtie displays what a company does to control those scenarios. The idea of the barriers is based on James Reason's Swiss Cheese model. Reason (1997) hypothesized that hazards are prevented from leading to losses by a series of barriers. According to him, these barriers are never 100% effective. Each barrier has unintended intermittent weaknesses and, when they line up, a hazard can lead to losses. Herein lies the reason for why multiple barriers are necessary.

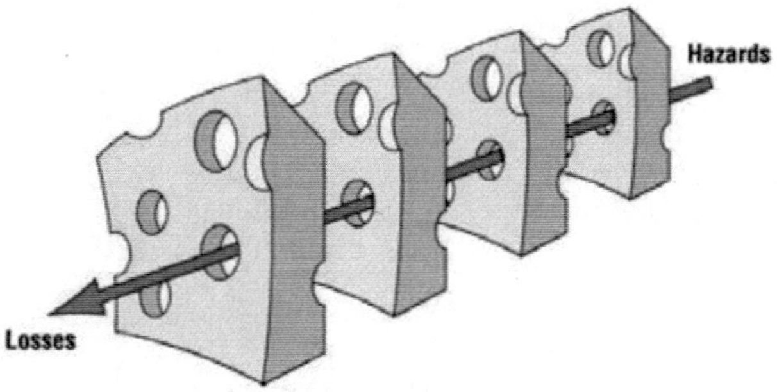

Figure 1: The Swiss Cheese Model

A metaphor commonly used to explain the barrier concept is to compare the barriers to slices of Swiss cheese with holes. The holes in the barriers are dynamic and continuously change in size and location. When holes in all the slices line up, the hazard (the arrow) can pass through these deficient barriers, leading to an accident (losses).

Bowtie elements

To answer the question of where human factors fit in the bowtie method, the bowtie terminology must be defined, and the bowtie elements must be examined more closely.

Hazard: a hazard is the process that an organization needs to do to run a business. If hazards are under control, they will not cause harm, but they introduce the potential for damage/injury.

Top Event: when control over a hazard is lost, it is often possible to identify the moment when a normal situation changes to an abnormal situation. The top event is not a catastrophe yet, but the company is now exposed to the potential harm of the hazard. It is still possible for the organization to control the situation.

It is important to realize that the top event is not an absolute event. Events are usually part of a longer chain of events: drinking and driving can lead to a loss of control over a vehicle, which can lead to crashing into open water, which can lead to occupants becoming entrapped in an enclosed space, which can lead to fatalities. Which event is appointed as the top event is important, because this choice dictates how the diagram will turn out. It is a very subjective choice that depends on the perspective taken. For example, an explosion can be a consequence for an organization, and a top event for a fire fighter. Sometimes a generic bowtie is created where the threats and consequences are specified in other bowties. In this case, a threat or consequence can become a new top event of another bowtie. In this manner, several bowties will be linked, creating a chain of events. This process is called bowtie chaining. Later in this paper (see 'Two visions') it is explained how this knowledge can be applied when including human factors in the bowtie model.

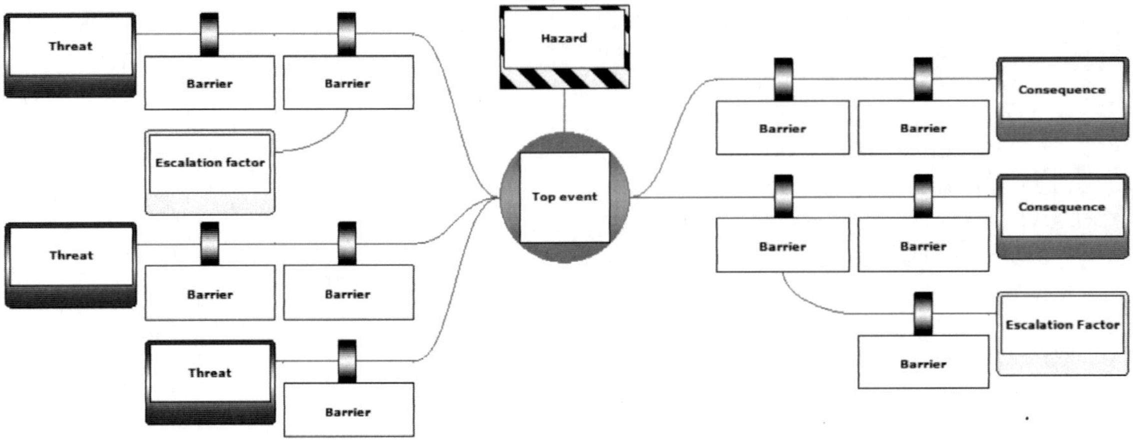

Figure 2: The skeleton of a bowtie

Barriers: risk management is about controlling risks. This is done by implementing barriers to prevent certain events from happening. A barrier (sometimes also called a control) can be any measure taken that acts against some undesirable force or intention, to maintain a desired state. Barriers are placed on both sides of the top event. Preventive barriers on the left side of the bowtie prevent the top event from happening. Recovery barriers on the right side of the bowtie either prevent the top event from resulting in unwanted consequences or mitigate the severity of the consequence. Barriers can be hardware systems, design aspects, human behavior and so on (for a more extensive analysis on barriers, see Sklet (2006).

Threats: threats are potential causes of the top event. They are located on the left-hand side of the top event. There can be multiple threats for one top event and each threat represents a single scenario that could lead to the top event directly and independently. Relevant for the human factors question is a common mistake that is often made: formulating a threat as the failure of a barrier. Barrier failures can be spotted by certain words such as 'lack of' ... or 'failure of' ..., for example 'failure of Anti-lock braking system'. Anti-lock braking systems do not cause loss of control over a vehicle; rather they are safety devices that allow the driver to remain in control over the vehicle during hard braking and slippery conditions. The described threat is in actual effect a failed barrier that, were it described positively, intends to prevent the original threat. In this case, the slippery conditions were the original threat with the anti-lock braking system as the barrier.

However, one exception to this is a primary process failure, which is a type of threat that is also described with these failure words while still being correct. If a piece of equipment that is part of the primary process fails (primary equipment such as an engine failure in a helicopter, or a pipeline integrity failure in an installation), this is an actual threat. Thus, a threat should be a force or condition that pushes an unwanted chain of events further. In the Chapter 'Two visions', it will be explained how this knowledge can be applied to include human factors in the bowtie model.

Consequences: consequences are unwanted scenarios that can be caused by the top event. Consequences are the incidents and accidents we want to prevent. Consequences are mainly unwanted because they will lead to loss or damage.

Escalation factors: once the control measures are identified, the bowtie method takes it one step further and identifies specific conditions or actions that make it more likely that a barrier will fail. An escalation factor can also be described as a failure mode of a barrier. It cannot directly cause an event but increases the chance that a certain threat or top event will lead to an event, by taking out a barrier. These are the holes in the 'Swiss Cheese Model'. Why it is agreed that human factors can be included in this bowtie element and how it can be included specifically, will be discussed later (see 'Differences').

Escalation factor barriers: similar to the barriers on the original threat line, there are barriers for escalation factors as well. These escalation factor barriers are more niche and protect the main barrier from an escalation factor.

Human error & human factors

Like the bowtie method, the topic of human factors has also become increasingly popular. More and more literature, research, books, magazines, newspaper articles, journals and conferences are available on the topic of human factors. The Ergonomics Society Conference in the UK and the Human Factors and Ergonomics Society Conference in the USA are two noteworthy institutions. Perhaps the best indicator of the rising popularity, though, is the number of human factors professionals being employed in many different industries (Jordan, 2002).

The study of human factors occupies itself with researching the interaction between the user and his/her working environment. The underlying philosophy to human factors, is that it is easier to change the working environment than it is to change the human condition (Reason, 1997). It therefore tries to design the working environment according to the physical and cognitive needs and limitations of the human mind. The goal of human factors is to optimize the performance, safety and usability of products, processes and systems.

Safety is of interest here for the purpose of this white paper. The study of human factors tries to optimize safety by designing environments in such a way that the possibility of human error is reduced as much as possible. Human error is an umbrella term that can be better analyzed by looking at the different types of errors that can be made: slips, lapses and mistakes.

Slips: This subcategory of errors are classified as visible execution failures (Reason, 1990, 1997). The plan is correct but the execution deviates from the original plan, it fails. An example is driving on the highway, but having to slow down due to traffic, thus changing gears but changing from fifth gear to third, accidentally skipping fourth gear.

Lapses: This subcategory of errors are less visible than slips as they are more 'in the mind' and are related to failures of memory (Reason, 1990, 1997). Like slips, lapses are about having the right plan, but again, the execution deviates from the original plan, it fails. Lapses involve not doing an action. An example is waiting for a red light with the handbrake on, the light turns green and you try to pull up but forgot to take the handbrake off.

Both slips and lapses involve accidentally activating the wrong routine (Reason, 1990, 1997).

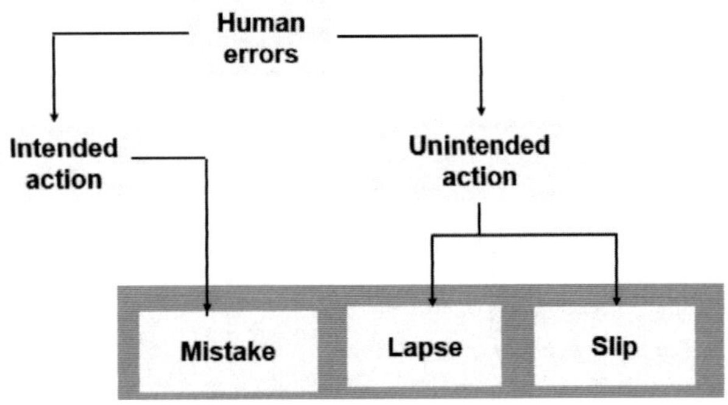

Figure 3. An abridged model of human error

Mistakes: Unlike slips and lapses, the failures of 'mistake errors' are that the plan itself is incorrect, but the execution is successful, it follows the original (wrong) plan. An example is driving on the highway and believing you need exit A and so you successfully take exit A only to find out later that you needed exit B (Reason, 1997).

More distinctions can be made in human error but are left out for the scope of this white paper. For other distinctions, please refer to James Reason, Human Error (1990 and 1997) and Rasmussen's Skill, Rules and Knowledge based model (1983) for more information.

To summarize: "Simply put, human factors can be thought of as conditions that if not managed correctly can lead to human error" (Edkins, 2016). Thus, a condition such as fatigue (human factors) could lead to inattention (human error, lapse).

Two visions

There is a clear distinction between two visions on how to include human factors into the bowtie method. Each can be related to what Hollnagel (2014) called either Safety-I or Safety- II thinking.

In short: *"A Safety-I approach presumes that things go wrong because of identifiable failures or malfunctions of specific components: technology, procedures, the human workers and the organizations in which they are embedded. Humans—acting alone or collectively—are therefore viewed predominantly as a liability or hazard, principally because they are the most variable of these components" (Hollnagel, 2015).*

Hollnagel believes safety management should move from *"ensuring that 'as few things as possible go wrong' to ensuring that 'as many things as possible go right'. We call this perspective Safety-II; it relates to the system's ability to succeed under varying conditions. A Safety-II approach assumes that everyday performance variability provides the adaptations that are needed to respond to varying conditions, and hence is the reason why things go right. Humans are consequently seen as a resource necessary for system flexibility and resilience" (Hollnagel, 2015).*

How do these visions translate to the bowtie method? Vision 1 (which is more closely related to Safety-I thinking) is in favor of the idea that human factors can be included as a threat as well as an escalation factor. Vision 2 (which is more closely related to Safety- II thinking) is in favor of the idea that human factors can only be included as an escalation factor.

Both similarities and differences between the visions will be explored in this article using the 'driving a vehicle' example. In a driving incident, imagine the driver was:

1. Speeding
2. Distracted, due to fatigue

assessment. Looking back, this guideline fits well with the paradigm of thinking about safety of that time (1960s-1980s), the Safety-I kind of thinking.

An argument in favor of human factors as a threat is that human factors can sometimes be viewed as being as vital to a process as

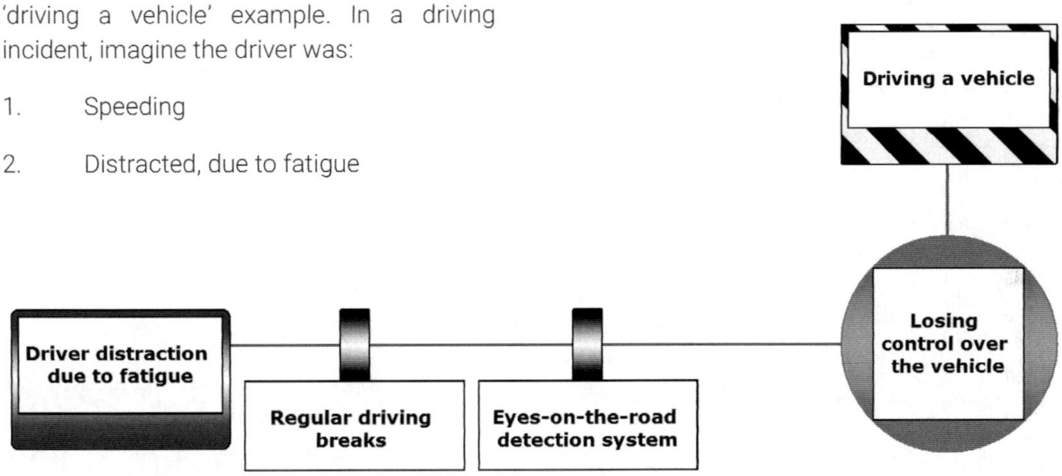

Figure 4: Including human factors in the threat

Differences in Safety-I and Safety-II

Vision 1 is in favor of the idea that 'Distracted due to fatigue' can be written down as a threat, as it can, and did, directly lead to the top event, 'losing control over the vehicle' (figure 4).

Human factors as a threat: A popular statistic is that human error represents a greater threat (80 - 90% of errors) to complex hazardous systems than technical failures do (Reason, 1997). The inclusion of human factors (or at the very least, human error) as a possible threat, therefore seems necessary to have a complete and thorough visual risk

primary equipment (see section 1.1). Many, if not all, organizations need humans to operate effectively. This holds true especially for the less automatable industries such as medicine and construction. Human factors as a threat should be a force or condition that pushes an unwanted chain of events further. Therefore, human factors appear as threats in a bowtie diagram when the human act has the 'causal power' to directly influence the hazardous operation to result in a loss of control. For example, 'Driver distraction due to fatigue' directly led to the top event of 'Losing control over the vehicle'. Therefore, before deciding to include human factors, it must be considered if the human is part of a safety (barrier) system, or part of the main process. In the latter case, an error could directly lead to loss of control. In the driving example, the driver can be considered as part of the primary process as the car would not run without the driver.

Human factors as an escalation factor: Vision 2 states that human factors can never be a threat and is always a failure mode on a barrier, i.e. an escalation factor (see Figure 5). Organizations such as Environmental Resources Management (ERM) and Center for Chemical Process Safety (CCPS) argue in favor of this vision, as they believe the impact of human factors solely affects the health of barriers (McLeod et al., 2016). This is more in line with Safety-II thinking. The popular '90% of errors are human related statistic' is looked at from a different angle. Despite the errors, organizations are still growing and staying alive. Therefore, the Safety-II vision argues, it must be safe to assume that people do things right more often than they do wrong. Indeed: ***"There is growing recognition that people are often the reason operations go well despite the upsets and the everyday variability that is normal to complex activities" (Hollnagel, 2014).***

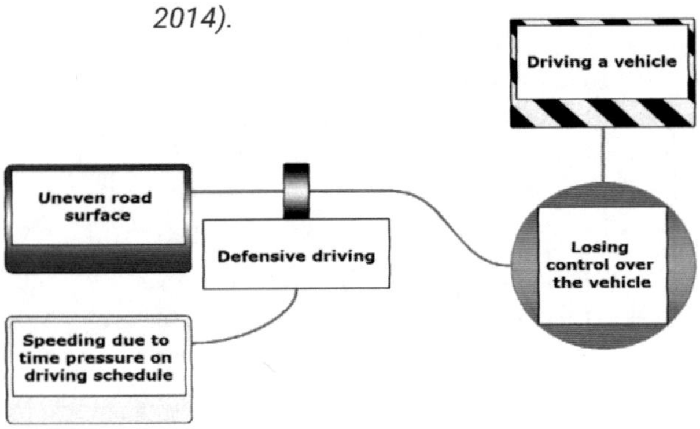

Figure 5: Human factors in an escalation factor

Furthermore, another possible negative side effect for putting in human factors as threats, as McLeod et al. (2016), argue is that, by focusing on human error as a threat, it can convey the wrong message to an organization's employees. If employees see that they are considered as threats in the organization's process, it can undermine their belief and faith in the organization and undermine the development of a strong safety culture. This runs counter to the point that human factors specialists try to make: the human is the strength of the system, not the weakness, i.e. humans are a solution to harness not a cause of failures.

Similarities

Beside the differences, there are also 4 parallels between the two visions.

Human factors must be included: both visions agree that every barrier is supported by the human element, whether it is a safety valve that needs to be maintained or an activity that requires certain competencies and/or training. Human factors are an integral part of risk management, and an organization needs to fully understand the impact these have on risks. Thus, excluding human factors from a bowtie means ignoring crucial information.

Non-generic wording to stimulate thought process: the way the threat or escalation factors should be described is not with a generic term like human error or human factor, but the specific results of those conditions, for example 'Speeding' or 'Distracted due to fatigue'. Wording the threat and escalation factor in this manner stimulates people's thought process and

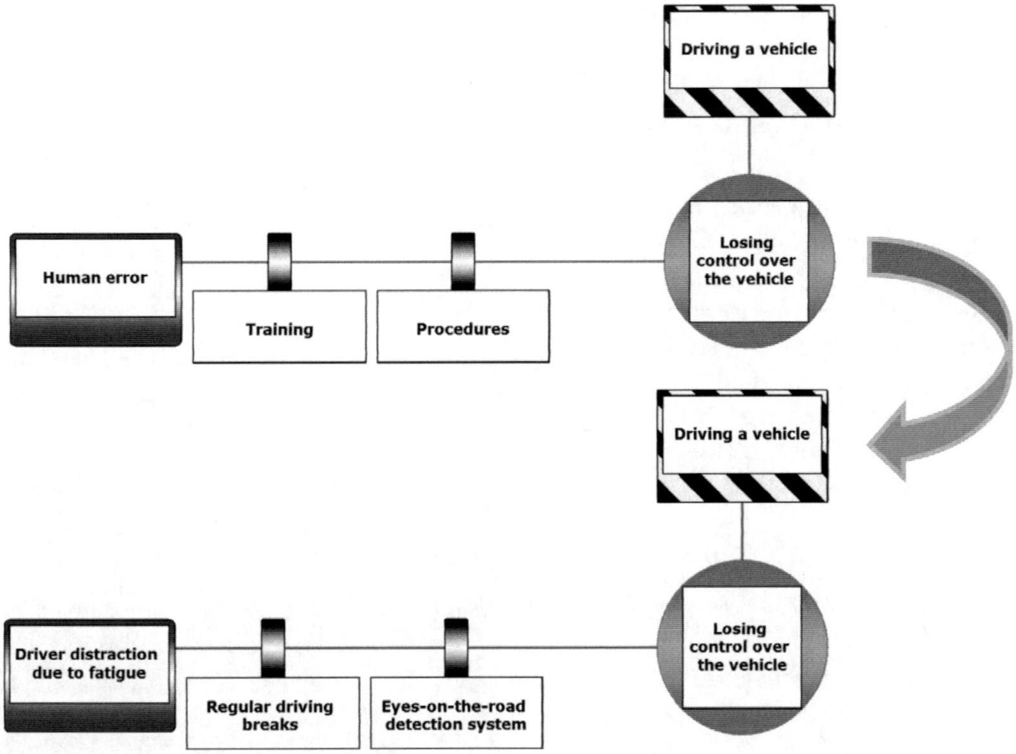

Figure 6: Improving the wording of human factors and human error

helps them complete the other steps in the bowtie method. It is much harder to come up with the barriers seen in figure 6, based on a generic wording like 'human error. 'Driver distraction due to fatigue' immediately focuses our thinking on both the result (distraction) as well as the cause (fatigue) so that we can come up with barriers that protect from either effect: regular driving breaks prevents fatigue, and an eyes-on-the-road detection system prevents distraction.

Escalation factors: Both visions agree that human factors are often part of the performance or failure modes of barriers. Failure modes are represented as escalation factors in the bowtie diagram. If the failure modes are understood, the activities a company needs to undertake to prevent these failure modes from compromising the barriers, can be known. This can be training, awareness campaigns, or it can go as far as organization development and culture transformation.

In the driving example, the driver was speeding. 'Speeding due to time pressure on driving schedule' would be an escalation factor to the barrier 'defensive driving', against the threat of the country's uneven road surfaces, see figure 5. Speeding, here, would be the human error, more specifically, a mistake (or 'violation', see James Reason, 1997). Whilst 'time pressure on driving schedule' refers to the driver's hurry to go on vacation, and therefore falls under the category of human factors. This view is in line with other organizations such as ERM and CCPS. Indeed, it can be called a standard.

A notable exception: bowtie chaining: both visions agree that it is possible to make a bowtie focused completely on human factors, with the top event usually being related to human factors by making use of bowtie chaining as described earlier in this paper (see "A' bowtie method?'). In such bowties, both visions agree that human factors can be put on the threat level to get the right level of detail and causation of the human factors.

In the driving example, 'driver distraction due to fatigue' can be chained into multiple bowties (see figure 7). In figure 7, 'driver distraction due to fatigue' has been phrased as 'Driver loss of attention (due to...)' in the bowtie with top event 'losing control over the vehicle'. Before the driver lost his attention, there were other threats that likely caused this to happen, so the threat of 'driver loss of attention' becomes the top event for the previous bowtie.

The right vision?

Perhaps the reason for the existing disagreement, is that the bowtie is, first and foremost a communicative tool. It is highly flexible and is being applied in several industries. Each industry has a different context and a different level of safety, and thus needs different guidelines for human factors and bowties. Maybe a third vision needs to be adopted, one that is as flexible as the bowtie method itself and combines the merits of both visions: For industries such as chemical, oil and gas or aviation, most processes work automatically, and their performance is protected by automatic barriers. Therefore, it is more likely that primary equipment failures involve mechanical solutions rather than human solutions. The human factor can reduce the effectiveness of the barriers but can never be a direct threat leading to the top event. For these industries, the Safety-II way of thinking supported by ERM/CCPS might be preferred, and thus human factors should only be included as escalation factors.

For the more manual labor-intensive industries such as medicine or construction, it is very difficult to automate processes and the human role is more present. In these industries, top events can directly be caused by human factor failures, as the human can be viewed as primary equipment. Therefore, as people are essential in such industries, the older style, Safety-I way of thinking might be preferred, and thus human factors can be considered both as a threat as well as an escalation factor.

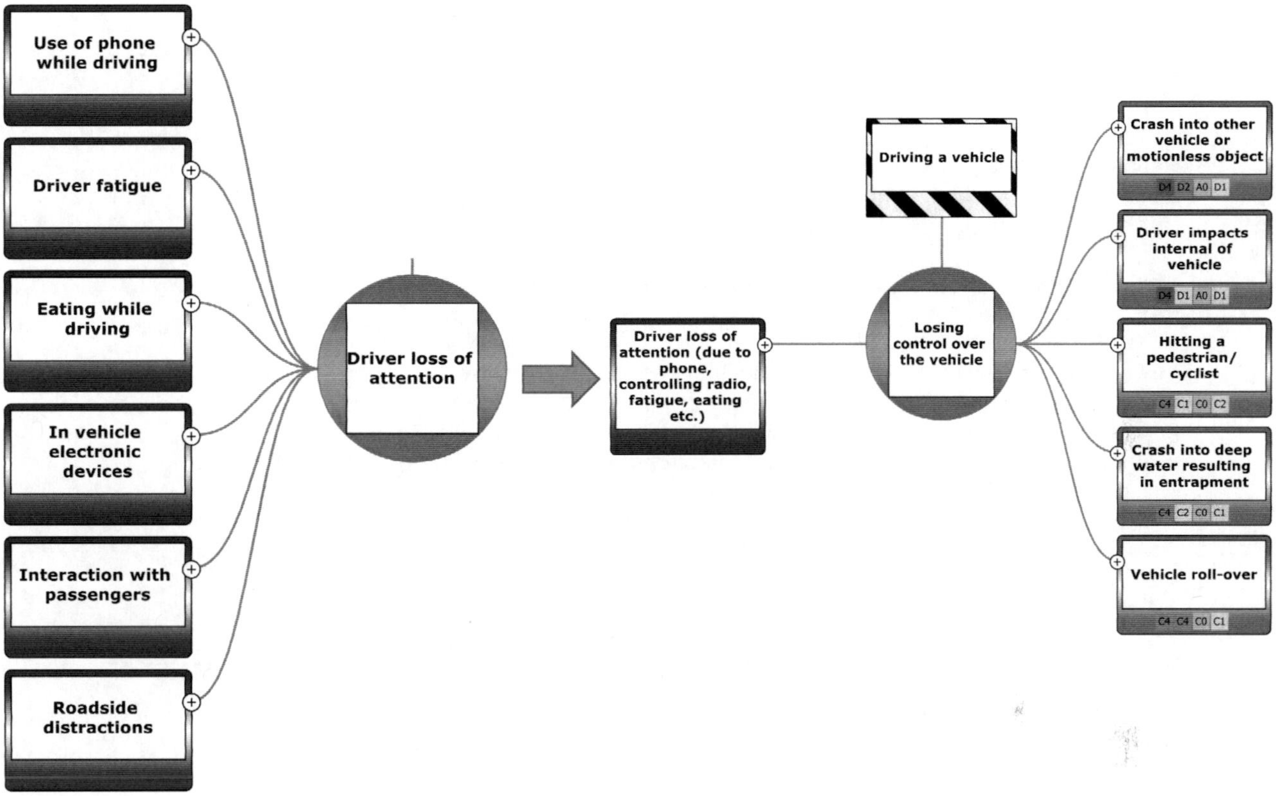

Figure 7: Bowtie chaining example of human factors

References

Edkins, G. (2016) Human Factors, Human Error & The Role of Bad Luck in Incident Investigations [Blog Post]. Retrieved from https://www.safetywise.com/single-post/2016/08/30/Human-Factors-Human-Error-The-Role-of-Bad-Luck-in-Incident-Investigations

Hollnagel, E. (2014) Safety-I and Safety-II: The past and future of safety management. Ashgate. Hopkins, A.

Hollnagel, E., Wears, R. L., & Braithwaite, J. (2015). From Safety-I to Safety-II: a white paper. The resilient health care net: published simultaneously by the University of Southern Denmark, University of Florida, USA, and Macquarie University, Australia.

Jordan, P. W. (2002). *Designing pleasurable products: An introduction to the new human factors*. CRC press.

McLeod, R., Randle, I., Miles, R., Hamilton, I., Wilkinson, J., Tomlinson, C., Thomas Jun, G., Wynn, T. (2016). *Human factors in barrier management: A White Paper by the Chartered Institute of Ergonomics & Human Factors* [White paper]. Retrieved August 17, 2018, via https://www.ergonomics.org.uk/Public/Resources/Publications/Barrier_Management.aspx

Reason, J. (1990) *Human Error*. Cambridge University Press.

Reason, J. (1997) Managing the Risks of Organizational Accidents. Ashgate
referentie in tekst: complex hazardous systems than technical failures do)

de Ruijter, A., & Guldenmund, F. (2016). The bowtie method: A review. *Safety science*, *88*, 211-218.

Sklet, S. (2006). Safety barriers: Definition, classification, and performance. *Journal of loss prevention in the process industries*, *19*(5), 494-506.

583490 Api's Process Safety Assessment Program - 8 Years Later - Learnings & Changes

Abstract:

In the past year, PSSAP has implemented two new protocols, developed PSSAP Focused (a slimmer version better suited to small sites/companies and petrochemicals) and rolled out a new pricing structure. This presentation will be an update on all these items, including learnings from the traditional 7 protocols, as well as the two new protocols (Incident Learning and Product Storage & Transfer).

Improving Process Safety Reporting from an Insurance Perspective

Nigel Cairns CEng FIChemE
Professional Process Safety Engineer
Aon Global Risk Consulting, The Aon Centre, The Leadenhall Building,
122 Leadenhall Street, London EC3V 4AN
Nigel.cairns@aon.co.uk

Prepared for Presentation at
American Institute of Chemical Engineers
2020 Spring Meeting and 16th Global Congress on Process Safety
Houston, TX
March 30 – April 1, 2020

Improving Process Safety Reporting from an Insurance Perspective

Nigel Cairns CEng FIChemE
Professional Process Safety Engineer
Aon Global Risk Consulting, The Aon Centre, The Leadenhall Building,
122 Leadenhall Street, London EC3V 4AN
Nigel.cairns@aon.co.uk

Keywords: insurance, bow tie, process safety barrier, losses, risk rating

Abstract

Energy sites viewed and assessed by insurance risk engineers continue to have losses, seemingly independent of their reported risk quality. Aon proposes a significant change in risk quality assessment using a safety barrier model to highlight strengths, weaknesses and exposures, and give a greater focus to learning from industry losses.

1 Introduction

Energy insurance risk engineers conduct hundreds of underwriting surveys globally each year, across upstream, midstream and downstream assets. Most of these surveys are conducted by risk engineers working for (or on behalf of) insurance brokers for the benefit of insurers and insureds.

Site operators are extensively audited by stakeholders and regulatory bodies, yet we know anecdotally that insurance underwriting surveys are highly regarded for bringing independent third party and industry good practice advice to sites. However, to some operators the process can often boil down to one question: "What will this mean for my insurance premium?"

The main output from the survey process, the insurance underwriting report, is written to give evidence-based opinion of insurable risk quality to underwriters and insureds. This is, in part, used to help set insurance premiums and policy conditions, and whilst there is insurance-industry guidance on what the content of underwriting reports should be, all insurance surveying organisations have their own method of rating the risk quality of a site, against recognised good engineering and management practices.

However, the fact that the industry as a whole continues to see losses begs the question whether insurance underwriting surveys are sufficiently objective, and suitably structured and executed in order to understand and address the root causes of losses so as to prevent (or reduce) reoccurrence.

The purpose of this paper therefore is to propose a more effective process for assessing insurable risk quality within the confines of an insurance underwriting survey, using a barrier-based methodology, whilst noting that within the framework proposed, the detailed assessment of individual barrier detail is still a work in progress by the author.

2 The origins of insurance risk engineering

The origins of energy insurance go back to 1972 when the insurance mutual Oil Insurance Limited (OIL) was formed by 16 energy companies in response to two large-scale industry losses in the late 1960s. The combination of an oil spill in Santa Barbara, California and a refinery explosion in Lake Charles, Louisiana highlighted the need for specialist energy insurance, beyond that available at the time from established property insurance markets. But it was the Flixborough disaster of June 1974 which ultimately led to the formation of International Oil Insurers (IOI), and to the recruitment of risk engineers into insurance from industry in an attempt to better differentiate risks. This led directly to the insurance survey process we see today, and the associated methods of risk rating, benchmarking, establishing Estimated Maximum Losses (EMLs), and risk quality improvement through targeted risk improvement recommendations.

Early reports would review site operations, document safety culture, and attempt to give some sort of risk differentiation, but process safety as a concept was in its infancy, and there was more of a focus on developing EMLs so that underwriters and insureds could understand their maximum exposures. However, industry losses continued.

The focus started to shift more towards process safety and away from occupational safety reporting following significant losses at Phillips Pasadena (1989) and particularly after BP Texas City (2005). However, progress has been slow, as insurance has traditionally been a conservative profession, and risk engineering surveys and reports have typically lagged behind the industry-wide push for better process safety management and reporting.

The objectivity of underwriting reports varies when it comes to defining risk quality and differentiating between risks. Anecdotally, there is also a widely-held view that reports generated from some global insurance hubs still over-emphasise the importance of 'after the fact' loss mitigation, rather than scrutinising loss prevention barriers. Overall therefore, the view of much of the insurance underwriting community is that although reports generally give good risk description and information, they do not give underwriters clear guidance on which insured assets are more likely to give them a claim by policy year end.

3 Asset risk rating

To try and counter this lack of direction, report authors have devised risk rating or risk ranking tools to try to benchmark sites against Recognised And Generally Accepted Good Engineering Practices (RAGAGEP). Sites are typically scored against a number of factors, with a variety of weightings applied to distil a few dozen or few hundred individual scores down to a single indicator, within a range of (for example) 'poor' to 'excellent'.

However, this raises a number of questions, such as does the rating process:

- define the health of a site's safety barriers?
- give meaningful guidance to insureds as to where risk reduction improvements could be made?
- give a rating that merely satisfies our appetite for measurable data, regardless of how useful those data actually are?
- differentiate between risks that indicate which are the sites or the insureds most likely to have a loss? This last point is perhaps the most important to underwriters.

The reality is that the industry continues to have a significant number of losses year-on-year, and these losses are happening across all perceived risk qualities. Often the problem is that without sufficient differentiation within the risk rating methodology, ineffective risk-reduction features can get 'normalised' out, masking defective safety barriers. Too many sites which are positively rated by underwriting surveys have property-damage losses, such as fires or vapour cloud explosions (VCEs), and there appears to be no direct correlation between loss-likelihood and risk rating score: in the words of one underwriter "too many good sites are having losses".

Is loss propensity truly independent of risk quality? Or are we using the wrong tools?

Part of the problem is that the risk rating processes in common use by insurance risk engineers are often used as the 'go to' tool for a variety of conflicting purposes, for example, as a benchmarking tool for comparing individual sites or regions, or even as a means for insureds to set Senior Management bonus levels based on performance at the last insurance survey. Consequently, the number of metrics typically measured and scored has gone up rather than down, become less rather than more focused, with the result that rating quality has often been 'averaged out' or normalised into the middle of the bell curve distribution, reducing differentiation. Despite attempts by some organisations to address this by greater weighting of more critical topics, the greater number of topics being scored masks poor barrier performance, and the combining of independent loss control barriers results in them losing their individual identity as performance metrics.

For example, a low score for a poorly-resourced Inspection department could be compensated for by an adequate hazard study process, or low levels of operator competence could be offset by a high level of predictive maintenance. The evidence however is that independent barriers such as asset integrity or operator competence by themselves are key factors when looking at what causes losses and, therefore, must not be 'swapped out' for other barriers or layers of protection.

Therefore, before we can define how we can more effectively measure a site's propensity to have a loss, and thus better define its risk quality from an insurance perspective, we need to understand what actually causes losses.

4 What causes losses?

Anecdotally, insurers have known for some time what causes losses, but the first formal presentation of data was by Jarvis in September 2016 (Reference 1) on behalf of the Lloyd's Market Association (LMA). This document reviewed 100 major onshore oil, gas and petrochemical property damage losses in terms of monetary value over a 20 year period from 1996 to 2015, excluding natural catastrophe (Nat Cat) events; it was recently updated to include an additional 37 losses between 2015 and 2019.

Major losses occur because of failures of both loss prevention and mitigation barriers in response to an initiating event. The LMA data highlighted the main initiating events which led to the majority of the considered losses, as well as those loss prevention barriers which most frequently failed:

- 38% of the losses were due to mechanical integrity failure; with 71% of these due to either internal or external piping corrosion (noting the higher propensity for this in refining over other occupancies such as gas plants or terminal operations).
- 20% of the losses occurred during maintenance activity, typically due to inadequate control of work.
- Of the remaining 57 losses, 86% occurred during transient (non-routine, infrequent, abnormal or unplanned) operation.

Note that the key messages from the LMA data are supported by evidence from underwriters' own loss data and claims analysis, which typically reference a much larger industry dataset. They suggest that rating asset quality from an insurance perspective would be more effective if focussed on the health of those barriers which line up to prevent or limit losses associated with a range of initiating events.

Of course, this has the potential to significantly change the insurance underwriting survey process as we know it. Insurance surveys are currently organised by discipline; meetings are organised by department, together with a tour of the site being surveyed. Typically, this means spending a couple of hours each with the Engineering, Inspection and Operations teams, among others, with each assessed against recognised good engineering and management practices. But if we are now to focus on the effectiveness of individual barriers in response to defined initiating events, then this is drawing us down a path of site interviews by barrier rather than by discipline – of course the reality is that we'll be talking to the same people, only with greater focus on the specific barriers which stand between a site and a major property damage / business interruption (PD / BI) loss.

And if the survey and rating processes need to change, then so does the reporting process, from discipline to barrier focus.

5 The insurance underwriting report

Underwriting reports have been based on much the same format for the last 15-20 years. In recent times, some have become shorter, but only because the same information has been presented more efficiently, rather than because of a fundamental shift in what is being reported. Underwriters and their risk engineers rarely have time to read a plethora of 100-page reports, and even if they did, do the reports give a true insight into the health and status of critical safety barriers or the likelihood of a loss? The message from the insurance markets is a resounding 'no'.

Therefore, to better define and report insurable risk quality to markets and insureds, the insurance industry needs to move away from a highly descriptive report style to one which focuses more clearly on the health and quality of defined process safety barriers; one way of doing this is to adopt the look and feel of a "bow tie diagram" (see Figure 1) to give a simple, clear representation of site risk quality to underwriters to help their insurance decision-making processes.

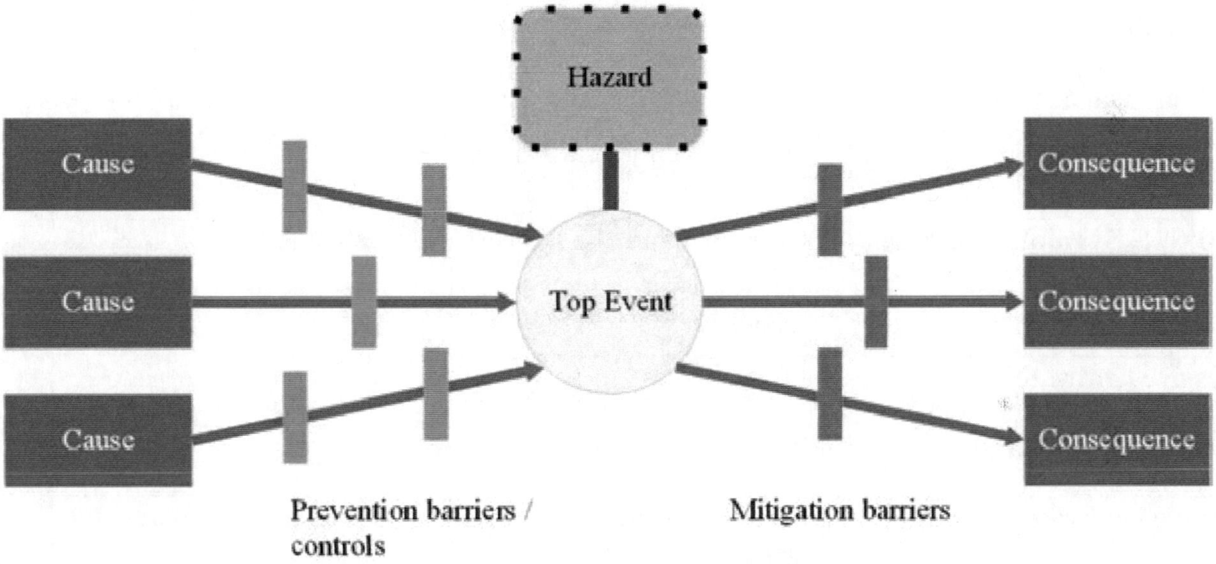

Figure 1. Typical "bow tie" diagram used for process safety reporting

The model should be simple and clear enough to effectively report on barrier health for a range of initiating events and causes, yet flexible enough to accommodate a variety of asset types and a range of review frequencies. If the model is too specific, the process becomes restrictive, cumbersome and quickly out-of-date as sites change – this is an important point for sites which will typically be surveyed by insurance risk engineers on a three or four-year cycle.

6 Adopting a barrier model

Adopting a barrier-based methodology begins with trying to build a process that focuses on the barriers which prevent what have been historically the key initiating events that lead to significant financial loss for insureds. This process can then be updated as loss initiators or barriers themselves change; for example, improved technology giving more inherently safe designs, global economics driving a shift in feedstock slates or changing demographics placing a greater strain on human resources.

Having defined the events or initiating causes which pose the greatest threats, our hazard is invariably hydrocarbons under pressure. Of course, extending the analysis to third party liability effects may require an extension of the methodology to non-hydrocarbons, for example to review the potential effects of the release of a typical refinery toxin such as hydrofluoric acid. For the purpose of the process however, the top event would not be a loss of primary containment (LOPC), but that event with which insurers are directly concerned, namely the fire or explosion which results in property damage. This is an important point, as it clearly defines those barriers which are preventing exposure and those which mitigate; it may also result in some barriers taking on a slightly different position within the model compared to a more traditional model based on an LOPC as the top event. In following this methodology through, no 'preventative' credit is taken for ignition source control on the left hand side (LHS) of our model; this is a deliberate stance from an insurance perspective to assume that a flammable cloud would invariably escalate to a VCE.

Such an analysis includes a range of barrier types and associated degradation controls (NB. passive barriers are typically stronger than active barriers) with barriers relying on human intervention typically among the weakest; refer to Table 1:

Table 1. Example of common barriers used in the process industries

Barrier type	**Example**
Passive hardware	Firewall
Continuous hardware	Active corrosion protection system
Active hardware	Safety Instrumented System (SIS) / automatic emergency shutdown system (ESD)
Active hardware + human	Operator-activated ESD
Active human	Visual fire detection and evacuation

The simplest and most effective way of reporting barrier health and giving a snapshot of performance (covering both the degree to which the barrier is present, and the quality of its implementation / maintenance) would be through the use of 'traffic light' colour coding as demonstrated in Table 2:

Table 2. Use of colour coding to define barrier status

Condition	Description
Effective	Barrier performing at design/intended effectiveness
Slightly degraded	Barrier performing slightly below design/intended effectiveness
Significantly degraded	Barrier performing significantly below design/intended effectiveness
Out of service	Barrier is not in place, turned-off, deactivated, fully degraded
Not applicable	Barrier not deemed applicable for that asset (for example, insurers wouldn't typically expect a remote terminal to maintain a full-time professional fire brigade, but might do so for a world-scale refinery)

So, for risk engineering reports, this might be applied to the first significant initiating event as follows.

A fire / VCE from the mechanical integrity failure of piping

For this scenario, Table 3 suggests examples of the key barriers and controls on the LHS of the model for preventing a loss caused by piping corrosion (NB. the constituent component detail of each barrier is pending continuing review by the author):

Table 3. Possible key preventative barriers and controls in a piping corrosion-initiated loss scenario

Barrier / control	Barrier / control type	Comment
Inspection competence	Degradation control	
Inspection programme philosophy	Degradation control / Preventative barrier	Identification and management of corrosion and damage mechanisms, use of appropriate NDT, etc.
Integrity Operating Windows (IOWs)	Degradation control	
Material selection	Preventative barrier	Including QA / QC

In terms of the survey process, all of these would currently be discussed with a site's Inspection department, and findings reported in the Inspection section of the underwriting report. In future, the robustness of the site's asset integrity management as a function of its defined barriers and degradation controls would be reported.

Moving from left to right along the model, further barriers and controls then play their part in preventing or mitigating a loss as a result of this threat; a selection is shown in Table 4:

Table 4. Possible key preventative and mitigating barriers in a piping corrosion-initiated loss scenario

Barrier / control	Barrier / control type	Comment
Engineering standards	Active hardware / preventative barrier	LHS of model
Gas detection	Active hardware / human	LHS – detection prompting action to prevent escalation of an LOPC to a VCE as the top event
ESD	Active hardware	LHS
Unit layout	Passive hardware / mitigating barrier	RHS of model – good layout features will reduce the impact of the top event
Fire detection	Active hardware / human	RHS – detection prompting action to mitigate the effects of the top event
Emergency response	Active human	RHS – typically the 'last line of defence' after the top event to limit escalation

This starts to build up a picture of how a new risk quality rating process might work, by identifying key barriers and degradation controls that prevent initiating events and threats becoming consequences through the top event.

A fire / VCE as a result of a transient operation

For this second initiating event, Table 5 suggests examples of those barriers and controls specific to the site's Operations team that should be in place to protect against unforeseen or transient events, such as the loss of a plant utility (e.g. power or cooling water). Again, note that the constituent component detail of each barrier is pending continuing review by the author:

Table 5. Possible key preventative barriers and controls in a transient operation-initiated loss scenario

Barrier / control	Barrier / control type	Comment
Operator competence assurance	Degradation control / preventative barrier	Includes training and recertification
Shift handover	Degradation control	
Operating procedures	Degradation control / active human	Includes Emergency Operating Procedures (EOPs), and requires action by the operator

There would also be a number of hardware controls upstream of the top event, and beyond this, the mitigation barriers on the RHS would typically be the same as in the earlier piping corrosion case.

A fire / VCE as a result of a maintenance activity

In this example, key barriers and controls on the LHS might include those in Table 6:

Table 6. Possible key preventative barriers and controls in a maintenance activity-initiated loss scenario

Barrier / control	Barrier / control type	Comment
Permit to Work (PTW) management system	Degradation control	
Training and certification of permit authorities	Degradation control	For both issuing and receiving permits
Isolation management procedures and practices	Degradation control	

Shift handover	Degradation control	

Again, the mitigation barriers on the RHS would be essentially the same as in the first initiating cause.

Aside from the specific barriers and controls postulated above for the three key loss-initiating events as identified by the LMA, incident investigations carried out by regulatory or industry bodies such as the UK's Health and Safety Executive (HSE), or the US Chemical Safety Board (CSB) have identified other key barriers whose documented absence or degradation has contributed directly to a loss.

There are many published examples of where poor or absent stewardship of key process safety elements has been the direct cause of a significant loss across a variety of initiating events and site operating modes. Examples of such elements include Process Hazard Analysis (PHA), Management of Change (MoC), Process Safety Information (PSI), process safety audits, and incident investigation.

It is worth noting that on an insurance survey, such topics don't always have a natural home for discussion, because from site to site, the processes may be owned by a variety of disciplines or functions. Not all sites have a dedicated Process Safety function managing process safety element responsibilities and ownership, so these topics may lie with any of the Engineering, HSE or even the Operations departments. Hence, they can easily be overlooked if they form part of a 'shopping list' of questions on an agenda for a particular department during a survey; or worse, the person responsible is not on site. Treating each topic as a barrier in its own right, with dedicated discussion time, could provide extra focus during a survey.

Other loss types

Adding the process safety barriers to the event-specific ones discussed earlier adds to the picture of what needs to be investigated and documented in order to determine a site's susceptibility to a large property damage (PD) loss resulting from a fire or VCE.

Of course, there are other types of losses than PD that underwriters pay out for, or that give increased exposure to insureds or clients. These may or may not have the same level of impact financially as a single catastrophic event, but could result in more frequent attritional losses that still significantly affect the bottom line. Then of course there is the Business Interruption (BI) coverage, which is triggered once the initial loss has occurred, to recompense the insured's lost income.

<u>Machinery Breakdown losses</u>

Machinery Breakdown (MB) insurance covers sudden and unforeseen losses to rotating and fixed equipment (i.e. not as a result of inappropriate levels of preventative maintenance, poor operation or corrosion over time), and is typically included on a property damage policy.

Therefore, the preventative barriers and degradation controls on the LHS of the model would be similar to those protecting against abnormal or transient plant operating conditions, as they demand a similar decisive response from plant operators. However, a degradation control covering the health of preventative and predictive maintenance schedules should also be included on the LHS to cover specific reference to this type of loss. Mitigation barriers on the RHS of the top event would be similar to those discussed earlier.

Nat Cat losses

Like MB, Nat Cat losses are typically covered by a standard property damage policy, subject to specific limits and exclusions, if the insured is operating in an area of known exposures, such as the Gulf of Mexico (windstorm). It would be unusual for a Nat Cat-related loss to be the defining maximum loss scenario for a single asset, though a Nat Cat event could be an initiator to a greater process safety loss. What tends to impact underwriters however is the aggregation of a number of insured assets in the same area. What stands out is that, aside from a decisive planned response from the site in reacting to the transient condition, the other main barrier that can be assessed is the suitability of the design and engineering codes employed during the site's construction, and the quality of the build.

Cyber event

This is only briefly mentioned here to acknowledge its existence as a threat that can lead to a fire / VCE top event, and that there are barriers that can be put in place and assessed. However, this is currently typically outwith the scope of a typical risk engineering survey and covered as part of a separate insurance programme. This could change in the future.

A note on Business Interruption

Business Interruption coverage is typically triggered by any of the above loss types. As a rule of thumb, across the Energy sector, for every £1 paid out by underwriters for a PD loss, a further £2 is paid out for lost revenues, so underwriters will look for evidence of any mitigating barriers that help keep this figure down (i.e. they will look at preventative barriers and controls to stop the loss happening in the first place). The mitigating barriers will typically be a well understood and rehearsed Business Continuity Plan (BCP), and redundancy in plant, site, or international operations. However, individual sites may not be aware of wider corporate BCPs, which suggests that a fundamental review is needed of how the insurance industry assesses those barriers that are most effective in mitigating BI exposures.

Bringing the barriers together

So far, this paper has identified the critical barriers and controls in this model that need to be examined during insurance surveys in order to try and redefine 'insurable risk quality', in response to the most common threats faced by operating sites.

For the survey process to be effective, the barriers should be grouped together to facilitate effective discussions at site level. This can then define the survey agenda based on the barriers to be assessed.

However, to define the health of each barrier, each one has to be broken down into measurable components or elements for use in site discussions.

For each element, there are aspects of recognised, good engineering practice that must be present before that element can be said to have been achieved at its designed and intended level. If it is not at the required level, then a means for scoring how near it is to the required level must be in place, not just in terms of how site procedures state it should work, but how effective its implementation has been, and what the evidence of its effectiveness is on site. As an example, in order to assess the health of the site's PHA barrier, we would need to consider at least the following elements:

- Site PHA policy and schedule
- The suitability of techniques used
- The provision for the review of process safety incidents as part of the study
- Quality of reference information such as P&IDs
- Quality of the review of abnormal operating modes
- Quality of actions generated etc.

Some barriers and controls will have fewer elements than others, but the overall health of the barrier must be no greater than its weakest element – this is a noticeable change in philosophy from current 'risk rating' systems which aggregate element or component scores, thus diluting the presence of poor performers. Scoring of each individual barrier's health would be done using the colour coding system defined earlier in this paper.

A key feature therefore of this process would be to demonstrate individual barrier health and effectiveness by a clear, colour coded status report, to define where the asset's strengths and weaknesses lie, and how likely they are to have a robust response to the main threats that cause losses.

What will the barrier model look like?

An example of what such a model would look like can be given using the first threat mentioned earlier, that of pipe failure due to corrosion. Figure 2 shows a number of directly applicable preventative and mitigating barriers, with those additional barriers relating to process safety shown in Figure 3, and for emergency response in Figure 4 (NB. the condition of each barrier has been randomly assigned merely to highlight the 'traffic light' nature of effectiveness reporting).

Table 7 gives the barriers under current consideration by the author relating to Figure 2:

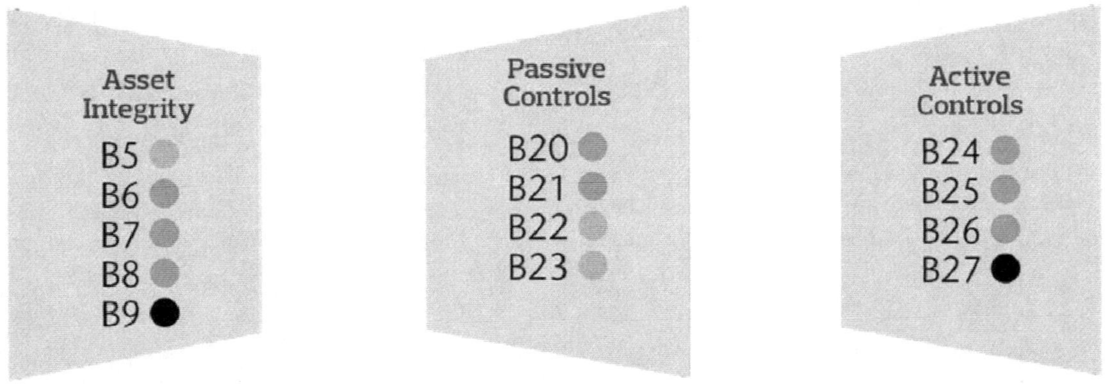

Figure 2. Asset Integrity (LHS) and loss mitigation (RHS) barriers for Passive Controls and Active Controls

Table 7. Relevant barriers under current consideration for the first threat

Barrier / control	**Barrier / control**	**Barrier / control**
B5 – Inspection competence assurance	B20 – Layout	B24 – Remote isolation
B6 – Inspection philosophy	B21 – Construction	B25 – Fire detection
B7 – IOWs	B22 – Passive protection	B26 – Emergency planning
B8 – Materials management	B23 – Inherent redundancy	B27 – Business continuity planning
B9 – Preventative maintenance		

In considering the second and third threats, the model then incorporates barriers specific to the site's Operations team; Figure 3 and Table 8 also include additional barriers from the review of learning from global process safety incidents and losses:

Figure 3. Representation of Operations and Process Safety barriers (LHS)

Table 8. Operational and process safety barriers for the second and third threats

Barrier / control	**Barrier / control**
B1 – Operator competence assurance	B10 – Process Safety Management (PSM)
B2 – Shift handover	B11 – Process Hazard Analysis (PHA)
B3 – Operating documentation	B12 – Management of Change (MoC)
B4 – Control of Work	B13 – Process safety auditing
	B14 – Incident investigation

Finally, Figure 4 and Table 9 demonstrate additional preventative controls and emergency response barriers which might be common across a range of initiating causes:

Figure 4. Representation of preventative (active and passive) controls (LHS) and mitigating (RHS) barriers

Table 9. Suggested further preventative and mitigating barriers across all threats

Barrier / control	Barrier / control
B15 – Engineering standards	B28 – Firewater system
B16 – Controls and safeguards	B29 – Active protection
B17 – Gas detection	B30 – Fire brigades and mutual aid
B18 – ESD	
B19 – Relief systems	

The ratings for the barriers above can be combined to give a summary representation of overall site risk quality, when measured against those threats known to cause energy losses (Figure 5).

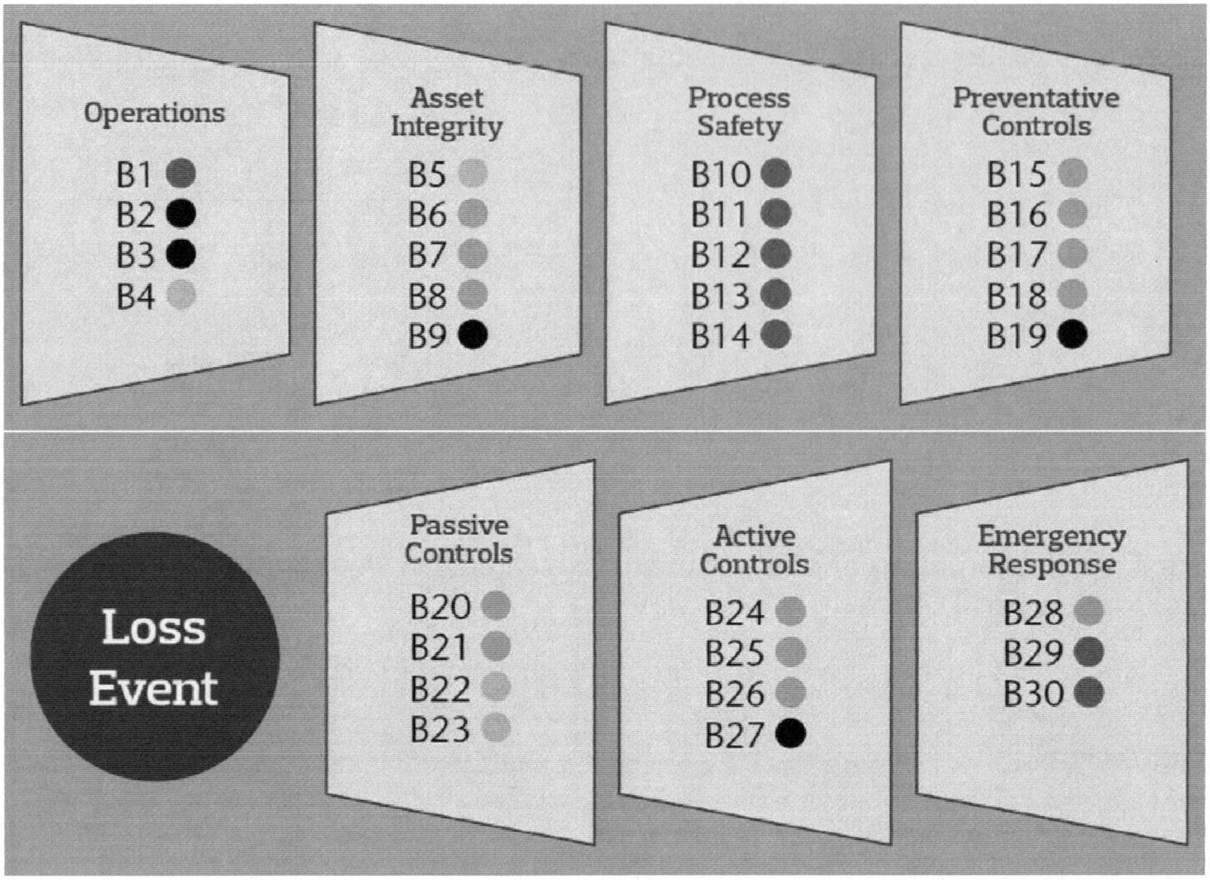

Figure 5. How a combined 'barrier diagram' might look in an Insurance Underwriting Report

Were this for an operating asset, a summary similar to Figure 5 at the front of an insurance underwriting report would quickly flag up concerns about the site's approach to process safety. Those reading the summary would be alerted immediately that the site's preventative maintenance programmes needed improvement (which would be a particular concern for underwriters were the insured purchasing Machinery Breakdown coverage). Based on the above summary, markets and the insured's senior management might also be concerned about the site's ability to respond to an abnormal or transient operation (such as loss of a utility), and depending on certain specific findings, would expect to see a number of recommendations from the survey team to address these concerns. The lack of business continuity planning (B27) might be a concern to insurance markets depending on the level of Business Interruption cover purchased by the insured.

Figure 6 shows another potential summary of findings for a site:

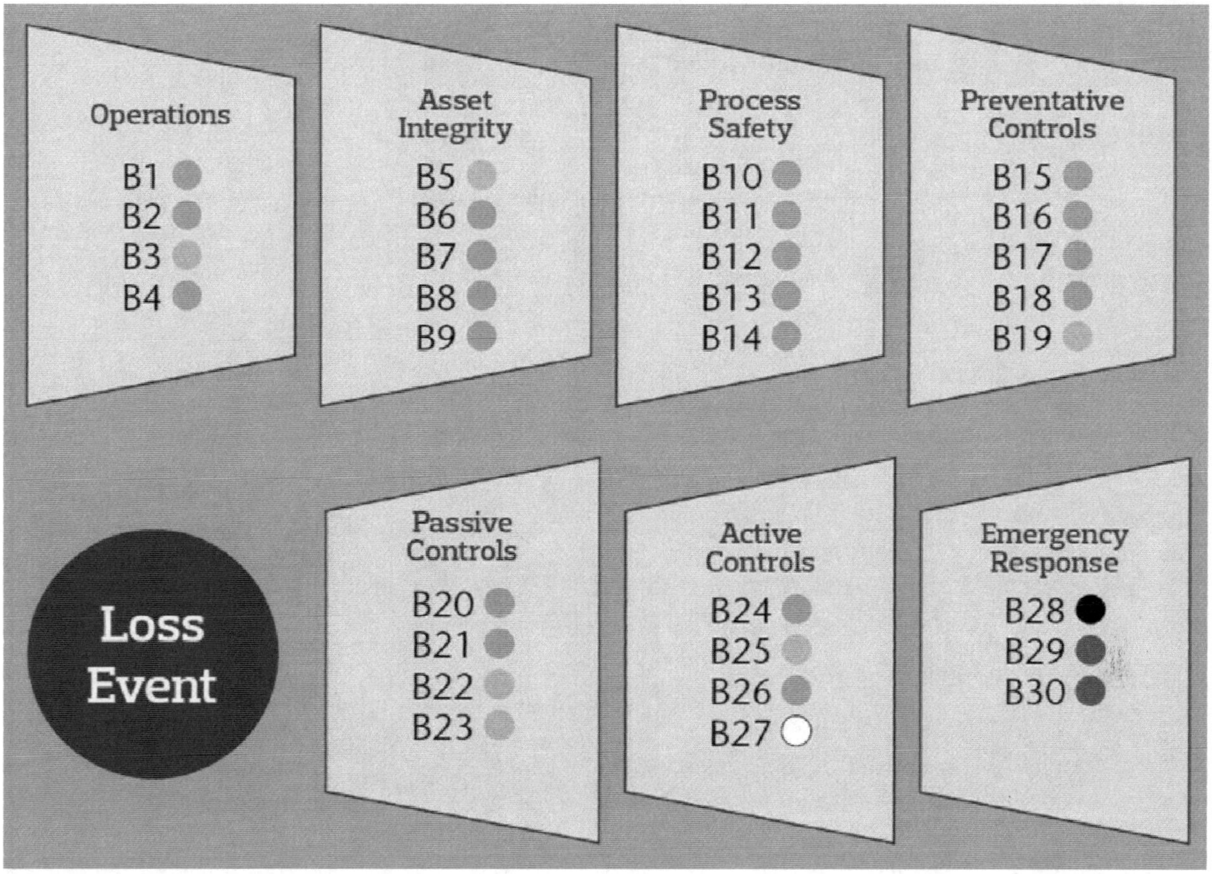

Figure 6. A further example of a possible 'barrier diagram' in an Insurance Underwriting Report

In this example, both the underwriter and insured could take comfort from the site's positive approach to installing and maintaining preventative safety barriers; if BI coverage had not been purchased in this case (note B27 is 'not applicable'), this would likely make this risk more attractive to insurance markets than that represented in Figure 5. Given the strong 'preventative' LHS of the diagram, the underwriter might not be concerned about the failings of the emergency response systems; this would depend however on underwriter 'loss appetite' and on recent loss history.

Conclusion

Aon Global Risk Consulting is proposing a significant change in the way that the energy insurance industry measures and reports risk quality in its insurance underwriting reports, giving a clear summary of the strengths, weaknesses and risks associated with each site, a greater focus on learning the lessons from industry losses, and a clear indicator of the strength of critical safety barriers at sites visited during insurance underwriting surveys for site operators, insureds and underwriters alike.

We believe that this methodology will give a more effective process to give better differentiation of risk quality across those energy assets surveyed.

7 References

[1] Jarvis, R and Goddard, A, September 2016, An analysis of common causes of major losses in the onshore oil, gas & petrochemical industries, Implications for insurance risk engineering surveys, Lloyd's Market Association (LMA).

Integration of Hydraulic Studies in Evaluating Abnormal Plant Conditions during Commissioning and Startup, Maintenance, and Intermittent Operations

Ritche Niño Li
SABIC
Petrokemya, Jubail Industrial City, Saudi Arabia
lirn@petrokemya.sabic.com
ritche.nino@gmail.com

Prepared for Presentation at
American Institute of Chemical Engineers
2020Spring Meeting and 16thGlobal Congress on Process Safety
August 17 – August 21, 2020

Integration of Hydraulic Studies in Evaluating Abnormal Plant Conditions during Commissioning and Startup, Maintenance, and Intermittent Operations

Ritche Niño Li
SABIC
Petrokemya, Jubail Industrial City, Saudi Arabia
lirn@petrokemya.sabic.com
ritche.nino@gmail.com

Keywords: hydraulic calculation, incompressible flow, transient operation, abnormal operation, management of change, pressure safety valve, preventive maintenance, commissioning, startup.

Abstract

Multiple abnormal operating circumstances are encountered in various phases of plant life cycle. During commissioning, maintenance, and irregular batch operations, challenges arise which may prevent in achieving an objective in differing modes of operation. Underestimated more often than not, a common operating problem boils down to issues of fluid transfer of process systems.

In aid of continued operation, hydraulic calculations were performed to attend on familiar dilemmas in a petrochemical plant. In the first case, simulation was utilized to determine viability of a proposed relief alternative pathway during a preventive-maintenance removal of a pressure safety valve (PSV). On the other hand, a separate calculation was also carried out to evaluate a proposal to remove a blower from a series of two units during commissioning of a project-delayed combustible dusts venting system. Lastly, a hydraulic analysis addressed an abnormal high pressure build-up and high flow in an intermittent recipe requirement of a corrosive chemical in a batch process.

The three scenarios are illustrations of process modifications in transient operating mode and abnormal operating conditions. If not decided with a sound technical judgment, process changes may induce operational and safety consequences affecting property and people. The study demonstrates the effectiveness of conducting hydraulic checks to recommend solutions or to assess proposal as part of the management of change proceedings. Based on the calculation summary, key recommendations were generated and implemented accordingly to assist requirements in varying plant conditions.

1 Introduction

The refining and petrochemical process industries have registered more than half of process safety incidents during non-routine and abnormal operations. Even though only 10% of the plant lifecycle is dedicated to transient operations, the bias of recorded process related incidents towards infrequent planned and unplanned operations is a point of concern for operators worldwide (Ostrowski& Keim, 2010; Duguid, 2008).

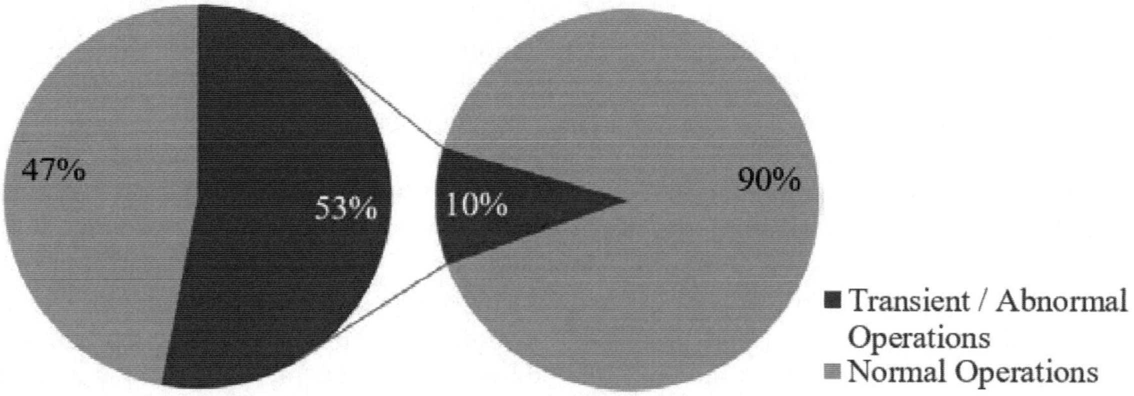

Figure 1. Percentage of Incidents Share Between Transient and Normal Operations

Figure 2. Percentage of Dedicated Time in Plant Life Cycle (Ostrowski & Keim, 2008)

A newly constructed petrochemical plant, which is the facility subject of this research, has encountered transient operations after its mechanical completion. The plant has undergone various types of operating mode from the inception of its commissioning prior to startup, then bridging to a sustainable operation, and up to halting for planned maintenance. Throughout these varying periods, a typical process plant requires to exert extra effort to overcome difficulties and to complete safely.

During the transitory phase and in some cases, changes to process units are inevitable and necessary. An example of this is providing an alternative pathway for pressure/vacuum relief to a storage vessel in the course of the temporary relief valve outage while in operation. In general, modifications in the process, if not managed properly and not critically checked, may impose hazard to the plant rather than providing a solution. One of the familiar incidents mostly associate with hazards of process modification is the Flixborough disaster. Accordingly, the introduction of uncontrolled process change to the reactor system led to the plant's demise.

With this, combining the ideas of process modifications while undergoing transient operations magnifies hazards inherent to this mode of operation. This poses a threat against the aim to preserve a safe condition for the plant, the environment, and its people.

A comprehensive evaluation of process modification is then mandatory to ensure all aspects of the change will fit the requirement of the facility. Through the process of change management, process safety information should go through a recheck.

A way of implementing confirmatory checks during management of change is by performing hydraulic studies, especially utilized to determine anomaly and solution about topics of fluid transfer. This paper promotes the use of the technique to evaluate process related inquiry. The underlying principle is to verify the compatibility of the change proposal to the system that will be modified, and to ensure potential hazards are addressed before the modification.

Moreover, hydraulic analysis can be an inexpensive alternative in answering inquiries to process systems that ranges from complex process system to hazardous materials. It is a powerful and handy tool generally used for line and pump sizing, and various piping component design.

The integration of hydraulic simulation as a tool for a safe mode of operation transition is demonstrated in this paper through the three case studies, differentiated through a define lifecycle of the plant such as commissioning and startup, maintenance, and intermittent operations. The cases are further categorized within the two types of transient operations namely non-routine planned operations, and abnormal or unplanned operations. Depending on the simulation results, a process modification (temporary or permanent) is either supported for implementation or declined to proceed.

2 Review of Literature

2.1 Transient Operations: Non-routine Planned, and Abnormal Operations

Vaughen (2020) defines transient operation as mode of operation wherein the process is undergoing a transition phase and is not at its designed normal operating mode.

Related to this operating mode, Ostrowski & Keim (2010) divided transient mode into two types: non-routine operations, and abnormal operations. Non-routine operations or planned operations are operating events that infrequently occur and have the potential for a loss of primary containment leading to a high consequence incident (Ostrowski & Keim, 2008). Some instances of non-routine operations include startup or shutdown of major units. Non-routine operations may also include continued operations in an equipment configuration beyond the intended design such as non-availability of a major pump, absence of a blower, and out of service of a boiler or a compressor (Ostrowski & Keim, 2008).

On the other hand, abnormal or unplanned operations are operating events that happen within the normal operating mode brought about by process upsets, and process parameter deviation from the normal operating limits (Vaughen, 2020). Abnormal operations have the potential for a loss of primary containment leading to a high consequence incident. Some examples of the unplanned operations involve operating instances beyond equipment's design limit or safe operating limit (SOL), responding to safety critical alarms, and loss of containment (Ostrowski& Keim, 2008).

As an illustration, the content of this paper incorporates case studies for both types of transient operations. These operating events include removal of safety equipment during preventive maintenance, blower outage during commissioning and prior to startup, and abnormal operation during an overpressure scenario in a batch process.

2.2 Commissioning

During commissioning executions, equipment is tested to check its capability to support varying operating conditions based on its design and intended function. Moreover, commissioning procedures are executed before the introduction of hazardous or process chemicals, and employs the use of safe chemicals. When executed well and along with its rigorous approach, commissioning phase is an ideal period to effectively identify design gaps or mishaps incurred during construction phase, and before startup with hazardous material.

Kletz (2009) demonstrated a near-miss incident during commissioning of a new unit. A vessel was isolated and was not pressure tested due to vessel service sensitivity to water as a test medium. The mishap led to a release of a dense vapor cloud. In spite of this, the operator was able to escape without sustaining injury (Kletz, 2009).

In general, a rightly executed commissioning plan involves the use of a disciplined, organized, and professional methodology, to bridge the newly constructed process plant into a fully integrated and operational unit. Ideally, commissioning revolves around the idea of getting it "right first time" in the most safe, efficient, cost-effective and timely manners to achieve startup and production goals (Killcross, 2012).

2.3 Startup

Process safety incidents during startups happen five times as frequent compared to events while at normal plant conditions. Interestingly in a study about a petrochemical facility, startups and other transient operations grab only a fraction of about <10% of the total operations time. The intense and stressful environment to push the plant to a start makes the startup prone to errors and incidents (CCPS, 1995; Ostrowski& Keim, 2008).

Startup may vary from initial startup, restart from a shutdown, startup after turnaround, and startup after extended outage. Initial startup aims to bring a newly constructed plant and equipment to its first operations (CCPS, 1995).

The non-conducive atmosphere of initial startup and commissioning is amplified in a case study included in this paper.

2.4 Maintenance

A huge portion of a plant's lifecycle is dedicated to operations along with maintenance. Preventive maintenance aims to ensure integrity and functionality of the assets, and is conducted on a predetermined time-based schedule (CCPS, 1995). Intuitively, about 50% of the process incidents were attributed to maintenance-related activities that took place during startups, shutdowns, maintenance, or abnormal operations (Kletz, 2009; Duguid, 2008).

Due to the significant amount of time dedicated during maintenance phase, the facility should be well prepared to determine hazards and be able to attend on it. A dilemma during online maintenance for a safety equipment is illustrated part of the research.

2.5 Batch Process

In a batch process, retention time in vessel and piping systems changes with time according to the particular processing requirement, product grade, or recipe. Time variations on reactant or monomer charging, reacting, transferring, recovering, and batch drop make it a challenging to check and troubleshoot processes (CCPS, 1999). A holistic approach integrating the operating philosophy of a batch process that is linked to a continuous process should be accounted as part of the process design. Otherwise, a mismatch of operating concept may lead to operability issues and safety concerns.

Kletz (2009) and Duguid (2008) have attributed 40% of process-related incidents to a less than adequate process design. In the report, safe design features were not applied in spite of its presence and availability industrywide.

This paper also reflects an instance when a batch process operates in interface with a conflicting design from a continuous process.

2.6 Hydraulic Calculation and Incompressible Fluid Flow

Hydraulic analysis can be an inexpensive alternative in answering inquiries to process systems that ranges from complex process system to hazardous materials. It is generally use for line sizing and determination of pump specification and equipment sizes as accurately as required for practical purposes (Moran, 2019).

Moreover, the simulation can also be used for troubleshooting issues in fluid networks. As an example, during its initial commissioning, once a pump is energized, its performance is tested to know its capability to deliver process parameters according its purpose. After individual run, its behavior will be tested with the piping systems and associated equipment. If unsatisfactory, troubleshooting onsite can be take place or a simulation can be performed if more practical. It is a powerful and handy tool in performing pipe and pump sizing, and various piping component design.

Nitrogen, air, and diluted potassium hydroxide solution, are the fluid separately involved in the three case studies. Intuitively, the aqueous solution of caustic potash was modeled through principles of incompressible flow.

During the simulation, the mach numbers, M, of nitrogen and air within their respective systems were found both below 0.3. When the mach number of a flowing fluid in any point is less than 0.3, the density variations due to pressure changes can be neglected, hence the gas density may be assumed as constant. As a general rule, gas flow with $M < 0.3$ is regarded as an incompressible flow (Pijush & Cohen, 2008).

In this paper and for calculation simplicity purposes, the flow of nitrogen and air are analyzed through incompressible flow calculations, therefore neglecting compressibility effects.

3 Hydraulic Analysis in Managing Modifications to the Process

3.1 Facility Description

Independent hydraulic analyses were carried out in response to operational requirement, troubleshooting activities, or improvement pursuit in one of the polymer plants that manufactures an engineering thermoplastic with various industrial and household applications. The facility is situated in Saudi Arabia, part of a petrochemical complex that manufactures olefins, basic petrochemicals, and other thermoplastic resins.

3.2 Problem Statements and Study Objectives

During the various plant lifecycle from commissioning and startup, operations until maintenance, the facility encountered varying challenges to run the plant in safe condition. Below case studies, describe the scenarios wherein hydraulic analyses were applied to facilitate decision-making.

3.2.1 Study 1: Maintenance

Scenario:

A conservation vent valve (PSV shown in Figure 3) was required to be removed during a preventive maintenance check. During the instance of removal, caustic storage tank was already filled-in with some level of NaOH solution.

The safety valve was installed for overpressure and vacuum protection of the NaOH solution storage tank. The governing cases of PSV by design are nitrogen blanketing control valve failure for overpressure scenario, and blocked discharge of inbreathing valve PCV_{in} for vacuum case.

Outage of the pressure safety valve for ex situ testing while operating the tank may impose lack of inbreathing and outbreathing protection for the tank. The removal of PSV may cause sustained tank damage in case of an overpressure or implosion scenario. In this paper, conservation vent valve is designated as pressure safety valve (PSV) for simplification and identification purpose.

Proposal:

During the preventive maintenance of the conservation vent valve, the breathing requirements of the tank should be fulfilled. To facilitate tank inbreathing and outbreathing conditions, continuous flow of nitrogen is introduced to the tank from PCV_{in} and released through the ¾" drain valve.

A hydraulic study was commissioned to determine the capability of the temporary arrangement to facilitate control valve failure and blocked discharge scenarios. From the PCV_{in} outlet, nitrogen supply is injected to the tank. For the blanketing outlet, nitrogen passes through antisiphon holes located at the top of the dip pipe and is ejected at the ¾" drain valve upstream of PCV_a.

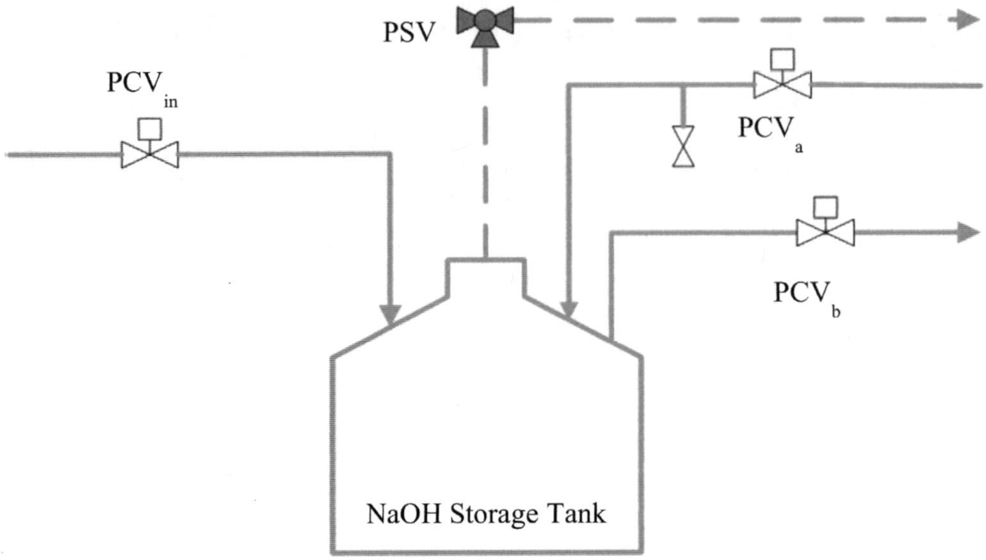

Figure 3. Conservation vent valve (PSV) will be taken for preventive maintenance.

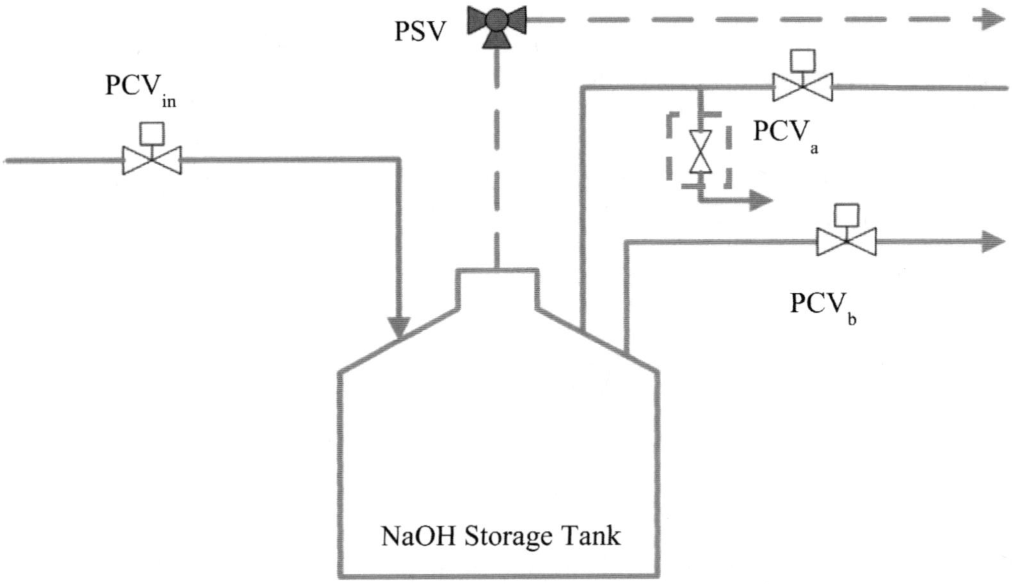

Figure 4. To facilitate tank inbreathing and outbreathing requirements, continuous flow of nitrogen will be introduced to the tank from PCV_{in} to the drain valve.

3.2.2 Study 2: Commissioning and Startup

Scenario:

During the commissioning of a vent system, Blower 1 sustained damage and required overhauling. The timeline of this event happened after the mechanical completion, part of the construction phase, bridging to commissioning phase.

Proposal:

A proposal was brought out to install bypass line across the centrifugal exhaust fan to run on a single Blower 2. If successful, the proposition will save the plant from a delayed startup.

However, the technical background of the proposal was challenged hence a confirmatory simulation was required to support the contradicting side. Since the nature of the inquiry was about fluid movement, hydraulic simulation was conducted to aid in decision-making. If not proven sufficient, the plant was required to wait until the new centrifugal exhaust fan / blower arrives.

By design, Blower 1 and Blower 2 convey combustible powder from the dryer to separator wherein wet air is also separated. Conveying air is further treated in the process.

Safety Consideration:

At low airflow, bigger particle sizes will settle in the piping vent systems. Relatively large dust particles introduce explosion risk when it adhered to surfaces and dryer internals. By design, there should be no settling of dust particles in the piping system.

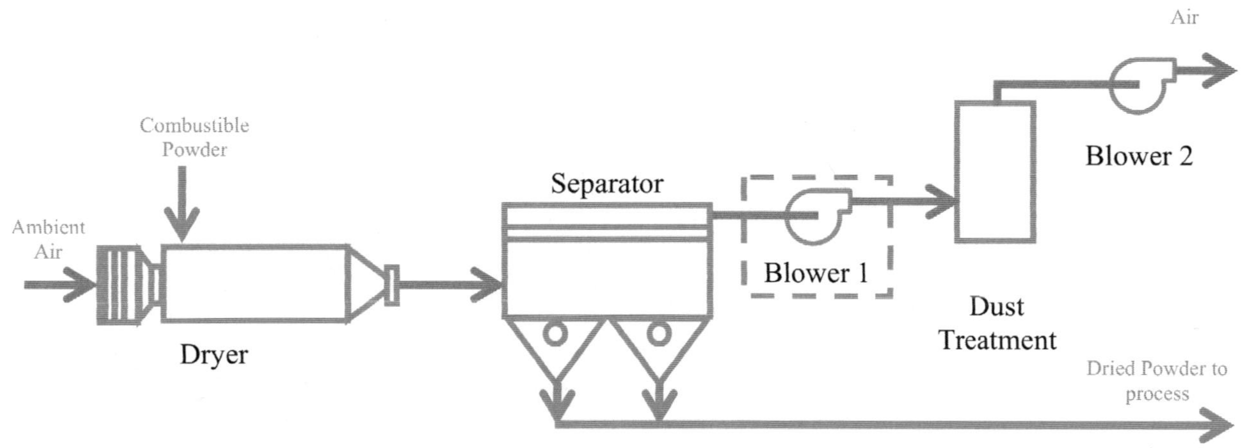

Figure 5. Schematic Diagram of Vent System by Design

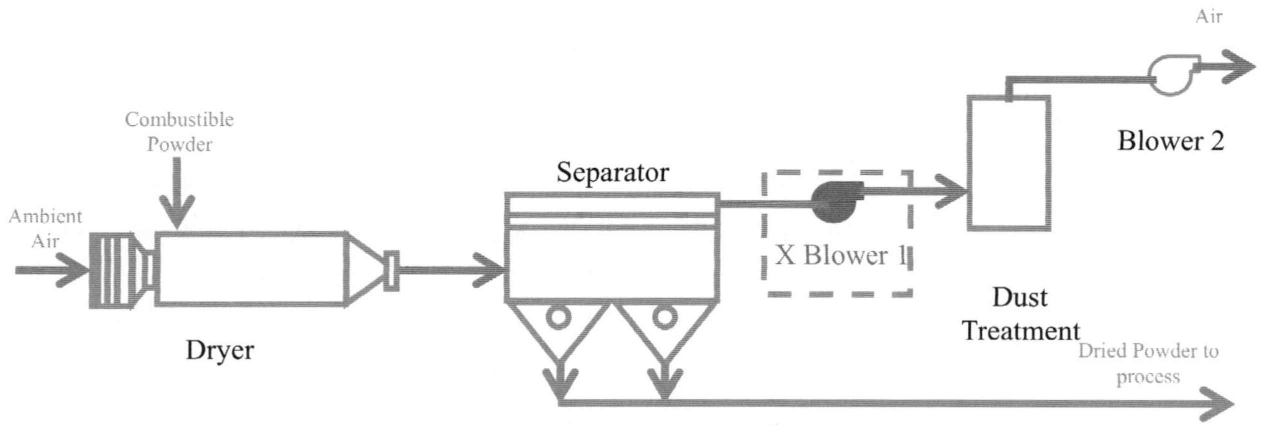

Figure 6. Proposed Schematic Diagram of Vent System wherein Blower 1 will be Bypassed

3.2.3 Study 3: Intermittent operation

Scenario:

Transfer pump delivers potassium hydroxide solution to reactors, solution make-up tank, feed tank, and storage tank. From the slightly pressurized and nitrogen blanketed storage tank, the KOH solution is delivered to the users via the pump, filters and main header. The recirculation line passes through a restriction orifice back to the KOH storage tank.

However, the system experienced excessive flow to one of the users when a batch was initially started leading to operational upset downstream. A probable flow and pressure imbalances in the system was the point of inquiry of the hydraulic simulation.

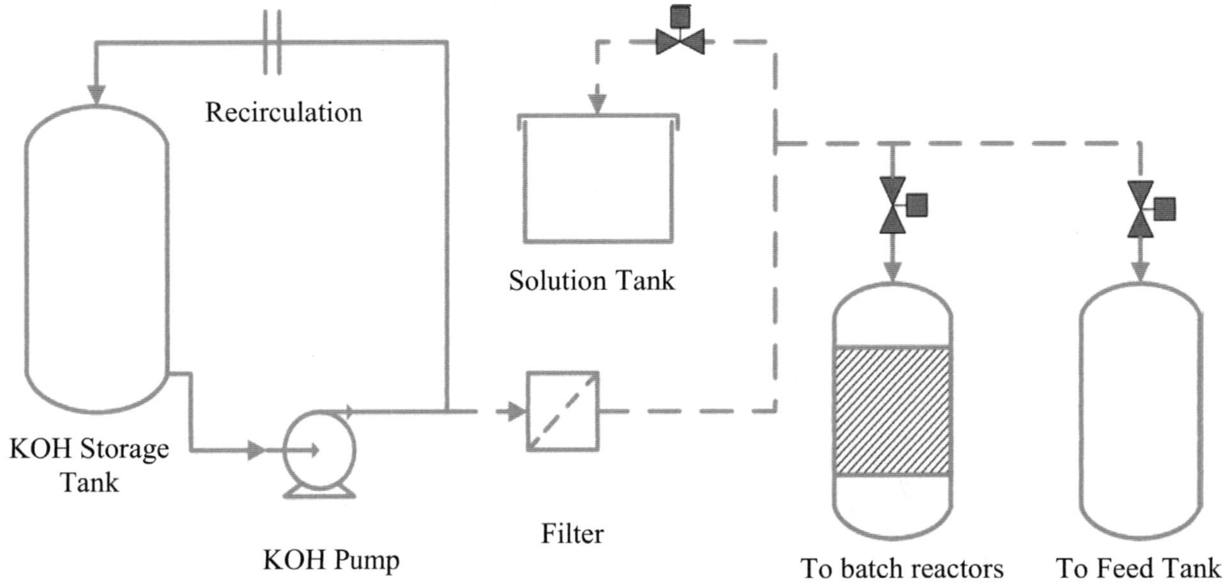

Figure 7. Schematic Diagram of KOH solution system wherein the header encounters blocked discharge when there is no demand during intermittent operation.

3.3 Problem Statements Summary

Table 1 below shows a brief summary of the background of the cases in varying plant conditions during maintenance, commissioning and startup, and batch operations including the respective problem statement or event drivers. The tandem of initiating cause (e.g., unavailability of vent system) happening at a specific operating phase of the plant (e.g., during plant initial startup) subjects the plant into an abnormal operation that requires acknowledgement to address the associated hazards.

Table 1. Problem Statement Summary

Description	Case Study 1	Case Study 2	Case Study 3
Plant Timeline / Status(A)	Online Maintenance	Commissioning and Startup	Normal Operation / Intermittent Operation
Initiating Cause(B)	Removal of conservation vent valve	Unavailability of damaged Blower 1	Closed shutoff valves to all users
Problem Statement(AxB)	Unavailability of PSV during ex situ test when the tank is in service.	Unavailability of vent system will prevent startup. Vent system is a critical path to plant initial startup.	Overpressure to KOH header due to blocked discharge, or no demand from users.
Related Asset / System	Caustic tank, inbreathing and outbreathing system	Vent system and Centrifugal Exhaust Fan 1	KOH header to users
Safety Consideration	PSV is used for tank overpressure and vacuum protection	At low airflow, bigger particle sizes will settle. Settled particles can disintegrate and initiate explosion.	The service is corrosive if exposed to personnel during relief through the PSV.
Potential Consequence	The removal of PSV may cause tank damage due to overpressure or implosion leading to potential financial impact and personnel injury.	Internal dust explosion may occur from accumulation of unstable powder due to low flow leading to potential fatality.	Overpressure in the piping may lead to release of corrosive chemical and upset to users hence potential personnel injury.

4 Hydraulic Calculations

4.1 Fulfilling Overpressure and Vacuum Requirements of NaOH Storage Tank during PSV Outage

4.1.1 Objective

Determine if ¾" drain gate valve downstream of PCV_a is enough for inbreathing and outbreathing requirement of NaOH storage tank during the temporary out of service of PSV.

4.1.2 Hydraulic Calculation Summary

Table 2 below shows the summary of the input and output of the hydraulic analysis. The hypothesis that was tested, whether a ¾" drain can be a temporary substitute for the venting requirements of the tank, was proven acceptable. The tradeoff to this arrangement is a possible dust contamination or moisture ingress into the caustic tank; hence, a quality issue may arise.

Table 2. Hydraulic Calculation Sheet – Case 1

Topic	Fulfilling Overpressure and Vacuum Requirement of NaOH Storage Tank on PSV Outage
Service	Nitrogen/Air Inbreathing and Outbreathing to/from NaOH Storage Tank
A. Objective	Determine if 3/4" drain gate valve downstream of PCV_a is enough for inbreathing and outbreathing requirement of NaOH Storage Tank during the temporary out of service of PSV.
B. Basis, Data and Assumptions	1. PSV Relief Calculation a. Overpressure Governing Case: Control valve failure (PCV_{in}) Maximum Relief Flowrate = approx. 100 kg/h Inlet Pressure = approx. 500 kPa_g b. Vacuum Governing Case: Blocked discharge (PCV_{in}) leading to vacuum Maximum Relief Flowrate = approx. 100 kg/h Inlet Pressure = atm 2. NaOH Storage Tank Nozzle Design Nozzle N3 Pump minimum flow nozzle has a dip tube with antisiphon holes. 3. Fluid Properties Air @ 25°C Nitrogen @ 25°C
C. Procedure	1. Determine if piping hydraulic profile including 3/4" gate valve as discharge has no significant pressure losses to be overcome for safe relieving and venting during overpressure and vacuum.
	2. Determine if equivalent area of antisiphon holes, will be sufficient as relief outlet/inlet.
D. Results and Conclusions	1. The maximum relief flow the piping can accommodate (based on the piping isometric) is about 200 kg/h during depressurization. This is greater than the maximum relief rate of both overpressure and vacuum scenario. Because of this, there is no significant piping and fittings pressure losses that can restrict safe relief venting using the drain downstream of PCV_a.

	2. Antisiphon at nozzle N3 has combined area equivalent to a 1 in. pipe. No significant impact on pressure losses brought about by antisiphon holes at dip tube that can restrict safe relief venting.
E. Recommendation	1. During this particular PSV outage, ensure 3/4" drain gate valve downstream of PCV_a is not blocked and fully open for safe relieving.
	2. Vent to safe and open area to avoid hazard of asphyxiation due to continued venting of nitrogen.
	3. Ensure antisiphon holes are not blocked.
	4. Reinstate PSV and piping as per P&ID immediately after calibration activity.

4.2 Testing the Capability of Vent System to Fulfill Duty while Running on a Single Blower

4.2.1 Objective

Determine if Blower 2 can safely handle the dust/air venting requirement when Blower 1 will be remove from the service.

4.2.2 Hydraulic Calculation Summary

Table 3. Hydraulic Calculation Sheet – Case 2

Table 3 below shows the summary of the input and output of the hydraulic simulation. The tested hypothesis, whether Blower 2 can permanently support the air movement requirement of the vent system, was not acceptable. The proposal may impose explosion hazard to the piping system causing damage to property and injury to personnel.

Topic	**Evaluating a Proposal to Remove Blower 1 from a series of two fans**
Service	Air, water vapor, and dust movement from dryer to dust treatment via Blower 1 and Blower 2
A. Objective	Determine if Blower 2 can safely handle the dust/air venting requirement when Blower 1 will be remove from the service.
B. Basis, Data and Assumptions	1. Fluid Properties Air, Water Vapor, combustible powder Normal flow @ 60°C 2. Inlet Pressure = atm 3. Elevation based on site data 4. Relatively large dust particle introduces explosion risk when it

	adheres to surfaces and dryer internals. By design there should be no settling of dust particles in the piping system.
C. Procedure	1. Determine the flowrate across the system when both Blower 1 and Blower 2 are running at normal operating condition. a. Set the inlet pressure = atm, at respective destination pressure, utilize the pump curves and piping systems.
	2. Determine the flowrate across the system when only Blower 2 is running. a. Set the inlet pressure = atm, utilize the pump curve and piping systems.
D. Results and Conclusions	1. When two blowers are running at normal operating conditions of: inlet pressure = atm to respective destination pressure the normal air flowrate is acceptable i.e., dust particles will not settle.
	2. However, when operating at a single Blower 2 (with bypassed Blower 1) at normal operating conditions of: inlet pressure = atm to respective destination pressure the normal air flowrate is not acceptable i.e., dust particles will not settle based on empirical data.
E. Recommendation	1. Neither Blower 1 nor Blower 2 can withstand total systems losses alone at normal condition. Two blowers in series must be in place to support safe movement of air, water vapor, and combustible dust.
	2. Centrifugal Exhaust Fan / Blower 1 shall be returned on service before initial startup of the vent system.

4.3 Preventing Overpressure in an Intermittent Process from KOH Header to a Batch Reactor

4.3.1 Objective

Determine the reason of buildup pressure in the KOH system. Check whether the design of recirculation line is adequate to prevent overpressure in the line.

4.3.2 Hydraulic Calculation Summary

Table 4. Hydraulic Calculation Sheet – Case 3

The following Table 4 shows the summary of the input and output of the hydraulic study. The restriction orifice located in the recirculation line back to the KOH tank was found underrated.

Topic	**Preventing KOH Header Overpressure in a Batch Process to Reactors and Makeup Tank**
Service	Transfer KOH solution from storage tank to users like batch reactors, KOH feed tank and solution tank via KOH pump.
A. Objective	Determine the reason of buildup pressure or flow imbalance in the KOH system. Check whether the design of recirculation line is adequate to prevent overpressure in the line.
B. Basis, Data and Assumptions	1. Fluid Properties KOH solution @ 25°C 2. Inlet Pressure = atm 3. Destination Pressure: KOH Feed Tank = atm Solution makeup tank = atm Storage tanks = slightly above atm Batch reactors = slightly above atm 4. Elevation based on site data
C. Procedure	1. Determine the required flowrate across the recirculation line during the normal operations. a. Set the inlet pressure, destination pressures accordingly, utilize the pump curves, piping systems, and restriction orifice size.
	2. Determine the adequacy of recirculation line to accommodate pressure buildup in the KOH header when there are no demands from the users. a. Set the inlet pressure, shutoff valves to users in closed position, utilize the pump curves, piping systems, and restriction orifice size.
D. Results and Conclusions	1. The required flowrate across the restriction orifice during normal operation is about 2 tons/h.
	2. When there is no demand from KOH header the minimum recirculation flowrate should be adjusted to approx. 5 tons/h.

E. Recommendation	Adjust the recirculation line to approx. 5 tons/h by resizing the orifice size to a bigger opening. A control valve in the recirculation line can be a long-term solution to prevent the pressure build up in the header.

5 Results and Discussions

5.1 Hydraulic Calculation Results

Table 5 exemplifies a summary of the simulation studies in each case study.

Table 5. Problem Statement and Recommendations Summary

Description	Study 1	Study 2	Study 3
Plant Lifecycle	Maintenance	Commissioning and Startup	Normal Operation / Intermittent Operation
Transient Operation	Non-routine Planned Operations	Non-routine Planned Operations	Abnormal or Unplanned Operations
Problem	Unavailability of conservation vent valve	Removal of Blower 1 from a series of two blowers	Design Inadequacy
Hydraulic Calculation Recommendation	Provide alternative overpressure / vacuum protection pathway. The use of ¾" drain valve can be used as relief outlet. **Accepted.**	Decline the proposition. The centrifugal fan must be returned on service before initial startup. **Rejected.**	Improve the process by increasing the size of restriction orifice. **Accepted.**
Type of Change	Temporary	Permanent	Permanent

5.2 Discussions: Transient Operations, Management of Change, and Hydraulic Studies

Restating the facts, almost half of the process incidents accounts for maintenance-related activities emerge during startups, shutdowns, maintenance, or abnormal operations (Kletz, 2009; Duguid, 2008). In a bigger picture, more than half of process safety incidents happen during non-routine and abnormal operations (Ostrowski & Keim, 2008; Duguid, 2008). Therefore, the statistics reiterates that a greater concern is posted by the concept of transition period operation alone.

By injecting process modification during the transient period, the risk is magnified bringing in the consequences mentioned in Table 1 Problem Statement Summary that ranges from potential operational upsets, financial impact, personnel injury, and fatality.

The recognition of hazard is critical in the course of action of the change review process. If by any chance that safety consideration was not credited in the decision-making, a catastrophic event may have happened. To demonstrate, Case Study 2 during commissioning and startup is the most critical of the three case studies because of the potential fatality consequence from an internal dust explosion in the dryer. The elements that can play around during this period include the naturally stressful commissioning environment to startup the plant, delayed commissioning schedule, inherently hazardous and combustible service, and continued operations in anon-standard equipment configuration beyond the design intention (Blower 1 removal). After the hazard was recognized, hydraulic calculation was promoted to determine whether removing Blower 1 was a safe proposal to be implemented. The result of the simulation rejected the idea (Table 5) and recommended to wait for a new centrifugal fan to arrive even in the expense of a further delayed schedule.

5.3 Effectiveness of Hydraulic Studies in Aid of Process Modifications

By performing hydraulic studies, hypothesis and suggestions were validated in solving or troubleshooting a particular process issue. This paper promotes the use of process engineering tools such as hydraulic simulation as a way to check proposals as part of management of change proceedings (MOC). The multidisciplinary team participating in the MOC must be able to recognize the underlying problem, hypothesize, and test the proposition by conducting simulation studies as deem required.

6 Conclusion

By injecting process modification during the transient period, the risk in the process is amplified bringing in the consequences that ranges from potential operational upsets, financial impact, personnel injury, and fatality.

As a resolution, hazard recognition is critical in the course of action of the change review process. If by any chance that safety consideration was not credited in the decision-making, a catastrophic event may have happened.

Moreover, conducting hydraulic simulations is a basic engineering capability that will able to support varying plant operating conditions. While this paper aims to demonstrate the efficacy of hydraulic calculations in solving various engineering questions, the study particularly focused on solving problem related to fluid transfer, a common issue across the lifecycle of the plant. The study exemplified the use of this engineering capability during transient operations from commissioning and startup, preventive maintenance, and abnormal operations from a batch process.

Finally, the criticality of conducting a thorough study before acting on a potential change cannot be overemphasized, especially during transient operation wherein an imbalance of process safety incidents reside during this period.

7 References

[1] Ostrowski, S. W. & Keim, K. K. (2010, June 23). Tame Your Transient Operations: Use a special method to identify and address potential hazards. *Chemical Processing*. Retrieved from https://www.chemicalprocessing.com/articles/2010/123/.

[2] Duguid, I. M. (1998). Analysis of past accidents in the process industries. *Loss Prevention Bulletin*. No. 142, No. 143, and No. 144. Institution of Chemical Engineers.

[3] Ostrowski, S. W. & Keim, K. K. (2008, October 28). A HAZOP methodology for transient operations. Paper Presented at the Mary Kay O'Connor Process Safety Symposium, College Station, Texas. Retrieved from http://pscmembers.tamu.edu/wp-content/uploads/15-ostrowski.pdf.

[4] Vaughen, B. K. (2020, March 14). Understanding and managing the risk during transient operations. *Process Safety Progress*. Retrieved from https://doi.org/10.1002/prs.12146.

[5] Kletz, T. (2009). *What went Wrong? - Case Histories of Process Plant Disasters and How They Could Have Been Avoided (5th Edition)*. Elsevier. Retrieved fromhttps://app.knovel.com/hotlink/toc/id:kpWWCHPPD1/what-went-wrong-case/what-went-wrong-case

[6] Killcross, M. (2012). *Chemical and Process Plant Commissioning Handbook: A Practical Guide to Plant System and Equipment Installation and Commissioning*. Oxford: Butterworth-Heinemann.

[7] CCPS. (1995). *Guidelines for Safe Process Operations and Maintenance*. Center for Chemical Process Safety, American Institute of Chemical Engineers, New York.

[8] CCPS. (1999). *Guidelines for Process Safety in Batch Reaction Systems*. Center for Chemical Process Safety, American Institute of Chemical Engineers, New York.

[9] Moran, S. (2019). *Applied Guide to Process and Plant Design* (2nd Edition). Elsevier.

[10] Pijush, P. K. & Cohen, I. M. (2008). *Fluid Mechanics* (4th Edition) - 16.1.3 Useful Thermodynamic Relations. (pp. 714-716). Elsevier.

[11] Nakayama, Y. (2018). *Introduction to Fluid Mechanics* (2nd Edition) - 13.3 Mach Number. (pp. 261). Elsevier.

[12] Shaughnessy Jr., E. J., Katz, I. M. & Schaffer, J. P. (2005). *Introduction to Fluid Mechanics* (OUP Edition) - 10.10.2 Incompressible Fluid and Incompressible Flow. (pp. 640). Oxford University Press.

586242 Occurrence of Black Powder in Gas Lines and Its Control

Abstract:

Occurrence of Black Powder in Gas Lines and its Control

Objective / Scope:

Black powder contamination in gas stream has been a matter of great nuisance. Its occurrence causes operational, maintenance, HSE, reliability & integrity issues and various undesirable consequences like product contamination, production loss, economic loss, and sustainability concerns (e.g. due to flaring), for both the operators and the downstream customers. The objective of this paper is to demonstrate the success of immediate / short-term measures taken to minimize its occurrence and subsequently its removal.

Methods, Procedures, Process:

Several technical reports and research works are found in the literature pertaining to black powder, still, need of more awareness has been felt by researchers / industry. In order to contribute for such cause, the present paper attempts to review the problem of occurrence of black powder and various strategies that are adopted to handle the associated issues, in general, and shares experience of minimizing the occurrence of black powder by revisiting the chemical injection dosages, frequent pigging operations and other methods at one of the oilfield areas in Kuwait, in particular classified as short-term / immediate measures.

Results, Observations, Conclusions:

Pipelines and connected systems are designed considering solid / liquid contaminations expected in the gas stream; but such provisions can become insufficient if occurrence of black powder starts, later, during operation. In view of involved complexity, problematic feature & uncertainties, operating companies may implement a black powder management system incorporating prevention, mitigation and removal strategies. Accordingly, when occurrence of black powder, along with liquid carry-over, in sour gas streams at one of the oilfield areas in Kuwait was observed, various short-term strategies in hand were reviewed such as adjusting chemical injection dosages, frequent pigging operations. This results in remarkable reduction in the occurrence of black powder and accelerating its removal from gas lines. On the basis of the results it is observed / concluded that the immediate short-term methods are effective in minimizing the occurrence of black powder in gas lines and its removal.

Novel / Additive Information:

The paper is expected to create awareness and enrich knowledge base regarding the occurrence of black powder in the middle-east region and its immediate / short-term handling solutions. This shall facilitate the Oil & Gas industry to address such problem of gas contamination and, in turn, handling its associated problems, like gas flaring, by attaining enhanced solutions based on such information.

"We Must Be Safe, Right? We Had a PHA": Effectively Crafting and Implementing PHA Recommendations

Emma Qingyuan He
ACM Facility Safety Inc.
926 5th Ave SW, Calgary
ehe@acm.ca

Richard Carter
ACM Facility Safety Inc.
926 5th Ave SW, Calgary
rcarter@acm.ca

Prepared for Presentation at
American Institute of Chemical Engineers
2020 Spring Meeting and 16th Global Congress on Process Safety
Houston, TX
March 30 – April 1, 2020

"We Must Be Safe, Right? We Had a PHA": Effectively Crafting and Implementing PHA Recommendations

Emma Qingyuan He
ACM Facility Safety Inc.
926 5th Ave SW, Calgary
ehe@acm.ca

Richard Carter
ACM Facility Safety Inc.
926 5th Ave SW, Calgary
rcarter@acm.ca

Keywords: PHA Recommendation, SMART Recommendation, Prioritization, Process Hazards Analysis, Action Items

Abstract

Recommendations are the heart and soul of a Process Hazards Analysis (PHA). PHAs are only as effective as the action taken to implement the recommendations made during the study. They are the key action items made during the PHA session and usually are made in response to an identified risk that needs to be mitigated or investigated. Therefore, if they are not followed through on, the value of the PHA study is compromised, not to mention the safety of the facility.

Valuable information is contained in the PHA worksheets, but because they are long and difficult to read it is often only the action items list that people will see once the session has finished. Therefore, each action item needs to clearly describe the uncovered risk and how to approach mitigating it.

Time and resources are limited, however, so not all recommendations can be implemented. Sometimes further investigation can show that the recommendation is not required, the risk can be tolerated, or an alternative approach would be better. It is vital that recommendations are carefully worded so that they can be correctly followed up on after the PHA study, and management can make effective decisions. Recommendations that are poorly constructed or insufficiently backed up by the analysis may be discounted without due process, and those that are overly restrictive may constrain the close-out action to a sub-optimal solution.

This paper will discuss the ways in which PHA recommendations are used and how that is supported during the PHA, common pitfalls when generating and addressing recommendations, and best practices for crafting recommendations during the PHA to ensure the risk to workers, assets and the environment is as low as reasonably practicable.

1 Introduction

Recommendations are the heart and soul of a Process Hazards Analysis (PHA). They are intended to provide additional protection layers to mitigate risk, identify future opportunities to improve operability, and potentially make the processing unit inherently safer. A PHA recommendation is an action item that is generated during a PHA session to address a specific concern that was raised during the study, usually to mitigate the risk of a particular hazardous scenario for which the risk was deemed to be too high. However, recommendations are only valuable if they are implemented.

There are numerous reasons why it is difficult to implement all PHA recommendations after a PHA study is completed. First, PHA stakeholders often find the recommendation wording vague, unclear, and lacking sufficient information to explain the rationalization behind the recommendation. The missing information may be the detailed description of the hazardous scenario, the risk level associated with the hazardous scenario, or any existing safety protections that are available to mitigate the risk. Such detailed information is only accessible by reading the study worksheets, which may range from tens to hundreds of pages, and require an experienced PHA participant to be able to find the relevant scenarios.

The second biggest challenge when implementing recommendations is how to prioritize them; some PHA will generate hundreds of recommendations. Prioritization of these recommendations becomes crucial when considering the limitations inherent in availability of human resources, financial constraints, and the turnaround schedule of processing unit in a facility. However, the criteria of prioritization differ between PHA stakeholders due to the differences in individual perspective on what is most important and what is most urgent.

The third most common challenge is to maintain a record of the recommendations and keep track of their status and completion throughout the lifecycle of the facility. If there is no consistent and robust method of tracking the progress on recommendations, the individual lists of recommendations from each PHA are likely to be mis-placed or overlooked in the huge database of process safety information for the entire facility. During the typical five-year gap between two rounds of PHAs, an effective tracking system for recommendations will lead to a more effective revalidation PHA, and overall safer operation of the facility.

2 Crafting of PHA Recommendations

The quality of a PHA recommendation is largely affected by the amount of information it contains to support the PHA stakeholders in making the best decision of when, how, and whether it is to be implemented. A well-crafted recommendation should be documented using a modified version of the SMART goal-setting method:

- Stand-alone,
- Measurable,
- Assignable,
- Risk-reducing, and
- Trackable.

These aspects will be expanded on in this section.

2.1 Stand-alone

A recommendation is "stand-alone" when the wording used in the recommendation description contains adequate information to explain exactly what the potential hazardous scenario in question is, and the proposed path forward for mitigating or eliminating that scenario. In other words, it describes both "what" and "why." This supports anyone involved in decision making for the recommendation, or implementing the recommendation, by providing the information they need without having to go back to the PHA worksheets.

PHA recommendations are usually reviewed with the PHA stakeholders a few weeks after the session is completed. By that time, all the memories of the discussion about the intention of, and reason for, the recommendation are already faded. Without a stand-alone description of the recommendation, the following question is likely to be asked "which hazardous scenario is this recommendation trying to mitigate?"

Table 1 List of Not Stand-Alone Recommendations

Recommendations	Place(s) Used in HAZOP Consequence #	Drawing Number	Max Risk Level (Before Action)	Max Risk Level (After Action)
1 Ensure PSV-1001 on Stabilizer Feed Drum is sized for gas blowby.	1.9.1.1, 3.9.1.1, 29.9.1.1, 29.9.5.1, 29.9.6.1, 31.9.2.1, 36.9.2.1, 36.9.3.1	P&ID 1	High	Low
2 Relocate PT-1205 sensing point from V-1202 2nd Stage Suction Scrubber to V-1205 2nd Stage Suct. Dampener.	2.1.6.1, 2.1.7.1, 2.1.8.1, 2.7.1.1, 2.7.3.1, 2.7.4.1	P&ID 2	Medium	Medium

For example, by reading Recommendation #1 in Table 1, one will question why the PSV sizing case needs to be revisited on the Stabilizer Feed Drum. Recommendation #2's description is also not stand-alone: without looking up the study worksheets using the referenced HAZOP consequence number, one will not understand the purpose of relocating the PT-XXXX's sensing point.

Table 2 List of Stand-alone Recommendations

Recommendations	Place(s) Used in HAZOP Consequence #	Drawing Number	Max Risk Level (Before Action)	Max Risk Level (After Action)
1 Ensure PSV-1001 is sized to protect the V-1001 Stabilizer Feed Drum against gas blowby from the following vessels: - V-1000 Slug Catcher, V-1201 Inlet Compressor Suction Scrubber, V-2000 TEG Contactor.	1.9.1.1, 3.9.1.1, 29.9.1.1, 29.9.5.1, 29.9.6.1, 31.9.2.1, 36.9.2.1, 36.9.3.1	P&ID 1	High	Low
2 Relocate PT-1205 sensing point from V-1203 2nd Stage Suction Scrubber to V-1205 2nd Stage Suct. Dampener, such that the PT is able to read low suction pressure if the demister pad in V-1205 2nd Stage Suction Scrubber is plugged.	2.1.6.1, 2.1.7.1, 2.1.8.1, 2.7.1.1, 2.7.3.1, 2.7.4.1	P&ID 2	Medium	Medium

On the contrary, the versions of these recommendations shown in Table 2 include the hazardous scenario, or the specific intention of the recommendation, so that the assigned responsible personnel will understand why the recommendation is proposed, and what it is intended to do. This allows them to make better decisions on whether to implement the action as described, implement a modified version of the action, or reject it.

2.2 Measurable

A "measurable" recommendation is one that identifies exactly what it is achieved once the recommendation is implemented. To craft a measurable recommendation, the PHA team needs to be cautious to choose a strong action verb to start the recommendation description. See the example shown in Table 3.

Table 3 Example of Non-Measurable Recommendation

Recommendations	Place(s) Used in HAZOP Consequence #	Drawing Number	Max Risk Level (Before Action)	Max Risk Level (After Action)
3 Consider moving LSHH-2504 on V-2500 Inlet Separator back into service, to prevent gas blowby to V-5800 Produced Water Flash Drum.	50.9.1.1	P&ID 3	High	Low

Table 4 Example of Measurable Recommendation

Recommendations	Place(s) Used in HAZOP Consequence #	Drawing Number	Max Risk Level (Before Action)	Max Risk Level (After Action)
4 Add an additional check valve, dissimilar to the existing 1" check valve, to the sweet fuel gas purge line to V-005 Inlet Slug Catcher, F-250 Inlet Filter / Separator and H.P. Flare. In addition, ensure the sweet/sour spec break is upstream of the last manual valve.	1.10.1.1	P&ID 4	Medium	Low

Recommendation #3 is a typical example of non-measurable recommendation, as the result of this recommendation is a "consideration". A consideration is not a measurable action that can be reflected on any documentation. On the other hand, Recommendation #4's goal is an additional check valve, which can be reflected on Piping & Instrumentation Diagrams (P&IDs), datasheets, and all other applicable Process Safety Information (PSI).

Strong verbs such as "add", "provide", or "configure" provide better information to the recommendation reviewer on what is the result of this recommendation, where as soft verb such as "consider", "review", "evaluate" leaves less evidence in record whether the recommendation is implemented, and technically allows one to mark the recommendation complete when it has only been considered and no actual change has been implemented.

2.3 Assignable

An "assignable" recommendation means that the action has a clear indication of the responsible party who will be accountable for it. The assigned responsible personnel for this recommendation should be one of the PHA team members, who has participated in the discussion during recommendation crafting. Often during the review or implementation process of PHA recommendations, the assigned personnel of each action will have better understanding of the intention of the action, the associated hazardous scenario, and the final goal of the recommendation. The person responsible for the recommendation is not necessarily the one who executes the recommendation, but they will be held accountable for the action.

In some cases, PHA teams have found that action items have been assigned to a team member, who shortly thereafter left the position or company and the recommendations were lost or forgotten. This will be addressed in section 4, Tracking PHA Recommendations.

2.4 Risk Reducing

An effective PHA recommendation provides additional protection layers or design modifications to eliminate, mitigate or reduce the risk of hazardous scenarios. If documented properly, the risk reducing credit of the additional protection layer is clearer to the recommendation reviewer, which helps them making the decision on the prioritization of the recommendation and allocating resources into implementing it. See Table 5 for a demonstration of how a risk reducing recommendation is documented.

Table 5 List of Risk Reducing Recommendation

Recommendations	Place(s) Used in HAZOP Consequence #	Drawing Number	Max Risk Level (Before Action)	Max Risk Level (After Action)	Category	Max # Credits
5 Change the setpoint of PSV-3401 on V-3400 Amine Regenerator, such that the PSV will provide adequate overpressure protection for the Lean Amine Overhead Condenser, Lean Amine Regenerator Overhead Accumulator, and Acid Gas Separator in the event of blocked flow.	13.1.1.1, 13.1.2.1, 13.1.3.1, 13.1.4.1	P&ID 5	Very High	Medium	MEC	2

As shown in Table 5, Recommendation #5 it to provide 2 "credits" of likelihood reduction for the HAZOP consequences it is applied to (in this case, equivalent to a reduction in likelihood by a factor of 100) as a potential mechanical safeguard ("MEC"), which will reduce the risk level from "Very High" to "Medium". Without referencing the Study Worksheet, the PHA recommendation reviewer has a clear understanding of the benefit of implementing this recommendation, and is shown that it is believed that the facility would be safer if it is implemented.

2.5 Trackable

Once a PHA recommendation is created, it is crucial to keep a trackable record in a centralized database for the facility, such that the implementation of each recommendation is tracked through Management of Change (MOC) and falls under the overall facility's Process Safety Management

(PSM) program. One of the biggest challenges during a PHA for an individual processing unit within a larger-scale facility is that any action items that might affect adjacent processing units or the overall facility might be mis-understood or overlooked. This may be caused by the limited expertise of the PHA team for a single processing unit, or limited accountability by the responsible personnel for the PHA recommendation.

Another challenge that many facilities face during recommendation implementation is that the person the action item is assigned to leaves their current position, facility or company, and the recommendation is lost or forgotten. With a centralized tracking system, these recommendations can be identified and re-assigned appropriately.

A trackable PHA recommendation allows for different parties in a facility to monitor the status of the implementation. Whether a PHA recommendation is accepted or rejected, in process or completed, or waiting for management approval or turnaround schedule, the responsible personnel of the recommendation should update the status of the action as necessary. This will ensure that facility stakeholders are aware of any cases where the facility is operating with unmitigated high risk. In addition, a trackable record of recommendations minimizes the extra work during the next cycle of the PHA review. It will significantly reduce the effort and time to determine whether a past PHA recommendation is implemented or rejected, and it will minimize the chances of the new PHA team proposing the same recommendations.

3 Prioritization of PHA Recommendations

Once recommendations are generated from the PHA study, the implementation of them becomes the priority of the PHA stakeholders. It is unrealistic to implement all PHA recommendations in a short period of time, because of the limited availability of financial or human resources, or even by the required shutdown period of a processing unit. Sometimes it takes years to implement all the recommendations from a given PHA. Managing the priority of PHA recommendations often becomes the biggest challenge of PHA stakeholders after the session is completed. Two of the most common methods used to prioritize PHA recommendations are: by the availability of resources, or by the risk reducing contribution of each recommendation. The benefits and drawbacks of these methods are discussed below, followed by a discussion on how to optimize the prioritization of PHA recommendations.

3.1 Prioritizing Recommendations by Availability of Resources

Using availability of resources to prioritize the order recommendations are implement in, the PHA stakeholders can complete the maximum number of recommendations in the shortest period of time. These resources include:

- human resources such as working hours for engineers, operations, or instrumentation technician, or;
- financial resources such as available operating budget to implement recommendation, or;
- equipment resources such as potential downtime for equipment or processing unit.

However, with constraints of time, budget, and resources, and often with company directives or key performance indicators (KPI) which focus on implementing as many recommendations as possible [1], the disadvantage of using resource availability as the primary criteria for sorting recommendations is that some of the "cheaper" or "easier" recommendations may have a minimal or negligible impact on the overall risk of the facility. As a result, the facility may be facing unanticipated risk exposure when those more complex, more expensive, or more resource-consuming recommendations remain unimplemented or set at low priority.

3.2 Prioritizing Recommendations by Risk Reduction Contribution

Using the risk reduction contribution to prioritize recommendations, the PHA stakeholders can theoretically achieve the maximum reduction in risk by implementing the minimum number of PHA recommendations. The risk reduction contribution can be determined by:

- The maximum risk level before implementing the recommendation, or;
- The number of potential safeguard credits by implementing the recommendation.

The amount of risk reduction offered by a safeguard is referred to as "credit".

Table 5 shows the list of recommendation that is prioritized by the maximum risk level before implementation.

Table 6 List of Recommendations in Order of Maximum Risk Level (Before Action)

Recommendations	Place(s) Used in HAZOP Consequence #	Drawing Number	Max Risk Level (Before Action)	Max Risk Level (After Action)	Category	Max # Credits
5 Change the setpoint of PSV-3401 on V-3400 Amine Regenerator, such that the PSV will provide adequate overpressure protection for the Lean Amine Overhead Condenser, Lean Amine Regenerator Overhead Accumulator, and Acid Gas Separator in the event of blocked flow.	13.1.1.1, 13.1.2.1, 13.1.3.1, 13.1.4.1	P&ID 5	Very High	Medium	MEC	2
1 Ensure PSV-1001 is sized to protect the V-1001 Stabilizer Feed Drum against gas blowby from the following vessels: - V-1000 Slug Catcher, V-1201 Inlet Compressor Suction Scrubber, V-2000 TEG Contactor.	1.9.1.1, 3.9.1.1, 29.9.1.1, 29.9.5.1, 29.9.6.1, 31.9.2.1, 36.9.2.1, 36.9.3.1	P&ID 1	High	Low	MEC	2
3 Consider moving LSHH-2504 on V-2500 Inlet Separator back into service, to prevent gas blowby to V-5800 Produced Water Flash Drum.	50.9.1.1	P&ID 3	High	Low	SDS	1
2 Relocate PT-1205 sensing point from V-1203 2nd Stage Suction Scrubber to V-1205 2nd Stage Suct. Dampener, such that the PT is able to read low suction pressure if the demister pad in V-1205 2nd Stage Suction Scrubber is plugged.	2.1.6.1, 2.1.7.1, 2.1.8.1, 2.7.1.1, 2.7.3.1, 2.7.4.1	P&ID 2	Medium	Medium	-	0
4 Add an additional check valve, dissimilar to the existing 1" check valve, to the sweet fuel gas purge line to V-005 Inlet Slug Catcher, F-250 Inlet Filter / Separator and H.P. Flare. In addition, ensure the sweet/sour spec break is upstream of the last manual valve.	1.10.1.1	P&ID 4	Medium	Low	MEC	1

Although theoretically achieving a safer operation of the facility by implementing the least number of recommendations, the PHA stakeholder is potentially facing a higher financial stress due to the complexity of the recommendation, or the potential process downtime to install additional equipment or instrumentation. Relying primarily on the risk reducing contribution will potentially delay the implementation of other PHA recommendations, which will expose the facility to unmitigated risks and bring negative impact on the overall safer operating condition of the facility.

Another drawback of this approach is that often management will rely solely on the risk level of the scenario, and eliminate all recommendations of a given risk ranking or lower. In the example above, it is tempting to accept all Medium scenarios as As Low As Reasonably Practicable (ALARP) without due investigation and review, and eliminate all recommendations that apply to Medium or Low scenarios in one stroke. Although it is possible that any given Medium risk scenario is indeed ALARP, it needs to be further investigated for each recommendation. It is also important to consider that when there are many Medium risk scenarios, together they may increase the facility's risk above a tolerable level.

3.3 Optimizing the Recommendation Prioritization

Optimizing the recommendation prioritization will require the consideration of both the resource availability and risk reducing contribution. In addition, the combined risk reducing contribution of a recommendation that applies to multiple hazardous scenarios may be larger than the recommendation that is only applicable to a single hazardous scenario. The reliability of proposed recommendation, along with the availability of other applicable safeguards and/or recommendations in the same hazardous scenario is also a contributing factor to priority of a certain recommendation.

Moreover, as recommendations are completed, it can affect the priority of any remaining recommendations, therefore it is important to maintain the recommendation priority as a dynamic list, and re-evaluate the prioritization order when any recommendation is implemented.

By improving the methodology with which recommendations are prioritized, the opportunity for unanticipated risk exposure can be minimized. By taking availability of resources, such as cost, time and effort, into account the financial and resource expenditures can also be optimized, while having a primary focus on reducing risk [2].

4 Tracking PHA Recommendations

In the previous section of "Crafting PHA Recommendations", one of the criteria for a SMART Recommendation is that it is trackable. A centralized database for all PHA recommendations in a facility is beneficial to track the status of recommendations, as well as planning and prioritizing the remaining recommendations. Some of the other benefits of tracking recommendations are:

- Optimize recommendation prioritization across the entire facility;
- Maximize visibility of recommendation status;
- Minimize the information gathering process during the preparation for the next round of PHA reviews;
- Minimize duplicate recommendations that may be applicable to multiple processing units.

There are many systems available for tracking these actions, and the specific one to be used depends on the requirements of the individual facility or operating company. At its simplest, a shared Microsoft Excel spreadsheet can be used. Whichever centralized recommendation tracking system is selected, it should include at least the following information to achieve the above-mentioned benefits:

- Affected processing units for the PHA recommendation;
- Risk Level of the associated hazardous scenario, before and after implementation;
- Responsible personnel and their department;
- The MOC tracking number (if an MOC is initiated);
- Recommendation Status;
- Proposed completion date;
- Actual completion date;
- Comments on the progress-to-date of implementation and explanation of the final decision to implement, reject or modify the recommendation.

Once a centralized recommendation tracking system is established, it should be embedded to the facility PSM program, which will benefit the facility throughout its entire lifecycle.

5 Conclusion

Although it is well understood by industry that the PHA recommendations are the most important information generated during a PHA study, very rarely are the recommendations interpreted effectively by the PHA stakeholders, nor are they implemented in the optimal manner. Crafting PHA recommendation and implementing them effectively ultimately determines the quality of the PHA study.

A well-crafted recommendation should follow the modified SMART recommendation criteria: Stand-alone, Measurable, Assignable, Risk-reducing, and Trackable. From a simple list of recommendations, the reviewer of the each recommendation should be able to understand the hazardous scenario it is associated with, the intention or final goal of the recommendation, the responsible party who is held accountable for the implementation, the potential risk reducing benefit of the recommendation, and have a method to track the status of the recommendation.

Implementing all PHA recommendations is never an easy task due to the inherent limitations on available human resources, financial resources, and equipment downtime. Optimization in prioritizing the PHA recommendations needs to consider not only the availability of the resources, but also consider the risk reducing contribution of each recommendation. In addition, the prioritization sequence needs to be re-evaluated and modified constantly to achieve the maximum optimization of the recommendation prioritization.

Tracking all PHA recommendations in a centralized database benefits not only the stakeholders of a single PHA study, but also the overall facility. It allows for interconnected processing units to share knowledge of common recommendations, maximizing the visibility of recommendation status from management, and minimizing excessive paperwork when preparing for the multiple future rounds of PHA review throughout the facility's lifecycle.

6 References

[1] AICHE. 2011. *Process Safety Leading and Lagging Metrics*. [ONLINE] Available at: https://www.aiche.org/sites/default/files/docs/pages/CCPS_ProcessSafety_Lagging_2011_2-24.pdf. [Accessed 20 October 2017].

[2] A. De Man and R. Mukherjee. "Recommendations and the Broken KPI, Optimizing PHA Recommendation Implementation Chemical Engineering Progress", AICHE 2018 Spring Meeting and 14th Global Congress on Process Safety, April 2018.

586097 An Integrated Approach for Risk Management of Transportation of Hazmat: Use of Quality Function Deployment and Risk Assessment

Abstract:

An integrated approach for risk management of transportation of HAZMAT: use of quality function deployment and risk assessment

Guldana Zhigerbayeva[1], Ming Yang[1,2]

[1] Safety and Risk Engineering Group (SREG), Department of Chemical and Materials Engineering, School of Engineering, Nazarbayev University, 53 Kabanbay Batyr, Nur-Sultan, Republic of Kazakhstan 010000

[2] The Environment and Resource Efficiency Cluster (EREC), Nazarbayev University, 53 Kabanbay Batyr, Nur-Sultan, Republic of Kazakhstan 010000

Abstract

Transportation of hazardous materials (HAZMAT) is inevitable in the process industries. The statistics show a significant number of accidents has occurred during the transportation of HAZMAT. This makes risk management of HAZMAT transportation an important topic. The tree-based methods including fault-trees, event-trees and cause-consequence analysis, and Bayesian network, have been applied to risk management of HAZAMT transportation. However, there is limited work on the development of a systematic approach. The existing approaches fail to build up the linkages between the regulatory requirements and the safety measures development. The analysis of historical data from the past accidents' report databases would limit our focus on the specific incidents and their specific causes. Thus, we may overlook some essential elements in the risk management, including regulatory compliance, field expert opinions and suggestions. A systematic approach is needed to translate the regulatory requirements of HAZMAT transportation into specified safety measures (both technical and administrative) to support the risk management process. This study aims to first adapt the House of Quality (HoQ) to House of Safety (HoS) and proposes a new approach- Safety Function Deployment (SFD). The results of SFD will be used in a multi-criteria decision-support system to develop find an optimal route for HazMats transportation. The proposed approach will be demonstrated through a hypothetical transporation case in Kazakhstan.

585664 Risk Management-Resilience-Business Continuity: What, Why and How

Abstract:

Since the founding of MKOPSC 25 years ago, process safety and risk management made much progress with respect to material hazards characterization and risk methodology. Regarding the latter, fundamental issues such as parameter variability, interdependency of event probabilities and expert elicitation, all contributing to uncertainty, have been tackled. Human and organizational factors also have been studied and related with safety management system performance indicators. Static risk assessment in the design stage and of ready built plant is moving to dynamic operational risk assessment making use of data and virtual simulation-based knowledge.

However, even despite projected future major improvements in hazard identification techniques, leaving HAZOP and FMEA far behind, major hazards may escape our sight. Also, if all potential technical plant mishaps would be under control, possible external factors, e.g., human intervention, cyber-attacks, or weather will reduce certainty. Apart from not being identified, cost considerations will ignore low frequency major risks, yet, these can still materialize albeit rarely, hence coverage can never be complete.

Around 1995 in studies why certain companies are better surviving than others, the concept of "resilience" emerged. Also, the term "resilience engineering" appeared in the psychological context of organizational performance. "Engineering" meant here analyzing and improving culture. MKOPSC, though, initiated a decade ago resilience research to involve the whole plant operation. This kind of resilience based on risk assessment and indicators with early warning signals, staff prepared attitude, restrained flexible thinking, effective emergency response and recovery contingency measures, e.g., regarding supplies and customer needs, staff and funds, will lower and shorten post-disaster production output dip. Digitization and data, e.g., of various kinds of sensors and on performance indicators, will further support dynamic operational risk management, including abnormal situations, improving resilience and business continuity.

Cyber-Physical Systems Based on Plant Models, Proactive and Safe Control, Toward Full Automating Operation

Makoto Nakaya
Yokogawa Electric Co. Marketing Headquarters Innovation Center
2-9-32 Nakacho, Musashino-shi Tokyo 180-8750 Japan
Makoto.Nakaya@jp.yokogawa.com

Tadaaki Ando
Yokogawa Corporation of America
12530 West Airport Blvd, Sugar Land, TX 77478, USA
Tadaaki.andou@us.yokogawa.com
(Tadaaki.andou@jp.yokogawa.com)

Prepared for Presentation at
American Institute of Chemical Engineers
2020 Spring Meeting and 16th Global Congress on Process Safety
Houston, TX
March 30 – April 1, 2020

Cyber-Physical System Based on Plant Models, Proactive and Safe Control, Toward Full Automating Operation

Makoto Nakaya
Yokogawa Electric Co. Marketing Headquarters Innovation Center
2-9-32 Nakacho, Musashino-shi Tokyo 180-8750 Japan
Makoto.Nakaya@jp.yokogawa.com

Tadaaki Ando
Tadaaki.andou@jp.yokogawa.com

Keywords: Computing and Systems Engineering, Process Design & Development and Safe Design

Abstract

Due to the rapid development of computing, sensing and network technologies, advanced Internet of Things (IoT) technologies enable the integration of the physical and the cyber world into a cyber-physical system. Although IoT is merely a gateway into and out of the real world and the cyber world, data and information from the physical world are analyzed and learned in the cyber world, and in which a model that describes behaviors in the physical world is created. Utilizing the cyber-physical system is expected to create new value to the physical world. In the cyber-physical system, the one of the major goals must be to realize automating plant operation system.

We have developed "Mirror Plant" which is an online dynamic plant simulator. Mirror Plant receives sensor-measured data directly form the plant for model calibration. The model parameters inside Mirror Plant are automatically adjusted in the real time. As the result, Mirror Plant can perfectly simulate the dynamic plant behavior of the actual plant. By making computer run faster than reality, Mirror Plant can accurately predict the near future plant behavior. When Mirror Plant tells an alarm will occur in 30minutes, the operator can make an action to prevent predicted accident before something happened. Mirror Plant realizes a proactive plant operation yet to be seen in the conventional plant operation and contributes to safer plant operation. Mirror Plant is used in commercial LNG and chemical plants by operators who are directly involved in production and technical staffs who support production operation.

The development of IoT technology is accelerating the digital transformation in the manufacturing site, where the physical world and the cyber world will interact closely with each other. It is expected that big data utilization and AI technology will lead to more sophisticated plant model like Mirror Plant in the cyber world. By using the high-fidelity plant models, the unattended operation at night and fully automated operation of plants will be realized in future.

In this paper, we introduce Economic Model Predictive Control (EMPC), which uses nonlinear plant models created in the cyber world and delivers optimal and safe control system design. We applied EMPC to a chemical batch process with strong nonlinearity. We introduce result of EMPC shows high production with high yield and energy saving compared to the conventional PID control. If a plant model is available, anyone can use EMPC and specify any production requirements, such as maximum productivity ensuring environmental operation.

1 Introduction

The rapid development of computing and network technologies is driving forward the transformation to a smart plant using Internet of Things (IoT). With the arrival of explosionproof wireless sensors, a larger number of plants store their operation data and use them to predictively diagnose equipment failures, and today, the focus of using IoT is shifting to the prediction of product quality. There are many terms such as IoT, big data utilization, cyber physical system (CPS), and AI. Although these terms are used as synonyms, their relationships can be organized as shown in Figure 1. IoT is a gateway into and out of the real world and the cyber world, and data from the real world is analyzed and learned in the cyber world and the results are fed back to the real world in the form of operation amount and new value.

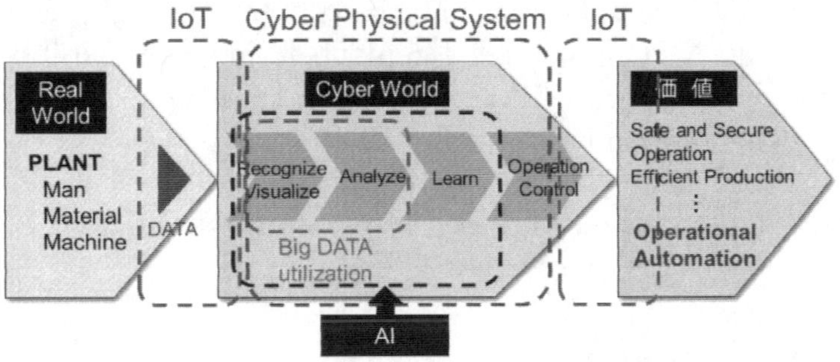

Figure 1. Relationship of IoT, Big DATA, CPS and AI

We have developed and released the Mirror Plant, which in the cyber world behaves in the same manner as a real plant. [1][2] The Mirror Plant is a dynamic simulator, which uses a physical model based on principles and rules and performs auto-tuning using the operation data supplied from the physical plant online. Plant operation is not determined by statistics and probability but determined by principles and rules. The Mirror Plant uses a physical model, since a model created by data provides good approximate accuracy only in the peripheral region where data exists.

The Mirror Plant that mirrors a real plant enables predicting plant operation in the near future by running the simulator in the cyber world faster than the real time. It can also provide a safe and secure operation environment by allowing operators to deal with alarms in advance that may occur in the future. It also allows for operators to search for the optimum operation conditions even during plant operation. The Mirror Plant, which faithfully models a plant, is widely used by operators who are directly involved in production and engineers who support the production operations. [3][4]

With the development of IoT, it is expected that a sophisticated plant model will be created in CPS as shown in Figure 1 using AI, Mirror Plant, and other technologies. For IoT, it is required to turn the results obtained in the cyber world into new value and deliver it to the real world through the Internet. Automated plant operation is one such value. There may be still a long way to go until we can use fully autonomous self-driving cars, but related technologies are being developed steadily. Meanwhile, automated plant operation requires various technical developments, such as creating a safe plant shutdown process without human intervention or measures to be taken by the operator for DCS in emergencies, as well as AI-based synthetic decision-making on operation. We are developing model predictive control using a sophisticated plant model generated by the Mirror Plant or other means in the plant control field. This paper introduces economic model predictive control (EMPC), which performs predictive control while performing an optimal computation without linearly approximating a non-linear plant model.

2 What is EMPC?

Most of the control used in plants is based on the feedback theory represented by PID control. Model predictive control (MPC) appeared in the 1980s. MPC predicts future behaviors using a mathematical plant model and calculates the operation amount while solving the optimum problem each time. Since MPC can perform advanced control for complex systems, it is used for a wide range of applications. MPC is also used in plants to increase stability against external turbulence or reduce the operation cost. EMPC described in this paper is MPC that belongs to the category of non-linear model predictive control.

Table1. Comparison between conventional MPC and EMPC

Ttem		Conventional MPC	EMPC
Prediction Model	Type	Linear	Non-linear
	Modeling	Created from operation data	Based on physical laws
	Application	only near operation point where model was created	a wide range
Optimaization	Objective Function	Minimizing the difference from target value of the prediced trajectory	Maximum yield and minimum energy consumption
	Structure	2 Layer-type (MPC+RTO)	1 Layer-type

2.1 Differences from Conventional Model Predictive Control

Table 1 summarizes the comparison between conventional MPC used in continuous plants and EMPC described herein. In model predictive control, the operation amount is obtained using a dynamic model of a plant to be controlled, so the quality of the model directly influences the control performance. In conventional MPC, the plant model is obtained by disconnecting the control loop at an operation point and conducting a step response test. Commercial MPC creates a linear model like “dead time primary delay approximation.” In MPC using conventional linear models, high control performance cannot be expected in some applications such as production output adjustment, in which the plant is operated at an operation point quite different from the operation point where the model was created. Meanwhile, EMPC uses non-linear physical models based on physical and chemical laws directly for control calculations. The operating range of the model is wide and there is the advantage that significant changes in the state can also be controlled. However, calculations are very complex since a non-linear model is run, so the calculation time is significantly longer than that in conventional MPC.

In optimization calculation, conventional MPC describes the following parameters in its objective functions: difference between the trajectory predicted from the model (or reference frame) and the target value, difference in the operation amount from the previous operation, and difference between the steady-state values for the optimization target amounts and the predicted values. The steady-state values for the optimization target amounts are generally given from the outside; for example, from the real time optimizer (RTO). Therefore, conventional MPC performs the optimization control calculation in two layers (MPC+RTO). In EMPC, economic functions (EF) such as maximum product yield or minimum energy consumption are described in the objective function. Different from MPC, EMPC is a single-layer type, which obtains the operation amount while performing optimization calculation.

2.2 EMPC Alorithm

EMPC is represented by formulas (1) to (8) below. [5]

$$\min_u \int_{t0}^{tf} EF(x(t), y(t), u(t))dt \tag{1}$$

Subject to:

$$\dot{x}(t) = f(x(t), y(t), u(t)) \tag{2}$$

$$g(x(t), y(t), u(t)) = 0 \tag{3}$$

$$h(x(t), y(t), u(t)) \leq 0 \tag{4}$$

$$x(0) = x_0 \tag{5}$$

$$xl \leq x(t) \leq xh \tag{6}$$

$$yl \leq y(t) \leq yh \tag{7}$$

$$ul \leq u(t) \leq uh \tag{8}$$

Formula (1) is the objective function of EMPC as described above. EF stands for Economic Function. Formulas (2) and (3) are plant model formulas. When the model formula of the Mirror Plant where the model is updated online is used, control can be maintained even if the process changes over time. Formulas (4) to (8) are constraint conditional formulas. x represents the state amount, y the plant output, and u the operation amount.

Since EMPC calculation uses nonlinear programming (NLP), there are problems regarding the convergence of a solution, stability, and real-time property. Accordingly, attention must be paid to the influence of the scale and the number of parameters of the target plant model.

3 Application of EMPC

The following reports on the results of applying EMPC to a continuous process and a batch process. In particular, the following shows how the production is changed by the setting of EF, which is a feature of EMPC.

3.1 Application to CSTR

We applied EMPC to a continuously stirred tank reactor (CSTR) process. [6] Figure 2 shows the model formula of CSTR and the parameters.

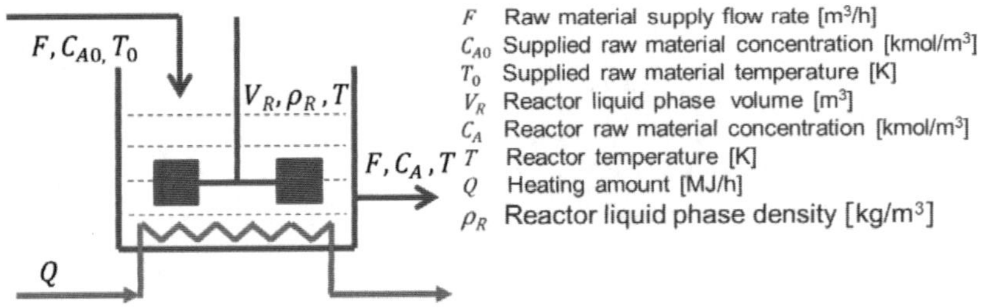

Plant model formula

$$\frac{dC_A}{dt} = \frac{F}{V_R}(C_{A0} - C_A) - k_0 e^{-E/RT} C_A^2$$

$$\frac{dT}{dt} = \frac{F}{V_R}(T_0 - T) - \frac{\Delta H k_0}{\rho_R C_p} e^{-E/RT} C_A^2 + \frac{Q}{\rho_R C_p V_R}$$

Figure 2. CSTR model

In this process, the density of the product generated in the reactor ($C_{A0} - C_A$) should be as high as possible, and heating amount Q to be input should be as small as possible in order to increase product yield. Accordingly, we set EF as shown in formula (9).

$$EF = -\alpha \cdot F(C_{A0} - C_A) + \beta \cdot Q \qquad (9)$$

"α" and "β" are factors used to adjust the priority between productivity and energy saving. Figure 3 shows the results. The blue line indicates the case where priority is placed on productivity, and the red line indicates the case where priority is placed on energy saving. These shows what production can be selected by setting EF.

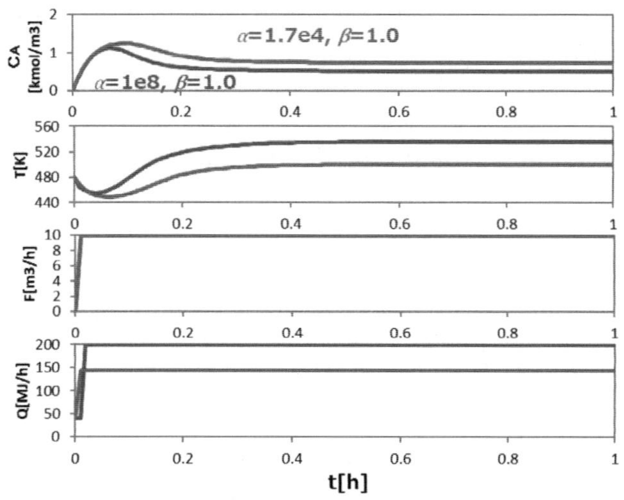

Figure 3. EMPC calculation results in CSTR

3.2 Application to a Batch Reaction Process

A batch process is used in many industries such as the fine chemical, pharmaceutical, and food industries that produce a variety of products in small quantities using the same equipment. The following describes an example of applying EMPC to a batch plant where the state keeps changing in contrast to a continuous process. [7] Figure 4 shows a batch reaction process considered in this paper.

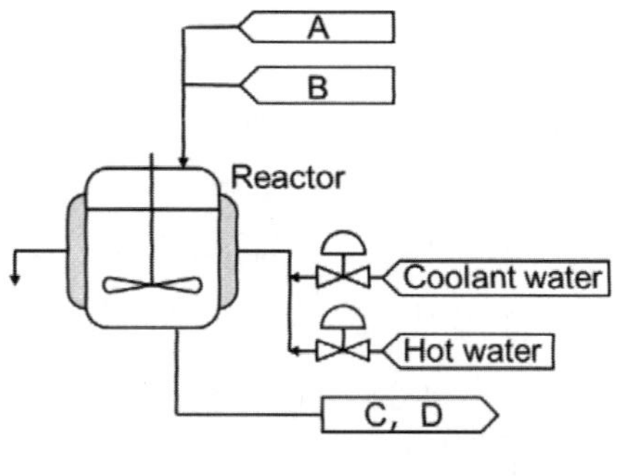

Figure 4. Batch reaction process

Raw materials A and B are input into the batch reactor, and after the reaction product C is output. A jacket is installed around the reactor to flow hot or cold water in order to regulate the temperature of the reactor. The concentration of components in the reactor is kept uniform by the agitator. Formula (10) represents a reversible reaction, and heat is generated by the forward reaction.

$$A + B \xleftrightarrow{k_1, k_2} C + D \tag{10}$$

$$C \xrightarrow{k_3} A + E \tag{11}$$

The reverse reaction in formula (10) reduces the production output of product C. There is another reaction that reduces the production output of product C, which is represented by formula (11).

The following shows model formulas that represent the balance between the materials and heat of this reaction process.

$$dC_A / dt = k_2 C_c C_D - k_1 C_A C_B + k_3 C_C^2 \quad (12)$$

$$dC_B / dt = k_2 C_c C_D - k_1 C_A C_B \quad (13)$$

$$dC_C / dt = k_1 C_A C_B - k_2 C_c C_D - k_3 C_C^2 \quad (14)$$

$$dC_D / dt = k_1 C_A C_B - k_2 C_c C_D \quad (15)$$

$$dT_r / dt = (Q_r + Q_j)/(C_r V_r C_{p_r}) \quad (16)$$

$$dT_j / dt = (T_j^{SP} - T_j)/\tau_j - Q_j/(C_j V_j C_{p_j}) \quad (17)$$

Where, C_A, C_B, C_C, and C_D represent the concentration of materials, Q_r the reaction heating amount, Q_j the heat exchange amount of the jacket, and C_{pr} and C_r are defined by formulas (20) and (21), respectively. T_j^{SP} represents the jacket entrance temperature and T_j the jacket temperature. The following shows model formulas that represent the heat quantity of the reactor and jacket.

$$Q_r = [k_1 C_A C_B * \Delta H_1 - k_2 C_C C_D * \Delta H_2 + k_3 C_C^2 * \Delta H_3] V \quad (18)$$

$$Q_j = UA(T_j - T_r) \quad (19)$$

$$C_{P_r} = (C_{P_A} * C_A + C_{P_B} * C_B + C_{P_C} * C_C + C_{P_D} * C_D) / C_r \quad (20)$$

$$C_r = C_A + C_B + C_c + C_D \quad (21)$$

EF in the batch reaction process was set as shown in formula (22) to maximize the yield.

Equation (22)

$$EF = -\alpha \cdot C_C + \beta \cdot \Delta u^2 \quad (22)$$

Figure 5 shows the results. PID control calculation results are also shown in Figure 5 for comparison with EMPC.

In PID control, the reactor temperature is kept at 50°C to prevent the influence of the reaction in formula (11). Meanwhile, EMPC sharply increases the temperature in the reactor at the start of reaction in order to give priority to the reaction speed. The temperature of the reactor is then reduced at the point when product C achieved the maximum production output in order to suppress the reverse reaction in formula (10) and the forward reaction in formula (11). As a result, product C can be produced in a shorter time compared to the PID control. The temperature of the fluid flowing in the jacket is also shown in Figure 5. Obviously, it is necessary to provide heating and cooling capabilities that quickly raise and reduce the temperature of the reactor in order to maximize the yield in the operation. EMPC uses the plant model at the plant design phase to estimate the equipment capabilities that are required to achieve target operation.

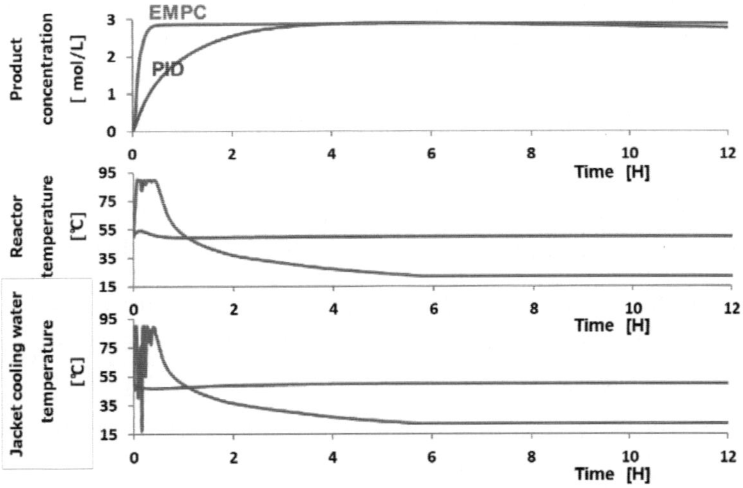

Figure 5. EMPC calculation results in batch process

4 Conclusion

The development of IoT technology is accelerating the transformation to a smart plant, where the physical world and the cyber world will interact closely with each other and efficient production and automation will be achieved. It is expected that the big data utilization and AI technology evolution will create a more sophisticated plant model in the cyber world, which will then be distributed to the real world. Use of this model opened an era in which our dreams of unmanned night operation and fully automated operation of a plant will come true.

In production at conventional plants, products were produced by controlling environmental variables in plants such as the temperature, pressure, and flow rate. In EMPC described herein, if a plant model is available, anyone can use it and specify any production requirements, such as productivity maximization and environmentally friendly operation.

With the current computing power, it is not possible to apply EMPC to the model of the whole plant and solve the overall maximization problem. Accordingly, the application of EMPC is limited to partial processes and devices. However, EMPC can also be used for designing and selecting plant equipment and devices for the optimal operation.

By leveraging the evolution of IoT which integrates the cyber world and the physical world at a high level, Yokogawa will develop and improve plant modeling and computing technologies.

5 References

[1] Makoto Nakaya, "MIRROR PLANT for Innovating Plant Operation," Yokogawa Technical Report English Edition, Vol. 56, No. 1, 2013, pp. 7–10

[2] Emiko Hatsugai, Naoto Takano, et al., "HMI of MIRROR PLANT for Facilitating Forecast of Plant Operation," Yokogawa Technical Report English Edition, Vol. 56, No. 1, 2013, pp.15–18

[3] Makoto Nakaya, Akira Yamada, et al., "Optimum Operation Supported by an On-line Simulator to Achieve Status Prediction," Instrumentation, Vol. 57, No. 6, 2014, pp.7–11

[4] K. Kato and T. Ito, "Active Use of Dynamic Plant Simulator in LNG Terminal Construction," speech draft for the 59th Automatic Control United Lecture Meeting, 2014, pp. 196–200

[5] M. Eills, H. Durand, et al., "A tutorial review of economic model predictive control methods," Journal of Process Control, Vol. 24, 2014, pp. 1156–1178

[6] X. Li, L. Zhang, et al., "Application of Economic MPC to a CSTR Process," Proceedings of 2016 IEEE Advanced Information Management, Communicates, Electronic and Automation Control Conference, 2016, pp. 685–690

[7] L. Zhang, M. Nakaya, et al., "Application of Cyber Physical Systems to Chemical Plants: Economic MPC in Batch Processes," Proceedings of the 13th IEEE Conference on Industrial Electronics and Applications, 2018, pp. 1265–1270

586511 Deployment of Process Safety Management & Risk Management for Integrated Steel Manufacturing Unit

Abstract:

This paper describes the challenges & leanings & progress made with regards to implementation of process safety management in a integrated steel manufacturing facility. The major Process risks involved are handling hot metal & toxic gases which had a risk of explosion, fire etc. A structured implementation was carried out through Center of Excellence exercise. This involved coaching leadership team , training line function on Process safety management. Apart from PSM culture development,Process Hazards were analysed using tools such as HAZOP, Consequence analysis, Bow tie , LOPA etc.Process risk was identified & communicated to management with necessary risk reduction measures. The Center of Excellence Model will cover the following activities.

Identification of High Hazard Operations

This activity focuses on the identifying process for which the system needs to be deployed, this is based on a pre-developed checklist focusing on Hazardous chemical properties such as flammability , toxicity etc.

Developing Process Design Basis

- Identification of Standard operating conditions & Safe Operating limits analyse their deviation & restoring the same to original condition.
- Identification of process safety critical equipment based on failure mechanism & developing Maintenance plan for critical equipment.

Process Hazard Analysis

- Appreciation of all process risks (Identification, Prioritization, Mitigation) using Process Hazard Analysis techniques such s *HAZOP for Batch & Continuous Processes.*
- Carrying out *consequence analysis* for toxic gas release, fires & *metal & water interaction explosions*.
- Developing *Bow ties* for the Major Accident Hazard Scenarios & identifying barriers.
- Determining the target mitigated risk through Layer of Protection Analysis.

Developing Operating Procedures & Emergency Response Plan

- Operating procedures for startup, shutdown etc were developed & operating personnel were trained.
- *Process Emergencies* were identified & response plans were developed. Control room operators were trained on handling process emergencies through *table top exercises.*

Building Process Safety Culture

PSM culture development activities such as PSM line walks focused on asset integrity, Process containment. Process safety performance Indicators were developed for monitoring the performance with a robust review mechanism.

This deployment has provided proven results in terms of achieving good operational control in the unit along with reduction in the Process Incidents.

Pressure Relief System and Safety Instrumented Systems - Common Weaknesses in Field Implementation

William Bridges
President
Process Improvement Institute, Inc.
1321 Waterside Lane,
Knoxville, TN 37922 USA
wbridges@piii.com

Stephen Bridges
Senior Process Engineer
Process Improvement Institute, Inc.
1321 Waterside Lane,
Knoxville, TN 37922 USA
wbridges@piii.com

Rashed Al-Zahrani
Manager, Process Safety
Jubail United Petrochemical Company (UNITED), a SABIC affiliate
ZahraniRA2@united.sabic.com

Prepared for Presentation at
American Institute of Chemical Engineers
2020 Spring Meeting and 16th Global Congress on Process Safety
Houston, TX
March 29 – April 2, 2019

Pressure Relief System and Safety Instrumented Systems - Common Weaknesses in Field Implementation

Presenters:

William Bridges
Process Improvement Institute, Inc. (PII)

Stephen Bridges
Process Improvement Institute, Inc. (PII)

Keywords: Human Factors, Human Error, Independent Protection Layers, IPLs, Integrity, Asset Integrity, Risk Based Process Safety, RBPS, PSV, Pressure Relief, Pressure safety Valve, Relief Valve, Safety Instrumented System, SIS, SIF, SIL, SIL Verification

Abstract

Pressure relief and safety instrumented systems are often the last line of defense before the occurrence of a catastrophic event. Their design, sizing, and reliability are hence of critical importance for the chemical process industries. Either are touted as capable of achieving a PFD = 0.01 or even 0.001. But, common weaknesses in implementation prevent this from being achieved. These include:

- Failing to take the PFD into account in the design of relief systems that require multiple relief valves to open to achieve the target capacity for the limiting case. Such systems should be designed as MooN instead of NooN.
- Failing to stagger maintenance for multiple channel systems (for multiple PSVs and for multiple sensors or multiple final element SIF). If maintenance or checks or calibrations are performed on the same day by the same person, then redundancy is lost as mistakes will be repeated on the other devices/channels.
- High error rate for control of block valves for relief valves or SIS root valves on taps and impulse lines
- Failure to account for the human error probability in the PFD estimates, resulting in expecting PFDs of 0.001 when only 0.01 is achievable long term.

Data, including from UNITED (a SABIC affiliate), are shared in this paper.

The major deficiencies found in implementation of PSVs an SIS stem from lack of understanding by design engineers of probability of failure on demand (PFD) and stems also from failure to understand and control human error probability in the inspection, test, and preventive maintenance (ITPM) programs and their practices in the field. This paper explains both issues and provides an approach to address these issues.

1 Design Deficiency in Relief Systems – PFD is not expressly considered in the design of relief systems

One major design deficiency seen across about 100 sites around the world is that the PFD of the relief system was not taken into account when designing the relief. Design engineers focus on sizing issues and setting issues such as:

- nature of the material being relieved (phase of the material, but also is it prone to cause plugging or corrosion)
- back-pressure issues in the tail pipes and discharge line/systems
- throughput necessary (and phase of the material, whether gas, vapor, two-phase, or liquid)
- design pressure (including allowable accumulation pressures)

However, very few relief system designers are trained in, or take into consideration, the PFD of the relief system. Because of this, designers specify multiple relief devices that all must open to relieve the throughput for the limiting design case. Instead of specifying a design that requires installed redundancy and instead of accounting for common cause error, they specify barely enough relief devices to me the limiting case throughput demand.

Take the case that a large petrochemical unit is required to relieve a large through of gas on the suction side of a compressor, if the compressor trips off or fails off; and further assume this release will be at relatively low pressure of about 1 atmosphere gauge pressure. For simplicity assume that the designers choosing 10 equally sized PSVs what when summed together will relieve the throughput for the limiting case (10 PSVs will be required to relieve the maximum throughput conceivable). So, if any One PSV does fails to open, then the design case throughput will not be provided; all 10 must open. If the designers do Not take into account the fault tree PFD calculation for the combined failures of the similar PSVs, then the PFD required for process safety of the relief system (all relief devices taken together as a relief system) will be too high, and the system will be under protected. This issue is summarized below:

- Target PFD for the relief system = 0.01
- PFD_1 of each relieve device (PSV) = about 0.01
- 10 PSVs required for the target throughput of the limiting case scenario, meaning if PSV_1 fails to open then we are unprotected from then limiting case, or if PSV_2 fails to open we are unprotected from the limiting case, or etc. for all 10 PSV_N. The overall system PFD is therefore calculated using fault tree as OR gates, which can be approximately reasonably well by summing the discreet PFD of each of the 10 PSVs.

- Therefore, the PFD_{SYS} of the relief system if all 10 PSVs (10oo10) must open for the limiting design case is given approximately as = $\sum$ (PFD_1 of single device) + β* PFD_1 = 10*(0.01) + 0.1*0.01= 0.101 (target PFD Not met)

  Where β is the fraction of failures due to common causes in the design of similar PSVs; this factor for PSVs is approximated here as β = 0.1 (10% fractional failure rate of a single device)

In this case, the set of NooN relief devices will be a factor of 10 less effective than the target PFD of the system (the PFD of the entire relief system is 10 times higher than the target PFD).

If the designers understand the how to estimate the PFD for a multiple device system, then they will instead include extra PSVs in the design, and achieve an MooN design, when M is the number of devices that must open to meet the capacity requirement and N is the total number of PSVs provided in the design. For such designs, then PFD of the system includes OR logic and AND logic, as the extra PSVs (those beyond what is minimally required for the throughput capacity alone) provide standby redundancy against the failure of other PSVs in the relief system. For this case, the PFD_{SYS} of the relief system of the N PSVs given MooN must open to reach the full throughput capacity of the limiting case can be approximated as:

$$PFD_{SYS} = (N!/((N-M+1)!*(M-1)!) * (2PFD_{1oo1})^{N-M+1})/(N-M+2) + \beta * PFD_{1oo1}$$

If MooN is 10oo11 (one additional PSV added to the design, and all 11 PSVs are designed to be online all of the time), then:

PFD_{SYS} (for 10oo11 case) = 0.0083 (target PFD met)

For this case the target PSV was met by adding one extra PSV and the designers simply need to account for the 11th PSV in the overall design of the hydraulics of the relief system.

2 Deficiency in the planning of ITPM activities – dependent human errors are not considered in the planning of tests or checks or PM of similar components in a system consisting of multiple key devices

Another major design deficiency seen across nearly ALL sites (hundreds) around the world is that the PFD of the relief or SIS system failed to account for dependent human errors when multiple similar devices are key to the PFD of the system. This is not a design error, per se, though designs can helped compensate for such errors. Instead this error in inherent in the programming / planning / scheduling of the ITPM activities related to a multiple channel system, as when there is more than one final element, such as multiple PSV or multiple safety block valve (XV), or more than one sensor channel in an SIS.

Probability of repeat errors (Coupled errors) – For many maintenance tasks, making a repeat error or "common cause error" or "dependent error" can lead to failures of all backup systems. Such errors have led to major process safety accidents as well as MANY airplane crashes.

Coupling represents the probability of repeating an error (or repeating success) on a second identical task, given that an error was made on the first task. The increased probability of failure on subsequent tasks given that an error has already been made is known as dependence. The list below provides some starting point guidance on values to use:

- 1/20 to1/90 – if the same tasks are separated in time and if visual cues are not present to re-enforce the mistake path. *This error rate assumes a baseline error rate of 1/100 with excellent human factors. If the baseline error is higher, then this rate will increase as well.*
- 1/2 – if the same two tasks are performed back-to-back but the worker is not in line of sight of the other component/device he/she already worked on, and if a mistake is made on the first step of two. *This error rate assumes a baseline error of 1/100 with excellent human factors. If there the baseline error is higher, then this rate will increase as well.*
- 8/10 to 10/10 – if the same two or three tasks are performed back-to-back and a strong visual cue is present (if you can clearly see the first devices you worked on), and assuming a mistake is made on the first of the two or more activities.
- Two or more people become the same as one person (with respect to counting of errors from the group), if people are working together for more than three days; this is due to the trust that can rapidly build. So, double-checking by someone you regularly work with is not a safeguard.

These factors are based on the relationships provided in NUREG-1278[1] and the related definitions of weak and strong coupling provided in the training course by Swain (1993) on the same topic [2], as shown here in Table 1. The following relationship is for errors of omission, such as failing to reopen a root valve or failing to return a safety instrumented function (SIF) to operation, after bypassing the SIF. The qualitative values in Table 1 are based jointly on Swain[1,2] (1983, 1993) and Gertman (SPAR-H, 2005 which is NUREG/CR-6883) [3].

One can readily conclude that staggering of maintenance tasks for different channels of the same SIF or for related SIFs will greatly reduce the level of dependent errors. Unfortunately, most sites PII visits do not stagger the inspection, test, or calibration of redundant channels of the same SIF or of similar SIFs; the reason they cite is the cost of staggering the staff. While there is a perceived short-term higher cost, the answer may be different when lifecycle costs are analyzed.

Simple Rule: Staggering of maintenance can prevent significant number of human errors in redundant channels. In fact, the US Federal Aviation Administration (FAA) requires staggering of maintenance for aircraft with multiple engines or multiple control systems (i.e., hydraulics) (*FAA Advisory Circular 120-42A*, as part of Extended Operations (ETOPS) approval).[4] (ETOPS is Extended-range Twin-engine Operational Performance Standards, a rule which permits twin engine aircraft to fly routes which, at some point, are more than 60 minutes flying time away from the nearest airport suitable for emergency landing.)

Table 1: Guideline for Assessing Dependence for a within-SIF Set of Identical Tasks (based partially on SPAR-H, 2005 [3] and partially on field observations by PII)

Level of Dependence	Same Person	Actions Close in time	Same Visual Frame of Reference (can see end point of prior task)	Worker Required to Write Something for Each Component
Zero (ZD)	No; the similar tasks are performed by different person/group	Either yes or no	Either yes or no	Either yes or no
Zero (ZD)	Yes	No; separated by several days	Either yes or no	Either yes or no
Low (LD)	Yes	Low; the similar tasks are performed on sequential days	No	Yes
Moderate (MD)	Yes	Moderate; the similar tasks are performed more than 4 hours apart	No	No
High (HD)	Yes	Yes; the similar tasks are performed within 2 hours	No	No
Complete (CD)	Yes	Yes; the similar tasks are performed within 2 hours	Yes	Either yes or no

The level of dependency is determined from Table 1 by assessing whether the same person is doing the tasks, the proximity of the actions in time, the proximity of the actions in space (same visual frame of reference), and whether the work is required to make a record for each component:

1. Read down the "Same Person" column and find the applicable row(s), then
2. Read down the "Actions Close in Time" column and find the applicable row,
3. Then check the two columns on the right for that row.
4. The Level of Dependence is shown in the left-most column for the applicable row.

Table 1 has two rows for Zero Dependency (ZD) because ZD can be achieved two different ways:

- Tasks done by different persons or groups (staggered people), or
- Tasks done by same persons or groups, but tasks are done several days apart (staggered times).

Once the level of dependence is known, the probability of either repeat success or repeating errors on identical tasks can be estimated. For these probabilities, we use Table 2, which is a re- typing of Table 20-17 from NUREG-1278[1] (and the similar table in SPAR-H [Gertman, 2005] [3]).

Table 2. Equations for Conditional Probabilities of Human Success or Failure on Task N, given probability of Success (x) or Failure (X) on Task N-1, for Different Levels of Dependence *Courtesy Process Improvement Institute, Inc., All Rights Reserved*

Level of Dependence	Repeating Success Equations (but shown as error probability)	Repeating Failure Equations
Zero (ZD)	$P_{Success@N} = x$	$P_{Failures@N} = X$
Low (LD)	$P_{Success@N} = (1+19x)/20$	$P_{Failures@N} = (1+19X)/20$
Moderate (MD)	$P_{Success@N} = (1+6x)/7$	$P_{Failures@N} = (1+6X)/7$
High (HD)	$P_{Success@N} = (1+x)/2$	$P_{Failures@N} = (1+X)/2$
Complete (CD)	$P_{Success@N} = 1.0$	$P_{Failures@N} = 1.0$

How to Avoid these Repeat Errors (dependent human error) – The straightforward solution is to follow the rules implemented by the US FAA for aircraft maintenance. This requires staggering the maintenance for each similar device in a set of devices that comprise the IPL.

- So, if there are 11 PSVs for the earlier case where 10oo11 PSVs must open for the design case, then set up the ITPM program such that either (1) there is 3 days between the servicing of each PSV in the set of 11 PSVs (if serviced by the same crew) or (2) have a different crew do the blocking, removing, checking, re-installing of each PSV that is checked on the same day.

- Similarly, if there are 3 instrument channels on a SIL 2 SIF, then either (1) ensure there are 3 days between the servicing of each channel in the SIF if performed by the same technician or (2) have a different technician do the testing, calibration, or checking of each channel that is checked on the same day.

To implement staggering, the site will need to track the ID number of each worker assigned to inspect or test or maintain devices that make up an IPL so that (1) staggering can be ensured up front and (2) staggering can be verified by auditing.

For more details on the major effect dependent errors can have on SIFs, read the earlier papers on the topic [5,6].

One means to improve the reliability and independence of the instrumented system is to use a smart sensor/transmitter for the LSH which will detect null movement of the sensor reading, indicating the valve is closed on the tap is plugged. Another possibility is to implement a limit switch (or captive key system) on the root valve. There is a probability that these safeguards against human error will also fail or be bypassed by the staff, but assuming the probability of that failure is the same as other human errors for this example, 0.02, then the systemic human error drops to about zero as the probability of leaving the root valve closed is now ANDed with the probability of smart sensor/transmitter or limit switch failing, [6].

3 Failing to Account for the Human Error Inherent in the positioning of Root Valves (block valves)

Another major deficiency in high integrity systems is failing to account for the relative probability of human error for such simple tasks like re-opening the block valves on either side of a PSV before the PSV is returned from servicing or similarly failing to re-open the root valve between the process and sensor of SIFs after servicing/checking. One common approach that used by industry is require a lock, or car seal (put in place by a human) on the block valve or root valve, and then administratively ensure the valve is indeed open by an checking the valve position and the placement of the Open Tag using someone other than the person who changed the position of the valve back to open when the maintenance activity was completed. The most common term for such administrative systems is a Car Sealed Open (CSO) or Locked Open (LO) or Chain Locked Open (CLO) system.

The problem observed with such systems, is when the checker is from the same site /local organization, then checking is not truly independent. During the writing of *Guidelines for Initiating Events and Independent Protection Layers*, 2015, CCPS/AICHE [7], the committee members stated that checkers from a site would report that all CSO or LO valves were found in the as-desired position (typically Open) but when an independent checker (from outside of the site) was used to do an unannounced check, they found 4/100 block valves in the wrong position. So, for a PSV, the impact on the system PFD is large:

- Target PFD for the relief system = 0.01
- PFD_I of each relieve device (PSV) = about 0.01
- Probability of leaving at least one block valve upstream or downstream of a PSV closed is about: P_{HUM_ERR} = about 0.04
- Therefore, the PFD_{SYS} of each relief system with a block valve upstream or downstream is approximately = $PFD_I + P_{HUM\_ERR} = 0.01 + 0.04 = 0.05$

In this case, the relief system for even a single PSV with block valves will be a factor of 5 less effective than the target PFD of the system (the PFD of the system is 5 times higher than the target PFD).

SIL 2 and SIL 3 SIFs are impacted to the same extent.

How to Avoid these Increases in the Overall PFD of the IPL – The straightforward solution is to not have block valves upstream or downstream of PSVs and not have root valves for SIL 2 or SIL 3 SIFs. But this solution is not always possible due to shutdown schedules for the plant (since if there are no isolation valves, maintenance can only be done when the process line is shutdown).

Another solution is having a truly independent check by an outside auditor about once per year, to find any valves in the incorrect position and to continually encourage accurate internal auditing of CSO, CSC, LO, LC block valves and any SIF root valve. This solution will cost about 1 to 2 staff-days of labor per 100 PSVs or 100 SIFs.

CASE STUDY on Third Party Inspection of CSO Valves for 448 PSV at UNITED (a SABIC Affiliate) in Jubail, Saudi Arabia:

PII was contracted by UNITED to check enough PSVs to determine the actual error rate at the site of PSV block valves being in the wrong position (and also CSO tags missing, though the tag does not affect the functioning of the PSV, obviously, although a missing tag implies that it is more likely in the future for the). There are approximately 2000 PSVs at the site and about 1423 in the 4 units that PII inspected. In these 4 units, 890 of these PSVs were inspected, but only 448 of the PSVs had a block-valve upstream, downstream, or both. (Since the likely error rate was expected to be between 0.005 to 0.05, PII chose to check about 500 PSVs with block valves.) One process safety engineer with 10+ years' experience was assigned to perform this check. Of the 448 PSVs (with block valves) that were checked, zero (0) were found to have a block valve upstream or downstream or both closed instead of open, without there being a valid reason (such as out-of-service for a valid reason); one (1) was found in the wrong position but there was a maintenance work-order for the related vessel. In addition, of the 840 block valves checked (counting both upstream and downstream), 79 were missing car seals, but the valve was in the proper position. The data from the inspection is summarized on the next page.

In this case, the positioning of the block valves were excellent, meaning that the PFD of the relief system is essentially the same as the PFD of the PSV alone, so the human error rate of mispositioning of the block valves are low enough to not effect the estimated PFD of the relief system at this site. But roughly 9% of the car seals were missing (mostly in one unit) which needs to be corrected. The site 50% converted to a more robust car seal mechanism, using uniquely printed metal seals, with plastic encases metal cables. They believe this system will greatly reduce the number of missing seals.

One last caveat on the data reported above: The inspection by a third party (PII) of the car seals was announced in October 2019, but the actual inspection by PII slipped until February 2020. It is possible, therefore, that the staff in the units took extra care in the 4 months to ensure the block valves were in the proper position. But there is no way to know for sure. But if special care was taken, then why were so many car seals missing?

RESULTS of THIRD PARTY INSPECTION of CSO Block Valves for PSVs at UNITED

			Upstream						Downstream					
Unit	PSV Total in Unit	PSV Total Sample	# BVs	# BVs wrong	# CSs on BV	# CS Missing	Fraction (Probability) BVs wrong	Fraction (Probability) Missing CS	# BVs	# BVs wrong	# CSs on BV	# CS Missing	Fraction (Probability) BVs wrong	Fraction (Probability) Missing CS
A	534	363	238	0	237	1	0	0.0042	154	0	153	1	0	0.007
B	65	49	35	0	35	0	0	0.0000	22	0	22	0	0	0.000
C	276	211	43	0	40	3	0	0.0698	154	0	7	0	0	0.000
D	548	267	132	0	89	43	0	0.3258	62	0	31	31	0	0.500
TOTAL	1423	**890**	**448**	**0**	**401**	**47**	**0**	**0.1000**	392	**0**	**213**	**32**	**0**	**0.127**

4 Failing to Account for the Human Error in the Baseline Estimation of the PFD.

The last major deficiency in high integrity systems to be covered in this paper is failing to account for the specific probability of human error in the PFD of the IPL, including for a single PSV installation, a multiple PSV set IPL, and SIL 2 or SIL 3 SIFs. A few of the specific human errors were discussed in sections 1-3 of this paper. For a SIF, below are the typical human errors of interest:

- Probability of leaving the root valve for the level switch (sensor/transmitter) closed
- Probability of leaving the entire SIF in BYPASS after maintenance or after some other human intervention (such as an inadvertent error or as a necessity during startup)
- Probability of miscalibration of the level transmitter/switch.

The new issue here applies to SIL 2 and SIL 3 SIFs more than to PSV or multiple PSV or single or multiple checkvalve IPLs. The reason is that for a SIF the user is required to a complete a SIL Verification calculation that states that the target PFD was achieved by the design. But, as explained in two earlier papers [5,6], the guidelines and standards for SIS from IEC and from ANSI/ISA do not require inclusion of the probability of human error (P_{HUM_ERR}) in SIL Verification. If the human error is not required to be in the estimate of the overall PFD of the SIF, then why try to control the human error? Further, leaving such terms out may give false hope that a high integrity IPL was actually achieved, whereas it nearly impossible to actually achieve a PFD of 0.001 for a SIL 3, due to the human error probability that cannot be lowered enough. A SIL 3 with a PFD of 0.001 is as rare as a Unicorn (they very likely do not exist) [6].

In this case, for any IPL with a target PFD of 0.001, the system will typically have a PFD that is 10 times greater than predicted, because the human error contribution was omitted from the PFD estimate. This can lead to the risk of a scenario being too high, and can lead to wasting resources on an IPL that the company believes is much more reliable than it actually is [6].

5 CLOSING

High integrity IPLs, such as PSVs, multiple PSVs with extra capacity, and SIL 2 and SIL 3, are highly vulnerable to human error and to mis-design. These errors have been observed across hundreds of facilities that use or make highly hazardous chemicals. But once these vulnerabilities are known, an organization can take measures to reduce the vulnerability. Not knowing about these vulnerabilities or not addressing them can lead to much higher risk of major accidents, or misallocation of funds and efforts on the wrong IPLs.

6 Acronyms Used

10oo10 ten out of ten voting architecture; all 10 devices must work for the safe state
10oo10 ten out of ten voting architecture; all 10 devices must work for the safe state
ANSI American National Standards Institute
CD Complete Dependence
CSC Car Sealed Closed
CSO Car Sealed Open
D Dangerous
ETOPS Extended-range Twin-engine Operational Performance Standards
FAA US Federal Aviation Administration
IPL Independent Protection Layer
ISA International Society of Automation
LC Locked Closed
LD Low Dependence
LO Locked Open
LOPA Layer of Protection Analysis
MooN M devices out of N total devices
NooN N devices out of N total devices
P Probability
PFD Probability of Failure (dangerous) on Demand
PII Process Improvement Institute, Inc.
PSV Pressure Safety Valve
SIF Safety Instrumented Function
SIL Safety Integrity Level
SIS Safety Instrumented System
SYS Systematic failure contributions to overall PFD
HUM_ERR Specific human errors and failures generated by human error
ZD Zero Dependence
CSO, CSC, LO, LC

7 References

References Cited

[1] A. Swain and H. Guttmann, *Handbook of Human Reliability Analysis with Emphasis on Nuclear Power Plant Applications, Sandia National Laboratories*, 1983 [this document became NUREG/CR-1278– The Human Reliability Handbook, guidelines from the US NRC on Human Reliability Analysis].

[2] A. Swain, *Human Reliability Analysis*, Training Course, ABS Consulting (formerly JBF Associates), 1993.

[3] D. Gertman, H. Blackman, J. Marble, J. Byers, and C. Smith, *The SPAR-H Human Reliability Analysis Method*, NUREG/CR-6883, U.S. Nuclear Regulatory

Commission, Office of Nuclear Regulatory Research, Washington, DC, August 2005.

[4] US Federal Aviation Administration, "AC 120-42B - Extended Operations (ETOPS and Polar Operations)", Washington, DC, June 13, 2008

[5] W.G. Bridges and H.W. Thomas, "Accounting for Human Error Probability in SIL Verification Calculations", 8th Global Congress on Process Safety, Houston, TX, April 1-4, 2012, American Institute of Chemical Engineers.

[6] A. Dowell III, W.G. Bridges, H.W. Thomas, and M. Massello, "SIL-3, SIL-2, and Unicorns", 15th Global Congress on Process Safety, New Orleans, LA, March 31 - April 3, 2019, American Institute of Chemical Engineers.

[7] CCPS, *Guidelines for Initiating Events and Independent Protection Layers in Layer of Protection Analysis*, Center for Chemical Process Safety, American Institute of Chemical Engineers, New York, NY, 2015.

Gas Flare Systems- Last Line of Defense

Rimon Vallavanatt, P.E.
Bechtel Oil, Gas & Chemicals
3000 Post Oak Blvd, Houston TX 77056
rvallava@bechtel.com

Freeman Self
Bechtel Oil, Gas & Chemicals
3000 Post Oak Blvd, Houston TX 77056
feself@bechtel.com

Prepared for Presentation at
American Institute of Chemical Engineers
2020 Spring Meeting and 16th Global Congress on Process Safety
Virtual Meeting
August 17 - 21, 2020

Gas Flare Systems - Last Line Defense

Rimon Vallavanatt, P.E.
Bechtel Oil, Gas & Chemicals
3000 Post Oak Blvd, Houston TX 77056
rvallava@bechtel.com

Freeman Self
Bechtel Oil, Gas & Chemicals
3000 Post Oak Blvd, Houston TX 77056
feself@bechtel.com

Keywords: flare, combustor, burner, smokeless, efficiency

Abstract

Process flares are considered a key equipment in a refinery or LNG plant for controlled destruction of unwanted gases during startup, shutdown, maloperation and maintenance at high destruction efficiency, but also producing some undesirable by-products including noise, smoke, heat radiation, light and pollutants. This paper will cover details of burners, structures, as well as operational factors, flaring regulations and flaring reduction. On the other hand, it will not discuss details well covered in the complementary API Standard 537 such as seals, pilots and flame monitoring.

1 Introduction

Gas flaring is a combustion process for flammable compounds used to dispose of gases that are wastes or generated under unsteady conditions of excess heat, pressure or concentrations.

Gas flaring, incineration and thermal oxidation are generic terms and often used interchangeably. In this paper, gas flaring will be used to describe the technology used for bulk combustion of streams composed mostly of flammables. Thermal oxidation is generally used for specialty combustion where gases may have a wide range of flammable concentrations. Incineration is a more general term for processes used for solids, liquids or gases.

This paper provides an overview of flare system to cover aspects that are not covered in detail in the complementary API Standard 537 "Flare Details for General Refinery and Petrochemical Service" [1]

- Flare Burners
- Flare Structures
- Smokeless flaring
- Operational factors such as flare efficiency and emissions, smoking, radiation and noise
- Flaring Reduction Initiatives
- Federal regulations

This paper will focus on the gas flaring and not liquid flaring. It will also not discuss specialty flares such as for ammonia, hydrogen sulfide, and high carbon dioxide, etc. Aspects that are well examined in major publications as API Standard 537 [1] and API standard 521 [2] will not be discussed. This includes flare headers, collection of liquids and knock-out drums, mitigation of air ingress and seals, pilots, flare pilot and flame monitoring and cases for cause a relief event.

2 Flare Burners

Burners are the heart of the flare designed to provide temperature, residence time, and turbulence which are the primary factors for efficient oxidation. The burner is a source of confusion since there are numerous types, specific uses, with various names for each burner type. This section will outline the types and characteristics. Burners can be classified in various categories.

- Pressure inlet to the burner. This paper will use this classification since it is becoming standard nomenclature for emission purposes. This classification is also important is the discussion of the structures for flares.
- Velocity through the burner. The three major subcategories are subsonic (not choked), sonic (choked) and supersonic. These are further described below.
- Smoking or non-smoking. These are discussed in a later section.
- Flame or flameless. All burners described exhibit flames except the mesh-burner.
- Size. The sizes are generally quoted on a duty basis (btu/hr) but also on gas mass or volumetric flow rate.

Figure 1. Courtesy of Escher. <Escher.nl/en-US/> Steam-assisted pipe burner with appurtenances.

2.1 Pipe Burners

Pipe burners, often called an "open-pipe-tip" or "open-pipe-flare OPF" and shortened to "tip", are basically the piece of pipe flanged to the stack (or seal), and outfitted with accessories such as wind shields, pilots, steam rings, etc. The diameters may be 6 feet or greater but are limited by the combustion efficiency. Typically pipe burners are supported on elevated structures to reduce radiation. They may be combined with pressure-assisted burners in the same structure.

2.1.1 Low-Pressure Pipe Burners

A typical tip is shown in Figure 1. The pressure at the inlet to the burner is just the backpressure across the tip, approximately five psig. Open pipe burners have a constant flow area so with a variable flow rate, the tip velocity, amount of air induced at the tip and mixing of air-gas are also variable. The tips are not inherently smokeless so assist gas is used for smokeless operation

2.2.2 Choked-Pressure Pipe Burners

Although low-pressure tips operating at less than choked conditions are prevalent, there are also large diameter tips operating at choked (sonic conditions). Refer to Figure 2. These are not used in the United States due to regulations on destruction efficiency and velocity limitations unless they can test to equivalent emissions discharge reduction. The sections on efficiency and regulations provide more information.

Figure 2. Two pipe tips (nozzle approximately 6" and 12" in diameter). Labeled "single point sonic pipe flare". Courtesy of <ArgoFlares.com>

2.2 Pressure-Assisted Burners

A pressure-assisted burner uses elevated pressure at the inlet to the burner to assist combustion. "Pressure-assisted burner" is the term frequently used by the EPA and other publications. If the pressure is sufficiently high so the gas flow is choked at the burner, they are sometimes called "sonic burners". Another characteristic is that the burners are available in many physical configurations and can be described by their shape as "bull-nose", "spider", coanda "tulip" or "furnace burner". Additionally, burners can be recognized by their trade name. Depending on the configuration, they can be oriented vertically or horizontally. Pressure-assisted burners are utilized in both elevated and non-elevated structures (such as ground flares or enclosed flares), and they may be combined with pipe burners in the same structure. Pressure-assisted burners are generally smokeless without assist mechanisms except at low-turn-down. The gas pressure and draft / residence time (with an enclosed flare) promotes mixing at the burner tip.

2.2.1 Choked Operation Burners (pressure-assisted)

- *Pipe Tips*

This type of pressure-assisted tip has a single opening and operates at choked flow. They may be designed as a venturi (high-pressure recovery) nozzle to enhance dispersion.

Figure 3. Multiple pipe tips arranged in a staged array. High-pressure gas permits choked flow in "NAO Jet Mix™". Courtesy of NAO, Inc <NAO.com>

These tips are generally small in diameter ranging up to approximately 10" depending on the vendor. Many tips are grouped into a "multi-tip" or "multi-nozzle" array with the nozzles at the end of arms which extend from a central core. Refer to Figure 3. However, larger tips are available but not used in the United States due to regulations on destruction efficiency and velocity limitations. Refer to Section 2.1 and Figure 2.

- *Coanda Tips*

The coanda flare tip has a profile like a tulip. The relief gas at high pressure (15 to 70 psig) is ejected through slots circling the outside of the tulip-shape cone. The gas adheres to the outside, drawing in air which mixes with the gas film adhering to the outside wall.

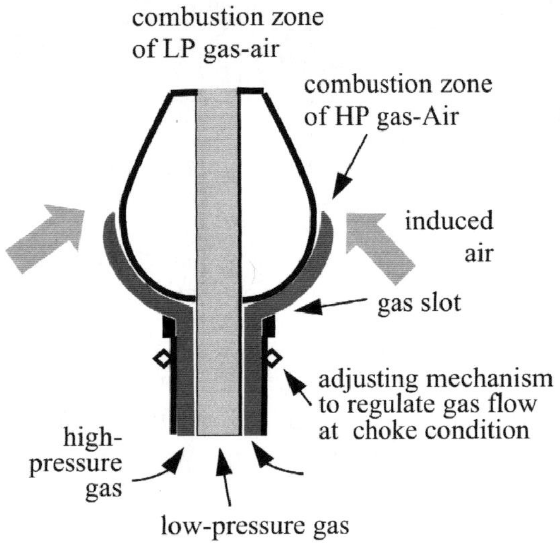

Figure 4 Coanda Tip

Figure 5. Courtesy of NAO, Inc. The burners are Coanda "Hotdog™". The high-pressure flow through the burner vent is choked. This array is staged.

Combustion occurs at the location of the cone maximum diameter. The slot opening may be adjusted for downturns, so that the flow always choked through opening slot thereby enhancing air-gas mixing. Refer to Figure 4. Sometimes steam is added to the relief gas to enhance combustion. Multiple smaller coander tips may be arranged in an array for better efficiency, Figure 5. The tips are understood to be smokeless, produce low radiation and can handle small amounts of liquids.

Figure 6. Courtesy of NAO, Inc. Spider burner tips have vanes with drilled holes where the high-pressure flow is choked. This array is staged in a ground flare.

- *Spider Burners*

Spider burners, also known as radial spoke burners, are arrangements of tubular/pipe/vanes burners that radiates from the center. Examples are shown in Figures 6 and 7.

• *Plate Fin Burners*
Fin plate burners are designed to distribute the gas into thin layers over a series of plates, which facilitate air and gas mixing and smokeless combustion. Refer to Figure 8.

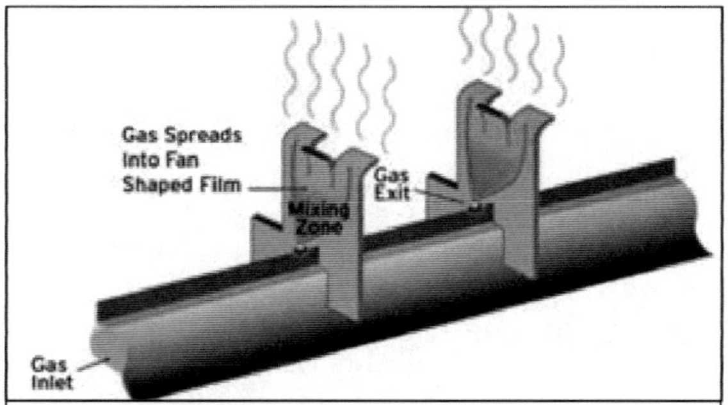

Figure 8. Fin plate burner. Courtesy of <johnzinchamworthy.com>

Figure 9. Pipe or tubular burner for non-choked flow

Figure 7. Courtesy of Zeeco, Inc. Spider burners in a ground flare

2.2.2 Non-Choked Burners (pressure-assisted)
For very low-pressure applications, the flow through the burner structure is not choked. These are primarily used in enclosed flares where the draft of the enclosure enhances combustion. Inlet pressure to the burner for small duties can be as low as 0.06 psig (1.8-inch water column-gage) (0.156 millibar-g) Larger units inlet pressure may be 0.25 psig (7-inch water column-gage) (17.2 millibar-g).

Low-pressure burners are variations of choked burners with the primary difference in the configuration is the size of the burner holes. A common type is the tubular or pipe burner, Figure 9. Since pressure is not available, smokeless operation may be provided typically using air, or steam or fuel gas.

2.3 Mesh Burners

A mesh burner is basically wire mesh. These are used for "flameless" flares, where the combustion process does not produce a flame since the gas combustion reaction occurs below the lower flammability limit. Since the flare rate is dependent on the area of the mesh, these flares have typically small capacity. Mesh burners are not common in the United States.

2.4 Super-sonic

A super-sonic burner is designed as a converging-diverging nozzle and is designed to operate above Mach 1, reputedly 1.5 to 2.0 times sonic. Interest was strong in the late 1990s but has declined. The advantage is that lower weight of the nozzle, higher air-gas mixing and lower radiation. However, turn-down is limited since at subsonic flows, the flares are less efficient, wind may influence the direction of the flame and increase local radiation levels and even produce flame impingement.

3 Flare Structures

Structures may be categorized several ways, but two major classification are elevated and ground flares.

3.1 Elevated Flare Design

Elevated flares are more common type but require more plot space for larger flares due to radiation impact. In general, they are less complex with fewer variations.

<u>3.1.1 Elevated Structures</u>
Structures are generally classified into four major types as shown in Figure 10. Stacks may be used with a pipe burner or pressure-assisted burner. They may also be equipped for smokeless flaring.

Self-Supported	Derrick	Guyed	Boom
• limit ~ 150 feet • small plot • lowest cost	• limit ~ 600 feet • medium plot • highest cost • may have 4 risers or more	• limit ~ 500 ft • medium plot • medium cost	• limit ~ 100 feet • small plot • medium cost

Figure 10. Types of Flare Structures. Courtesy of MRW <MRW-Tech.com> and NAO <NAO.com> for guyed stack

Since they are not shielded, the radiation levels at target (personnel or equipment) establishes distance from the target and the height resulting in potentially very tall stacks. The tall flares can laterally move up to 6" per 100 ft. due to wind load. Both guy wire and/or derrick supported takes the wind loads from the flare risers and results in a much more cost-effective design. In very tall stacks both derrick and guy wire supports are simultaneously used for cost-effective design. Elevated flares are generally chosen primarily for low cost, which can run about $65 to $85 per MM Btu.

Figure 11. Demountable derrick (segmented) with three risers. Courtesy Bechtel.

3.1.2 Operational Issues

The continuous flaring tip in an elevated flare requires maintenance at least once in every three to four years. Drones are used for visual verification of the condition of the tips. Maintenance of the flare tips becomes a major problem. Demountable flares are designed to lower the tip to grade level for replacement or maintenance while the other tips are operational. Demountable derrick flare with a spare riser can also be considered for added flexibility. Refer to Figure 11.

3.1.3 Burners for Elevated Flares

Both pipe burners and pressure-assisted burners may be used on elevated structures. And both types are commonly employed together on the same structure.

- Open pipe burners are more popular since they less expensive, radiation determination is simple, and are easy to maintain. Pipe burners are not inherently smokeless so assist gas is used for smokeless operation - refer Sections 2.1 and Figures 1 and 3.

- Pressure assisted burners have diverse choices. Combined in an array of multiple burners, in general the about 70% less radiation due to lower flame length and shielding of inner tips by the outer tips. Pressure-assisted burners are generally smokeless without assist mechanisms except at low-turn-down. Refer to Section 2.2 and Figures 5,6 and 12.

Figure 12. Multi-tip elevated flare with pressure-assisted (choked) burners. Courtesy MRW <MRW-Tech>

3.2 Ground Flare

In this paper, a ground flare is defined as utilizing a series of burners, separated into stages and is configured in one single open area surrounded by a wind and radiation fence. The base equipment cost of a ground flare can run about $120 to $150 per million Btu. The cost can vary based on the runner material of construction, retractable or stationary thermocouples, location of fabrication, and various other design variables.

A significant disadvantage with a ground flare is the potential accumulation of a vapor cloud in the event of a flare malfunction. As a result, special safety dispersion systems are included in the ground-flare system. For this reason, instrumentation for monitoring and controlling ground flares is typically more stringent than for an elevated system.

3.2.1. Installation and Layout - Ground Flare

For a large ground flare, the gases from the plant comes to the flare via large headers, often 20 to 60 inches in diameters, and supplied to many runners, installed in parallel, with staging valves. Each of these runners, ranging in sizes 12 to 24 inches in diameters, can have many burners depending on the flare vendor designs. On each runner, the number of burners is usually limited to fifty and the capacity of each of these burners can vary between 80 to 160 million Btu/hr. Depending on the number of runners, the flare consists of hundreds or even thousands of burners. Refer to Figures 13 and 14.

Figure 13. Ground Flare – view from top showing burner layout. This flare has a small flare release. Courtesy of Bechtel.

The burners and risers are spaced to prevent direct flame impingement and to allow air flow between risers (and/or burner flames). The runners are above ground to allow for calculated expansion and protected by stainless steel radiation shields.

Figure 14. Ground Flare – plan view showing open air slats in the wind fence. Courtesy of MRW <MRW-Tech>

3.2.2 Pressure and Staging - Ground Flare

A staged system is used for varying flow rates and turndown-ability. Additionally, staging provides higher inlet pressure to the burner which enhances the destruction efficiency and smokeless burning. Gas pressure is typically about 18 to 35 psig gas pressure at inlet of the burner. Low flows are flared in the first stage, which is open to atmosphere and protected with a velocity seal and purge to prevent air ingress. Depending on the vendor and burner type for the first stage, an air blower may be provided. Higher flows are routed to the other high-pressure stages as successive staging valves open.

In case a staging valve fails to open, a bypass is installed around the staging valve to maintain relief gas flow to the burners. The bypass is fitted with rupture pins, large specialty relief valves, or rupture disks that open on the appropriate pressure. Some installations will provide nitrogen to purge the burners after the bypass valve is closed.

The staging valves are located outside the structure where they are protected against high heat. The flame and flame field are monitored by pilot thermocouples, optical scanners and other devices. Since the pilot thermocouples and ignitors are inside the flame field, they are positioned so they are cooled by air flowing toward the burners. Additionally, they are designed to be retractable so they may be maintained on-line outside the structure. The retraction mechanism and thermocouple wires are protected by stainless steel radiation shields.

3.2.3 Smokeless Operation - Ground Flare
A ground flare will burn smokeless through the full capacity. For the high-pressure stages, the smokeless combustion is achieved using the kinetic energy of the high-pressure gas to entrain and mix air to pressure-assisted burners. The first low-pressure stage, which is open to atmosphere, will not smoke due to high residence time and forced air. Refer to Section 4 for more discussion on smoking reduction.

3.2.4. Radiation and Noise - Ground Flare
The flare field is surrounded by 45 to 60 ft tall wind fence, which allows entry of combustion air, protection from wind, and shielding of radiation. The height of the flame is kept below the height of the fence. The wind fence blocks the sight of the flame and may reduce the radiation exposure by approximately two-thirds. Still, the radiation adjacent to the fence is sufficiently high to require a personnel exclusion zone. Light is reduced but some is emitted vertically from the fence. However, unlike the enclosed flare, the wind fence does not offer significant noise attenuation. For large flare loads, the ground flare occupies less space than an elevated flare due to less radiation impacts.

3.3 Enclosed Flare

Enclosed ground flare is similar to the ground flare but is enclosed by a tall structure.

3.3.1. Installation and Layout - Enclosed Flare
The enclosure offers several advantages. First it essentially blocks all radiation and light and much of the noise. Secondly, the height supplies a large draft, which adds combustion air and turbulence. The structure also allows more residence time. These enhance destruction efficiency to 99.9% and higher. Another advantage of the enclosure is that the destruction efficiency can be measured to meet prove regulatory requirements.

Enclosed flare can range from a duty of 2 million btu/hour (approximately 80 lb/hr based on lower heating value) to more than 10 billion btu/hour (approximately 400,000 lb/hr). The large flares can reach 70 feet in diameter and 120 ft tall. The enclosure is internally insulated ceramic fiber blanket. In a large enclosed flare, the gas is fed via 8 to 20" diameter runners. Each runner will have many flare burners and each burner can support up to 50 million Btu/hr flaring. Even though the capacity of the typical large enclosed ground flare is limited to approximately 400,000 lb/hr of gas, larger capacity have been built. The equipment cost can run about $400 to $425 per million Btu for large flares. The cost can vary based on the runner material of construction, retractable or stationary thermocouples, location of fabrication, and various other design variables Refer to Figures 15 and 16.

Small enclosed flares of approximately150 million btu/hr (7,000 lb/hr) and 13 feet in diameter can be shop built and shipped on trucks to site. In special situations, the small enclosed flare may be

Figure 15. Enclosed Flare – side view

Figure 16. Enclosed Flare –top view showing burner layout

elevated on a structure, such a distillation tower to save piping costs. Shop built enclosed flares are competitive with elevated pipe flares at $100 per million BTU.

3.3.2 Pressure and Staging - Enclosed Flare

An enclosed flare utilizes a series of burners, separated into stages. For large flares, with high-pressure staging, the burners and staging valves are similar to the ground flare. A staged system is provided for turndown-ability. Additionally, staging provides higher inlet pressure to the burner which enhances the combustion efficiency and smokeless burning. Low flows are flared in the first stage, which is open to atmosphere and protected with a velocity seal to prevent air ingress. The first stage smoking is avoided by high residence time. Higher flows are routed to the other stages as successive staging valves open. After a staging valve closes, a small-volume short-term nitrogen purge is activated downstream of the staging valves to remove any flammable gas.

Section 2 on burners describes the range of pressures. Low-pressure small flares, in the 25 million btu/hr (1,000 lb/hr) range may be designed to 1.0 oz/in2 at the burner inlet while larger flares may be limited to minimum of 4 oz/in2.

3.3.3 Smokeless Operation, Radiation and Noise - Enclosed Flare

Because of the burners, an enclosed flare burns smokeless with high destruction efficiency. The enclosure essentially blocks all radiation and light and much of the noise. Refer to Section 4 for more discussion on smoking reduction.

3.4 Pit flares.

Pit flares are usually used in oil production flaring and in remote areas where smoke and sound are allowed by regulations or special permit. The flaring tip is exposed to a pit as shown in Figure

17. Pit flares generally do not meet the destructions efficiency or opacity smoking required by flares in the United States.

Figure 17. Pit flare firing horizontally. Left shows the frontal view of burners and right shows side view of operating flare. Courtesy of Zeeco, Inc

4 Smokeless Flaring

4.1 Smoke Formation

4.1.1 Mechanism of Smoking

Smoke is formed by unburnt hydrocarbon which have cooled to form soot particles or smoke. For smokeless combustion, sufficient air must be mixed with the fuel before combustion. In absence of ample air and high flame temperatures, the hydrocarbons crack, breaking carbon-carbon and carbon-hydrogen bonds. The breaking of bonds forms lower molecular weight compounds including carbon. If the carbon is not oxidized by the air drawn into the flame during combustion, smoke is formed.

The type of compound, the hydrogen-carbon (H-C) weight ratio and the hydrocarbon molecular structure, influences smoke formation. Methane (with a H/C weight ratio of 0.33) and carbon monoxide which oxidizes to carbon dioxide, do not smoke. Ammonia (NH4) and hydrogen with no carbon do not smoke. However, hydrocarbons with H-C weight ratios less than 0.33 will smoke. Branch chains, especially long branch chains, facilitate smoking since they can be easily cracked off. Compounds with ring structures and double-bonds tend smoke more than compounds with single-bonds.

Ringelmann and Opacity
Figure 18

Ringelmann 0	0 % Opacity (Clear)
Ringelmann 1	20 % Opacity
Ringelmann 2	40 % Opacity
Ringelmann 3	60 % Opacity
Ringelmann 4	80 % Opacity
Ringelmann 5	100 % Opacity (Black)

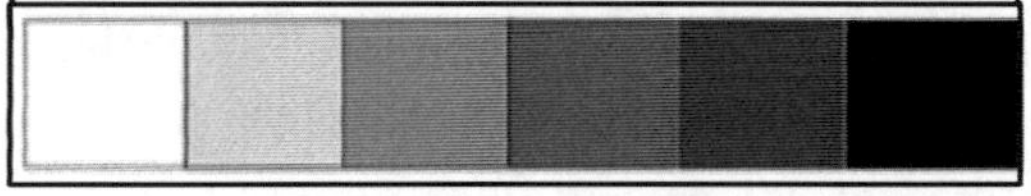

4.1.2 Measure of Smoking

Visible emissions or smoke is measured using opacity methods on the Ringelmann scale Figure 18. The current scale evolved from a method developed by Maximillian Ringelmann in the 1800s for coal-fired boilers. Federal opacity standards require the use of two procedures, either Method 9 or Method 22. These methods are found in the Code of Federal Regulations EPA 40 CFR part 60, Appendix A. State Implementation Plans (SIP) may also dictate other requirements. Observers responsible for measuring opacity attend visible emission (VE) training for certification.

4.2 Available Smokeless Methods

The primary methods to control smoke in heavy hydrocarbons are discussed below. The selection of which method is a function of several factors which can significantly affect the economics of the flare system.

- Utility requirements and availability. For example, facilities that do not have steam (or sufficient steam) must resort to other methods.
- Burner-design. Some burners inherently smoke less than others. For example, pressure assisted burners smoke less than open pipe flares OPF.
- Plot-space requirement i.e. elevated flares versus ground flares. The plot space may affect the selection of the flare type, which in turn affects the method to reduce smoking
- Flare operational considerations. The flow rates and duration of the flaring may be considerations, especially when assist gas is used to reduce smoking
- Associated maintenance issues due to additional equipment (e.g. motor / blower maintenance for air-assisted flare)

4.2.1. Steam Injection for Pipe Flare Tips

Pipe flare tips that use steam to control smoking are quite common. The basic mechanisms by which steam controls smoking include:

- Steam promotes turbulence and promotes air-fuel mixing.
- Steam reacts with compounds containing carbon to oxidize the carbon (e.g. to carbon monoxide) before the compound can crack and cool forming smoke.
- Steam dissociates to highly reactive free radicals H+ and OH- which lower the activation energies for the many intermediate reactions leading to the ultimate formation of CO2 and H2O. The enhancing of the overall combustion reaction lessens the opportunity for carbon cracking.

Steam is injected into the pipe flare flame zone to create turbulence and or aspirate air into the flame zone via the steam jets. The amount of steam required is primarily a function of the flare-gas composition, flow rate, steam pressure and location of injection (whether premixing with air or premixing with relief gas). Steam is normally provided from a supply header at a gauge pressure of 100 to 150 psig. API Standard 537 Table G.2. [1] provides an estimate for steam required for smokeless flaring. Systems are available to stage steam injection according to need.

4.2.2 Low-Pressure Forced Air (Air Assisted Flare) for Pipe Flare Tips

A low-pressure forced-air system is usually the first alternate if on-site utilities are not available to aid in producing a smokeless operation. The system is designed to create turbulence in the flame zone by injecting low-pressure air in large volumes. The air is supplied from blowers and is injected across the flare tip as the waste-gas is being ignited. This promotes even air distribution throughout the flames. The excess air required is normally 20% to 30% of stoichiometric air for saturated hydrocarbons, and 30% to 40% of stoichiometric air for unsaturated hydrocarbons. Air is supplied at a gauge pressure of 0.005 – 0.025 bar (2" – 10" wc). This system has a higher initial cost due to the requirement for a dual stack and air blowers. However, air assisted flaring has lower operating cost than a steam-assisted design since air assist requires just power for the blowers.

4.2.3. High Pressure Air for Pipe Flare Tips

High-pressure air can also be used to prevent smoke formation. This approach is less common because compressed air is usually more expensive than steam. Smoke abatement using high-pressure air is preferred in arctic or low-temperature applications where steam can freeze and plug the flare tip/stack. Other applications include desert or island installations where there is a shortage of water for steam, or where the waste-flare gas stream reacts with water. The air is usually provided at a gauge pressure of 6.8 bar (100 psi). The mass quantity required is approximately 1.0 lb of air per lb of gas flared. To cover the power outage flaring case, a diesel-driven compressor would be necessary in addition to a main electric compressor.

4.2.4. High Pressure Fuel Gas for Pipe Flare Tips

High-pressure fuel gas can also be used to prevent smoke formation. This approach is less common because fuel gas enhances air aspiration but it not as effective as high-pressure air. The situations used are similar to that for high-pressure air. If high-pressure fuel gas is available, then this option eliminates the need for an air compressor.

4.2.5. Water Injection for Pipe Flare Tips

High pressure water, while quite uncommon, is also used to control smoking. The mass quantity required is approximately 1.0 lb of water per lb of gas flared at 50 psig to 100 psig. 1 lbFreeze protection is required in cold climates and because of the difficulty in controlling the water line at low flaming rates, it is not so common.

4.2.6. Pressure Assisted Burners

Pressure assisted flares uses the flaring gas pressure to promote mixing at the burner tip. If enough gas pressure is available (15 psig or more, depending on the flare gas composition), these flares can be applied to streams previously requiring steam or air assist for smokeless operation and can be used in flares that are not enclosed. On the other hand, enclosed flares with sufficient draft and residence time, can burn smokeless at low pressures. Refer to Section 2 on burners and Section 3.3.2 on enclosed flares..

5 Efficiency of Flares

5.1 Efficiency Definition

Efficiency of flares is one of the primary concepts underlining emissions and air pollution regulations on flaring . Efficiency may be defined two ways: combustion efficiency or destruction efficiency.

Combustion efficiency (CE) is the percentage of hydrocarbon, carbon monoxide and soot that are completely oxidized to carbon dioxide. Combustion efficiency is 100% when all components, including CO are oxidized. (Water vapor is the other product of complete combustion.) Combustion Efficiency is also called Control Efficiency.

Combustion efficiency % = 100 * CO2 / (CO2 + CO + THC + Soot)

Destruction efficiency (DRE) is the percentage of hydrocarbon, carbon monoxide and soot that are oxidized to carbon dioxide and carbon monoxide. Destruction efficiency is greater than or equal to combustion efficiency since the definition of destruction efficiency does not include carbon monoxide oxidized to carbon dioxide. In general, a destruction efficiency of 98% is equivalent of a combustion efficiency of 96.5% in a typical pipe flare. On the other hand, for a pressure-assisted burner operating at high destruction efficiencies, the combustion efficiency is essentially the same as the destruction efficiency

Destruction efficiency % = 100 * (CO2 + CO) / (CO2 + CO + THC + Soot)

Where:

- CO2 = parts per million (volume) of carbon dioxide in flare plume
- CO = parts per million (volume) of carbon monoxide in flare plume
- THC = parts per million (volume) of total hydrocarbon as methane in flare plume
- Soot = parts per million (volume) of soot as carbon in flare plume. (Soot is nominally zero in many cases and often neglected in the formula.)

5.2 Open Pipe Flare Efficiency Studies

Flare combustion studies begin in earnest in the 1970s, increased with the major studies in the 1980s, and continue today, with over 200 published studies and models. Of the landmark flare tests performed in 1980's one of the most important was the "Flare Efficiency Study" in 1982 that led to the development of emission factors for open pipe flares [3] Refer to box.

> "Flare Efficiency Study", Marc McDaniel, EPA-600 / 2-83-052, July 1983, was conducted to determine the combustion efficiency and destruction efficiency of full-scale pipe flares (air-assisted, steam-assisted and non-assisted) over a wide operating range.
>
> Separate elements of the study were sponsored by the EPA and the Chemical Manufacturers Association (CMA) with other participants including John Zinc Company, Opti Metrics and Engineering Science, Inc.

Following the 1982 study, follow-up studies added more insight to pipe flare efficiency. These studies explored a multitude of topics including:

- Lower Flammability Limit (LFL) and Lower Heating Value (LHV) in the combustion zone of a steam-assisted flare
- wind effects on combustion efficiency
- flame shape and dimensions
- temperature in the combustion zone
- excess steam and air
- flame stability and lift-off
- sulfur dioxide flaring

5.2.1 High Combustion Efficiency

These other studies have determined that high combustion efficiency for pipe flares requires adequate residence time, turbulent fuel-air mixing and sufficient flame temperatures. Combustion efficiency is also improved for LHV greater than a limit, typically 270 BTU/SCF (in the combustion zone), with low to moderate exit velocities which facilitating a stable flame. Properly operated pipe flares, operating within the "stable flame envelope" have destruction efficiencies and combustion efficiencies 98%.

Combustion efficiency has been quantitatively correlated with many parameters. Figure 19 is the correlation for a steam-assisted pipe flare for lower flammability limit LHV as measured in the combustion zone. The LHV is one of the best predictors of combustion efficiency for steam assisted flares. This example is only appropriate for a steam-assisted flare. The different definitions of lower heating value are given in Figure 20.

With air-assisted flares, low excess air is one of the best predictors of high combustion efficiency.

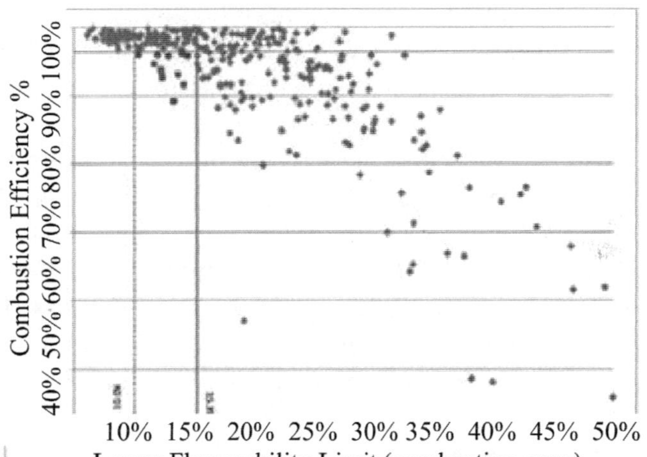

Figure 19

5.2.2 Poor Combustion Efficiency

On the other hand, poor combustion efficiency for pipe flares with steam assist and air-assist is promoted by many factors. Note there is still dispute of the degree of the following guidance. The guidance does not apply to flares utilizing "pressure-assisted burners" such as used ground flares or enclosed flares.

- "Over-Steaming" in steam assisted flares occurs near the region of flame instability and high-steam flows at low flare loads. In general, steam has minimal effect on combustion efficiency. But over-steaming dilutes the fuel-air mixture and cools the flame. As an approximation, over-steaming occurs when the steam-flared gas ratio exceeds 0.5 (weight basis). In contrast, insufficient steam facilitates poor fuel-air mixing and enhances smoke production, as explained in the section on smoking.
- Excess Air in air assisted flares cools the flame and dilutes the heat value. Excess air cools the flame and dilutes the heating value. As an approximation, excess air occurs when mass flow rate of assist air is approximately seven (7) to ten (10) times greater than the stoichiometric air mass flow rate. As a point of reference, assist air rates are generally less than half of stoichiometric air.

Lower Heating Value (LHV)
Combustion Zone or Relief Composition
Figure 20

The LHV lower heating value is one of the best predicators of combustion efficiency. However, there are two measures of LHV.

- LHV based on relief composition. This is the LHV of the relief stream to the flare. This is used in regulations before 2015.
- LHV in the combustion zone. This is defined as the gas composition in the combustion zone and includes the flow rates of the relief gas, pilot gas, and all steam just outside the flare tip, where combustion is supposed to take place. It is a better indicator of efficiency than the heating value of the relief gas. This definition was developed in several EPA enforcement settlements in the early 2010s.

- Wind. Low wind speeds do not significantly affect combustion efficiency. However, there are disagreements where wind speed impacts combustion efficiency which range from speeds over 22 MPH (32 ft/sec) to over 13 MPH (19 ft/sec).
- Flame Lift-Off for pipe flares. Flame liftoff occurs when flame (bottom of the flame) is physically located above the tip of the pipe flare. This can ultimately lead to flame extinguishment. Conditions for flame liftoff can be estimated from the flare gas velocity for pipe flares as compared to a lift-off criterion determined from tests. The criteria are a function of gas composition and flare gas and air velocities.

5.3 Pressure-Assisted Flare Efficiency Studies

Flare combustion studies and modeling for flares using pressure-assisted tips started in the 1980s and continue. Besides field measurements, computational fluid CFD dynamic models are used to extend results for mixtures and parameters outside test results.

Pressure assisted burners use high inlet pressure to the burner and the kinetic energy from the exit velocity to entrain and mix combustion air. This provides smokeless and high-efficiency combustion over a wide range of velocities up to sonic, for a range of gas compositions and Lower Heating Value LHV. Ground flare flares with pressure-assisted burners have nominal destruction efficiencies of approximately 99.5%. Enclosed flares with pressure-assisted burners have higher nominal destruction efficiencies of approximately 99.9%. Enclosed flares can obtain accurate measure of combustion efficiency because they have a tight enclosure. The EPA recommends using test data instead of emission factors for enclosed flares. (See Section 6 for emission factors.) Actual flare efficiencies differ between types of burners, types of enclosures and operational factors with each situation and design providing a different value. The above values are for general relative comparisons.

6 Emissions from Flares

Flare efficiency is closely related to emissions. Control of emissions are important for federal regulations and ambient air quality. These emissions include nitrogen oxides (NO_x), sulfur oxides (SO_x), carbon monoxides, particulate matter (PM_{10}), and volatile organic compounds, and unburnt hydrocarbons. Sulfur and nitrogen containing compounds may react with oxygen to form their respective oxides. However, nitrogen oxide (NO) may also be formed from the nitrogen in air by direct oxidation. Emissions from flares may be determined from the document "AP-42, Compilation of Air Pollutant Emission Factors", 5th Edition, EPA, 2018. [4] Refer to box.

6.1 Emissions from Pipe Flares

To estimate emissions from pipe flares, the U.S. EPA in 1991 updated AP-42 to provide emissions factors for industrial flares. The emissions factors were developed based the pipe flare efficiency study performed in 1982 and subsequent studies.

6.2 Emissions from Pressure-Assisted Flares

In the 1990s, it was recognized the emission factors were not accurate and more studies were performed. In February 2018, chapter 13.5 of AP-42 was revised with new emission factors for flares. One important change was to add emission factors for pressure-assisted flares as well as pipe flares. Another change in the 2018 edition was to revise the criterion for a well operated [pipe] flare to have 98% destruction efficiency (DRE) rather than 98% combustion efficiency (CE) as in the 1995 edition. The effect was to slightly lower the required efficiency since CE is lower than, or equal to, the DRE.

> The AP 42, Compilation of Air Pollutant Emission Factors, was first published by the US Public Health Service in 1968. Since 1972, AP-42 has compiled EPA's emissions factors. The 4th Edition, Supplement D, 1991, added a new chapter 11.5 Industrial Flares with flare emission factors. The 4th Edition, 1995 changed the industrial flare chapter to 13.5. In the 5th Edition, 2018, the flare emissions factors were revised in chapter 13.5, including pressure-assisted flares and pipe flares.

7 Radiation from Flares

The amount of radiation from an elevated flare (pipe flare or multi-tip-pressure-assisted flare) is one of the most important factors in siting the flare. On the other hand, the amount of radiation can be safely absorbed by the target is also important.

7.1 Radiation Generation

7.1.1 Pipe flares Radiation

There are many correlations based on field data to estimate radiation generated by the flare, most of them for pipe flares. API Standard 521-Annex C and F provides models that have been successfully used and are adequate in most situations. The traditional model is the "point source" where the flame is at a distance that is can be modeled as a single point. In actuality, the flame shape and length are not a point source, so the models use empirical functions to reduce the shape to a point source.

The models are calibrated against field data. Currently there is an assortment of high-quality experimental data to evaluate pipe flare radiation models or for developing correlating factors. This data allows a comparison of mathematical models for varied configurations: high to low flows; close-to-stack and far-from-stack; crosswind, up-wind and down-wind. To match the model with field data, most of the point source models use either two or three correlating factors. The correlating factors are specific to the model and field data. The correlating factors determined for one model should not be used with another model.

- The "F" factor is the fraction of the combusted relief flow that is radiated, as determined as the net heating value times the mass flow rate. The "F" factor has the most effect on the radiation since the value has an order of magnitude of about 0.3, depending on the model and data set.
- The "tau" or "τ" or "transmissivity" is the fraction of radiation that is absorbed by water vapor in the air. It has an order of magnitude of about 0.7.
- The "alpha" or "α" corrects the radiation for objects close to the flare.

7.1.2. Pressure-Assisted Burners Radiation

These burners are employed in flare fields that utilize multiple burners, such as enclosed flares, ground flares and elevated multi-point flares. The radiation is not amenable to simple analysis such as a source-point model since the burners shield radiation, the flame is shorter and stiffer, and the different amounts of air within flare field affect the combustion of individual burners. The calculation method of choice for the last several decades has been computational fluid dynamics (CFD). To a certain, extent, radiation may be shielded by the wind fence by approximately two-thirds. An enclosed flare will shield almost all direct radiation.

7.2 Target Radiation

7.2.1 Personnel Radiation

Radiation on personal is fully discussed in API Standard 521 among other publications, including whether to include the solar radiation as part of the total radiation on the target. API Standard 521 provides recommended thermal radiation for personnel.

7.2.2 Equipment Radiation

Radiation is important for design of equipment, control systems and structures. The following are some examples of damage effects from radiation. For damage effects based on temperature, consult API Standard 579 [7].

Effect	**Radiation Heat Flux (Btu/hr-ft2)**	**Reference**
Wood ignites with flame	4,000 after 10 minutes exposure (minimum flux)	Barry [5]
Fuel Oil ignites	5,500 at 1-minute exposure	SPE [6]
Cable Insulation Degrades	63,500 after 10 minutes exposure	Barry [5]]
Damage to process equipment (steel tanks, machinery, etc.). Steel columns - uninsulated and load bearing at 1,100°F	11,100 to 11,900 after 10 minutes exposure	Barry [5]

8 Flaring Reduction Initiatives

Reduction of flaring has been a concern for over a century. One of the first laws to regulate flaring in Texas was in 1899, when the Texas State Legislature declared that any gas well was to be "shut-in within ten days after its completion until such time as the gas produced therefrom is to be used for light, fuel or power purposes". After a series of rulings and three landmark flaring cases in 1947 and 1949, the Texas Supreme Court and Texas Court of Civil Appeals upheld regulations to prevent the "wasteful flaring of casinghead gas".[8]

In late 1999, officials of the U.S. Environmental Protection Agency (Region 6) began a series of meetings with eleven companies responsible for a large portion of unplanned releases in the region to share information on avoiding flaring. The Episodic Release Reduction Initiative (ERRI) was an outgrowth formed in September 1999 and included EPA (Region 6), Texas Natural Resource Commission (TNRCC), Louisiana Department of Environmental Quality, and thirteen petrochemical companies to "evaluate and implement methods for emissions reduction, for the measurement and reporting of progress and for information sharing".

Starting in 1999 and extending into the early 2000s, several Texas organizations began flare minimization outreach programs including Texas Natural Resource Conservation Commission (TNRCC), a local business group of manufacturing firms, and non-governmental organizations such as the Environmental Integrity Project. When the Texas legislature considered bills to reduce emissions, operating companies responded and their results were publicized in trade magazines and papers presented at meeting such as the AIChE and NPRA (National Petroleum Refiners Association - now American Fuel & Petrochemical Manufacturers AFPM).

On the other hand, the EPA recognized that flare minimization also led to flares operating at reduced capacity resulting in over-steaming, excess air and diminished flare efficiency. New flare performance standards (see Section 9) are intended to rectify this issue.

However, in the last few years, flaring has increased in the Permian Basin, and impacting the University of Texas's McDonald Observatory in west Texas. The Texas Railroad Commission, the Big Bend Conservation Alliance and other organizations are working to minimize light pollution from flaring.

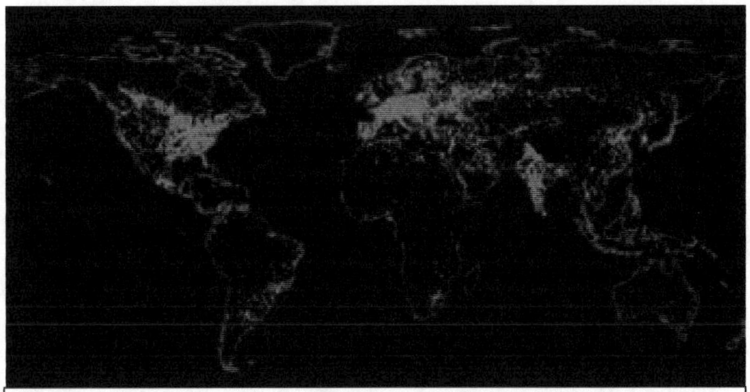

"City Lights and Gas Flares", 1994-1995 satellite data. Flares are RED and city lights are GREY (World Bank and GGRP, 2001). Other organizations such as NASA and US NOAA also publish data

Internationally, the "Global Gas Flaring Reduction Partnership" (GGFR), a public and private partnership created at the 2002 World Summit on Sustainable Development (Johannesburg), intends to improve energy efficiency, improve access to energy, reduce carbon emissions and decrease the environmental impact of gas flaring. [9] The partners include the World Bank, European Union, European Bank for Reconstruction and Development (EBRD), major petrochemical companies and governments (the US was a former member but has dropped out). The organization has been collecting flaring data (satellite and reporting), and analyzing the fundamental causes of flaring, regulatory actions and flaring reduction mitigation. The current program is "Zero Routine Flaring by 2030".

Satellite view of flares from US National Oceanic and Atmospheric Administration (NOAA). Courtesy of GGFR

There are other international state and non-state organizations supporting flaring reduction. Prominent among the oil and gas industry is the International Petroleum Industry Environmental Conservation

Association (IPIECA) which is an oil and gas industry association for environment and social issues . Established in 1974 in response to the United Nations Environment Program, its members include oil and gas companies and industry associations. The International Association of Oil and Gas Producers (IOGP), founded in 1974, is a global forum for the petroleum industry to share best practices. The organization collects and publishes data on flaring and gas emission. The International Energy Agency IEA and the US Energy Information Agency EIA also track emissions.

9 Flare Regulations

Flare regulations have an a long, complex and changing history. This section will provide an overview.

The regulatory implications for flare design and performance can be quite complicated. The following is only a high-level summary of the federal regulations regarding flaring, and does not address regional, state or local regulations regarding flare design and performance. The reader is encouraged to engage the services of specialty legal and regulatory consultants regarding regulatory requirements applicable to your specific process design and project location.

9.1 The U.S. Clean Air Act CAA

The U.S. Clean Air Act CAA stipulates that two groups of air pollutants have national ambient air quality standards (NAAQS).

9.1.1 Criteria Pollutants

Criteria pollutants are pollutants whose "criteria" were established from health and environmental damage studies. The six criteria pollutants are lead, particulate matter, carbon monoxide, sulfur dioxide, nitrogen dioxide, and ground-level ozone.

9.1.2 Hazardous Air Pollutants

Hazardous air pollutants (HAPs or Air Toxics) include a total of almost 200 compounds including asbestos, benzene, chlorine, etc. The list was first issued in the 1990 Clean Air Act Amendments Section 112(b). The industries cover refineries, oil and natural gas production facilities, synthetic organic chemical, polymer manufacturing and others. Every industry produces different types of hazardous air pollutants based on their product manufacturing. Therefore, to reduce HAP air emissions, the Environmental Protection Agency EPA developed standards known as "National Emission Standards for Hazardous Air Pollutants" (NESHAPs) and codified in the EPA Code of Federal Regulations 40 CFR Part 61. The subparts cover selected pollutants and their industries.

9.1.3 NSPS and MACT

To accomplish the goal of reducing hazardous air pollutants (HAPs), each industry must comply with two requirements.

- New industrial sources of emissions must conform to levels per NSPS or "New Source Performance Standards" published in 40CFR Part 60 "Standards of Performance for New Stationary Sources". Each industry or source has its own subpart with over 200 industries listed. For example, 40CFR Part 60-Subpart J -refers to the "Standards of Performance for Petroleum Refineries".

- Flares must meet a minimum standard known as MACT or "Maximum Achievable Control Technology" and codified in 40CFR Part 63 "National Emission Standards for Hazardous Air Pollutants (NESHAPs) for Source Categories". Each industry or source has its own subpart with about 300 industries listed. For example, 40CFR Part 63-Subpart CC -refers to the "NESHAPs from Petroleum Refineries". Similarly, 40CFR Part 63-Subpart UUU -refers to the "NESHAPs for Petroleum Refineries: Catalytic Cracking Units, Catalytic Reforming Units, and Sulfur Recovery Units".

9.2 General Duty Requirements to Minimize Emissions

The standards produced general duty clauses which encompassed the goal of emission reduction. For flares, the general duty clauses for both MACT and NSPS are quoted. They are very similar in concept but differ slightly in phrasing.

- <u>NSPS 40 CFR §60.11(d)</u>. The citation is 40 CFR Part 60 "Standards of Performance for New Stationary Sources". Subpart A, Section 60.11 "Compliance with standards and maintenance requirements", Clause (d). "At all times, including periods of startup, shutdown, and malfunction, owners and operators shall, to the extent practicable, maintain and operate any affected facility including associated air pollution control equipment in a manner consistent with good air pollution control practice for minimizing emissions.
- <u>MACT 40 CFR §63.6(e)(1)(i)</u>. The citation is 40 CFR Part 63 "Standards for Hazardous Air Pollutants For Source Categories", Subpart A, Section 63.6 "Compliance with Standards and Maintenance Requirements", Clause (e)(1)(i). "At all times, including periods of startup, shutdown, and malfunction, the owner or operator must operate and maintain any affected source, including associated air pollution control equipment and monitoring equipment, in a manner consistent with safety and good air pollution control practices for minimizing emissions."

9.3 Open Pipe Flares OPF - Performance Standards

Following the flare efficiency studies in the 1970s and predominately in the 1980s (see Section 5), the rules for minimizing emissions and sizing the open pipe flare tip were consolidated in NSPS 40CFR60.18 and MACT 40CFR63.11. Different industries have different standards.

<u>9.3.1 General Industry - Performance Standards for Flares</u>

The performance standards for most industries of concern are provided in the following for both MACT and NSPS. The two industry exceptions are the performance standards for the refining industry and the organic chemical manufacturing sector, which are discussed following.

- <u>NSPS 40 CFR §60.18(b)</u>. The citation is 40CFR Part 60 "Standards of Performance for New Stationary Sources". Subpart A "General Provisions", Section 60.18 "General control device and work practice requirements". Clause (b) "Flares". To minimize emissions, the pipe tip sizing is based on complying with 98% destruction efficiency. The rules apply to pipe tips that are air-assisted, steam-assisted, or non-assisted. Based on the data from the efficiency studies, the primary factors are lower heating value (LHV between 200 and 300 BTU/SCF), type of efficiency assist and velocity. The velocity is calculated at standard conditions (14.696 psia,

68°F, 385.33 SCF/mole) to not penalize a hot release. Hydrogen, since it enhances the combustion efficiency of pipe flares, has slightly different rules.

- MACT 40 CFR §63.11(b). The citation is 40CFR Part 63 "NESHAPs for Source Categories". Subpart A "General Provisions", Section 63.11 "Control device and work practice requirements". Clause (b) "Flares". This section duplicates the rules in 40CFR60.18 and is applicable for pipe tips that are air-assisted, steam-assisted, or non-assisted.

9.3.2 Refinery Industry - Performance Standards for Flares - MACT 40CFR63.670

The EPA, in an effort to reduce emissions from the refining industry, promogulated new regulations for refining pipe flares, which supersede those of NSPS 40CFR60.18 and MACT 40CFR63.11 for refinery flares. The rules are published in 40CFR Part 63-Subpart CC "NESHAPs from Petroleum Refineries", Section 63.670 "Requirements for flare control devices" and known as the Refining Sector Rules (RSR). The final rule was issued December 2015 with the latest amendment published in February 2020.

Like the current 40CFR60.18 and 40CFR63.11 statues, the 40CFR63.670 rules distinguish between air-assisted, steam-assisted, and non-assisted flares. Exit velocity are limited. However, two new Net Heating Value parameters are introduced: NHV of the combustion zone and NH-dilution. A major difference for refinery flares is that the owner has to define the maximum smoking capacity where emissions would not be exceeded without a violation. If the flowrate and emissions exceeded the maximum, then the facility had to perform actions to avoid a re-occurrence. Other requirements include monitoring of composition, flow rates, visible emissions, flame and elimination of the Startup-Shutdown & Malfunction (SSM) exemption; refer to Section 9.5

9.3.3 Organic Chemical Manufacturing - Performance Standards for Flares - MACT 40CFR63.2450(e)

The EPA, in an effort to reduce emissions from organic chemical manufacturing, proposed new regulations for pipe flares, which supersede those of NSPS 40CFR60.18 and MACT 40CFR63.11. The rules are similar to the Refinery Sector Rule and cross-references the refinery rule outlined in 40CFR63.670. The final rules will be published in 40CFR Part 63 Subpart FFFF "NESHAPs: Miscellaneous Organic Chemical Manufacturing", Section 63.2450(e) "Requirements for flare control devices". The final rule prepublication was issued 29 May 2020. Rules for other industries are expected.

9.4 Pressure Assisted Flares - Performance Standards

Until 2015, all performance standards were for open-pipe flares limited to a certain velocity (for comparison, approximately 0.3 Mach through the open tip). There were no specific performance standards for pressure-assisted flares, which generally are choked through the burner (Mach 1), although some studies on efficiency of pressure-assisted flares were performed in the 1980s. This issue has been acknowledged since pipe flare requirements were issued.

As experimental efficiency data became more prevalent, and pressure-assisted flares became more popular, the need for legislation on pressure-assisted flare was crucial. The Clean Air Act and rules in 40CFR 60 (NSPS), 40CFR61 (NESHAPs) and 40CFR63 (MACT) allow for an "Alternate

Means of Emission Limitations" (AMEL) for a flare to be permitted if the flare can demonstrate equivalent reduction in emissions. Tests had to prove equivalent destruction efficiency, no visible emissions over the full range of compositions (NHV / LFL), pressures, flow rates and tip velocity. A streamlined approval process for AMEL was finalized in the Federal Register on April 31, 2016 (81FR-page 23480). The rule using pressure-assisted burners has been applied in the refining sector and proposed for organic chemical manufacturing. It is expected that similar demonstrations can be developed for all flares (elevated, ground, enclosed) that use pressure-assisted-burners.

9.5 State Implementation Plans and Startup-Shutdown & Malfunction (SSM)

Startup-Shutdown & Malfunction refers to the rules for flare efficiency when flaring during these scenarios. The rules have a disorderly history, so a brief introduction is warranted. Startup and shutdown cases are generally specific to the operating unit and are defined by the regulation. A malfunction is breakdown of an operation. "SSM Events" are periods when emissions exceed limits during these scenarios.

Scheduled maintenance as well as Startup-Shutdown and are considered "planned" as part of normal operation and must meet emission limits. On the other hand, malfunctions, careless operation, poor maintenance and "unplanned startup-shutdown" events, such as during an accident, are not permitted. The facility may claim an "Affirmative Defense" to void monetary penalties. An example of affirmative defense is a failure which is infrequent, could not be reasonably prevented, quickly rectified and documented with a root cause analysis. Injunctive relief is the requirement for the facility to revise the process to prevent a re-occurrence.

"SIP Calls" refer to the instructions for revising a State Implementation Plan (SIP) when it is considered substantially inadequate to for national ambient air quality standards (NAAQS). Major calls have been issued every decade in 1980s. Following are some of the more interesting.

- 1982 - Malfunctions are not automatically exempted from emission standards although the use of affirmative defense is allowed.
- 1983 - Unavoidable excess emissions during Startup-Shutdown are exempted and not violations if infrequent and short
- 1999 - States allowed to address excess emissions during short and infrequent start-up and shutdown periods with a broad affirmative defense
- 2001 - Issued guidance document on SSM which would provide new rules to rectify inconsistent SIPs
- 2008 - Court decision - vacated SSM exemption, overturning 1983 guidance
- 2013 - 2015 Clarified planned and unplanned startup-shutdown rules with some states having to revise their State Implementation Plans (SIP) to remove automatic exemptions for SSM Events and eliminate the affirmative defense.

10 Conclusions

This paper provides an overview of flare system to cover aspects that are not covered in detail in the complementary API Standard 537 "Flare Details for General Refinery and Petrochemical Service", providing information on flare burners and structures, smokeless flaring, flare efficiency and emissions, flaring reduction initiatives and Federal regulations on flares.

11 References

Reference in order of first citation

[1] "Flare Details for General Refinery and Petrochemical Service", API Standard 537, 3rd Edition, 2017, with Addendum 1, June 2020

[2] "Pressure-relieving and Depressuring Systems", API Standard 521, 7th Edition, June 2020, American Petroleum Institute, Washington D.C.

[3] "Flare Efficiency Study", Marc McDaniel, EPA-600 / 2-83-052, July 1983

[4] "AP-42 Compilation of Air Pollutant Emission Factors, Volume 1: Stationary Point and Area Sources", Environmental Protection Agency EPA, 5th Edition, 2018

[5] "Fire Exposure Profile Modeling- Some Threshold Damage Limit (TDL) Data", T.F. Barry, 2003

[6] "Simplified Radiation Heat Transfer Calculations from large Open Hydrocarbon Fires', P.J. DeNenno, Air SFPE Technology Report TR82-9, Society Fire Protection Engineers, 1982

[7] "Fitness for Service", AI Standard 579, 3rd Edition, 2016, American Petroleum Institute, Washington D.C.

[8] "History of the Railroad Commission", State of Texas, <http://www.rrc.state.tx.us/about/history/index.php> Retrieved April 2020

[9] "Global Gas Flaring Reduction Partnership", World Bank, <www.worldbank.org/ggfr>. Retrieved April 2020.

584034 Methodology for Conducting Flare System Assessment in Petroleum Refinery

Abstract:

Presenter Information

Presenter Name:

Bejoy Bharatiya

Contact Information

Contact Name: Bejoy Bharatiya

Telephone No.:

+61 8 9440 0440

M: +966 553 543 149

Fax No.: +61 413 098 008

E-mail Address (Required):

pinkubejoy@iinet.net.au

Presenter Biography: (50 words Maximum)

Bejoy Bharatiya a Chartered Professional Engineer (CPEng), is Senior Process Safety Engineer at Saudi Aramco Shell Refinery (SASREF), and has over 30 years' of professional experience in the upstream oil & gas, refining, petrochemical, and mineral processing industry. Prior to joining SASREF, he was the Process Safety Consultant at SANTOS (South Australian Oil Search) Pty Ltd. Australia, Senior Operations Engineer with Saudi Aramco, and the Principal Process Engineer of Brown & Root Energy Services (currently, Kellogg, Brown & Root), Australia.

Paper/Presentation Title: Flare System Assessment-A Basic Approach

Abstract: (1500 words Maximum):

Methodology for Conducting Flare System Assessment in Petroleum Refinery

Bejoy K. Bharatiya

Saudi Aramco Shell Refinery

The petroleum industry has long since recognized certain risks are involved in the flare and relief system of a hydrocarbon processing facility. Over the years, the oil and gas industry has learned lessons from a number of process safety events and various hazards to flare systems such as loss of control and loss of containment top events that could result in: fires, vapour cloud explosions (VCE), toxic and asphyxiant exposure, which occurred in various parts of the world and are not limited to the country of origin. The design and operation of flare system s have been the subject of concern to the oil industry.

It has been the continuous effort of the industry to update the design and operating philosophies to improve the inherent safety of the flare and relief systems. The flare system is the last line of defense for various emergency scenarios. The present paper is concerned with defining the major safety considerations of international oil companies including Saudi Aramco and Shell, and International Industry Standards in designing and operating the flare systems. The flare system scope is defined in this paper as the piping and equipment downstream of the relief devices, control valves and equipment vents tied into the flare system all the way to the flare tip including knock out vessels and seal vessels.

These design considerations include basic design philosophies, and safety instrumentations in relief valve back pressure limitations and sizing of the discharge piping. Major design considerations and hazard assessment for flare system to be discussed in this paper are described below. Some oil companies use required relief valve capacity for sizing the closed discharge systems, rather than rated capacity which is mentioned in API 520 Part II. The following relief load reduction credits are used for designing the flare system:

1. Flare system design philosophies - Generally, there are three permitted methods typically adopted by some of the leading oil companies for flare load mitigation: Use of 100-50-50 method; Flare IPF (Instrumented Protective Functions) method; and Flare Bundle method.

a) Use of 100-50-50 methodology

· The impact of the common discharge system design produced by relief flows from the connected process units as a result of global utility failure is calculated by taking 100% of the quantities established for the relief and de-pressuring load from each unit in turn, together with 50% thereof for the other units.

· If it is determined that multiple trains will discharge simultaneously, the relief loads from those trains shall be taken at 100% (e.g. parallel trains that would simultaneously relieve or de-pressure during a utility failure).

· The individual relief valve backpressure shall be taken as the highest value from any of the cases mentioned above.

· 100-50-50 method shall not be combined with the other methods described below.

This methodology considers beneficial effects of dynamics, operator intervention, control and safeguarding systems without the need to specify these. The resulting design is ALARP based on the long operational experience demonstrated by the company.

b) Flare IPF method - This considers beneficial effects of all safeguarding systems on the flare loads. Additional inputs required are: mitigated relief loads (relief loads when the safeguarding is effective) and the SIL classification of the safeguarding systems.

• The flare IPF method shall not be combined with the 100-50-50 methodology to analyze the same common mode scenario.

• The system shall meet the tolerability requirements with any single IPF failing (not working as designed).

• The combination of up to three concurrent individual or partial IPF failures resulting in total flare loads exceeding the acceptance criteria be designed to the appropriate SIL level to meet the acceptance criteria.

• It results in an enumerated probability of exceeding design requirements such as backpressure, flare radiation etc. The resulting design is ALARP based on the application of SIL classification to manage the risk.

c) Flare Bundle method - This considers beneficial effects of safeguarding systems as well as basic process control and operator intervention. Required inputs are following parameters:

I. common mode relief loads for all utility failures

II. as I above. but mitigated relief loads (relief loads when the basic process control or operator is response is effective)

III. the probability of the mitigation of effective operator response (to be established and documented in a team review)

It results in a design with probability of exceeding design requirements being managed both by systems and by operator competencies. The resulting design is ALARP.

This method requires a commitment to establish systems capable to maintain each of the flare load management instrumentation barriers as completely independent of each other. This method requires lifetime commitment to maintain and assure the ongoing effectiveness of the barriers.

The advantage of this approach is that additional instrumentation (basic process control or IPF) barriers can be installed on a system to reclassify loads from the Likely Load, or Somewhat Likely categories, where statistically a fewer number of loads are required for the analysis.

The above methods (a, b, c) rely on the hydraulic model for the flare system network based on the accurate input data (relief loads and piping data). Generally, software/applications such as Aspen Flare System Analyzer or, Honeywell UniSim Flare is used for the network analysis. There are also other software available in the market for performing the hydraulic calculations.

1. **Hazard assessment** – Hazard assessment is based upon the hazard of the materials present in the processing facility. Some of the highest ranking included in the risk assessment (taken from facility Hazard Register) are:

Hazard	Top Event	Consequence
All Liquid Hydrocarbons (HC) including condensed flare gas	Loss of Containment	Fire
	Loss of Control or	

Toxics / H_2S / N_2	Containment	Personnel Exposure
All HC, including H_2 enriched HC	Loss of Containment	Generation of vapour cloud leading to VCE or fire
Heavier than air HC	Loss of Containment	Vapour cloud and/or Fire resulting from heaving than air HC

1. Some of the major **process threats** to the flare systems will be discussed are following:

I. O_2 Ingress into the Flare System

II. Loss of Pilots / Flame

III. Flare Line Entry / Live Flare Opening

IV.Overpressure due to Blocked Flow

V.Flare Header Liquid Hold-up / Slug Flow

VI.Flare Stack Structural Failure

VII.Hydrocarbon / H_2S Breakthrough to Sewer / Atmosphere

VIII.Backpressure Exceeding Flare System Design leading to Equipment Overpressure

IX.Flow induced vibration

X.Acoustic Fatigue

XI.Low-temperature effects/brittle fracture

XII.Radiation

Etc.

4. Flare Network Analysis – In the industry, there are many software that are being used to carry out the network analysis, viz. UniSim Flare and Aspen Flare System Analyzer.

Overview of our network analysis as an example will be presented to describe the effect of global power failure case (100%) and the results of high backpressure and/or high noise from certain relief valves in the process unit.

This paper discusses industry experience and summarizes the salient design considerations and risk assessment to improve safety integrity of the flare system. An effort has been made to highlight the major design approaches/ methodologies adopted by oil companies using company design standards, and industry standards like API. Emphasis has been made to increase the awareness among the process design engineers to develop safer designs of the flare systems.

586906 Develop a Hazard Index Using Machine Learning Approach for the Hazard Identification of Chemical Logistic Warehouses

Abstract:

With the rapid development of chemical process plants, the safe storage of hazardous chemicals become an essential topic. Several chemical warehouse incidents related to fire and explosion have been reported recently. Therefore, an accurate hazard identification method for the logistic warehouse is needed not only for the facility to develop a proper emergency response plan but also for the residents who live near the facility to have an effective hazard communication. Furthermore, the government can better allocate the resources for first responders to make fire protection strategies, and the stakeholders can lead to improved risk management. Hazard index is a helpful tool to identify and quantify the hazard in a facility or a process unit. The challenge for this research is to improve the current method with the novel technique to implement our purpose.

The first objective of this research is to develop a "Storage Hazard Factor" (SHF) to evaluate and rank the inherent hazards of chemicals stored in logistic warehouses. In the factor calculation, the inherent hazard of chemicals is determined by various parameters (*e.g.*, the NFPA rating, the flammability limit, and the protective action criteria values, *etc.*) and validated by the comparison with other indices. The current criteria for flammable hazard ratings are based on flash point, which is proved to be insufficient. Two machine learning based methods will be used for the classification of liquid flammability considering aerosolization based on DIPPR 801 database. Subsequently, SHF and other warehouse safety penalty factors (*e.g.*, the quantity of the chemicals, the distance to the nearest fire department, *etc.*) are utilized to identify the Logistic Warehouse Hazard Index (LWHI) of the facilities. Last but not least, this method is applied to real-time data from Houston Chronicle, and several statistical analyses are used to prove the hazard index is helpful for hazard identification to emergency responders and hazard communication to the public.

Use of Historical Data to Provide Process Leak Frequency Estimates

Brian Bain
DNV GL
Cromarty House, 67-72 Regent Quay, Aberdeen, AB11 5AR
brian.bain@dnvgl.com

Zoë Wattis
DNV GL
Ashby Road Holywell Park, Loughborough, LE11 3GR
zoe.wattis@dnvgl.com

Carol Humphreys
DNV GL
Ashby Road Holywell Park, Loughborough, LE11 3GR
Carol.humphreys@dnvgl.com

Prepared for Presentation at
American Institute of Chemical Engineers
2020 Spring Meeting and 16th Global Congress on Process Safety
Houston, TX
March 30 – April 1, 2020

Use of Historical Data to Provide Process Leak Frequency Estimates

Brian Bain
DNV GL
Cromarty House, 67-72 Regent Quay, Aberdeen, AB11 5AR
brian.bain@dnvgl.com

Zoë Wattis
DNV GL
Ashby Road Holywell Park, Loughborough, LE11 3GR
zoe.wattis@dnvgl.com

Carol Humphreys
DNV GL
Ashby Road Holywell Park, Loughborough, LE11 3GR
carol.humphreys@dnvgl.com

Keywords: Hydrocarbon, Leaks, Releases, Frequency

Abstract

A key driver in the risk assessment of an offshore installation or process plant is the leak frequency as this is combined with the consequence information to determine the process safety risks. Frequencies are normally determined by processing historic information on incidents.

The UK's Hydrocarbon Release Database (HCRD) captures the details of almost 5000 leaks for UK North Sea offshore installations from 1992 to the present day. It is the most commonly used source of leak frequency data both in the UK and the US. It is also applied as many other parts of the world as well as being used for onshore process plants and refineries.

Revised correlations for leak frequency and hole size distribution have been developed by DNV GL on behalf of the International Association of Oil and Gas Producers (IOGP). This analysis indicates a significant reduction in leak frequency which has implications for risk-based design and operational decision making. Alternative correlations produced by organizations in the Norwegian offshore industry show a similar significant reduction.

With such a high reliance on the use of the HCRD as a source, it is important that the reporting, recording, collation and interpretation of the data is thorough and technically sound. There are initiatives in place or planned to further improve the data collection and thereby the accuracy of the estimates produced.

This paper discusses how the IOGP leak frequency correlations were derived from the HCRD and the measures taken to ensure robustness of the conclusions. It also explores the impact of using the updated frequency correlations in quantitative risk assessments and future work investigating the technical quality of the HCRD.

1 Introduction

The oil and gas industry make considerable investments to reduce the likelihood and consequence of accidents. It is a complex task to assess the relative values of these investments to achieve an appropriate balance of financial expenditure and risk reduction.

While all hazards need to be considered in a full risk assessment, the focus of attention is often the hydrocarbon loss of containment events. Hence, one of the key inputs to a risk assessment is the estimate of frequency of accidental releases of different sizes. The leak frequency estimate is based on analysis of historical incidents. Of the various sources of data available, the UK Health and Safety Executive's (HSE) Hydrocarbon Release Database (HCRD) [1,2,3] has become established as the prime source of data for the development of leak correlations for used in risk assessments [4]. Correlations based on the HCRD are often used for onshore facilities as well as offshore even though the environmental and operational conditions may be rather different from the North Sea. Due to its maturity and comprehensiveness, the HCRD is the most significant source of information for use in the majority of risk assessment of petrochemical plants and offshore installations throughout the world. The risk is directly proportional to the leak frequency, hence the estimate of the leak frequency will directly affect risk based decisions on the provision of safety barriers and the costs of constructing and operating facilities.

In recent years, hydrocarbon leak events in the Norwegian Continental Shelf have been analyzed to provide an alternative set of correlations known as the Process Leak for Offshore Installations Frequency Assessment Model (PLOFAM) [5]. Both this and the most recent analysis of the HCRD indicate significantly lower frequencies than may have been used in the past and this will reduce the calculated risks when used in the analysis of a given installation.

Recent work by DNV GL identified some areas for improvement in the way in which information in the HCRD is collected, collated and analyzed. Some of these issues were addressed in the information provided on the HSE's web site [1,2,3] which covered the period from October 1992 until December 2015. A model for calculating suitable estimates for process equipment leak frequencies has been developed using this dataset and is discussed in this paper.

In September 2019, the International Association of Oil and Gas Producers (IOGP) revised their guidance on the calculation of process leak frequencies and used the analysis of the HCRD carried out by DNV GL as the primary source [6]. Details and parameters for the PLOFAM model are also included in the IOGP report for use in risk assessment of Norwegian offshore installations.

2 The UK HSE Hydrocarbon Release Database

2.1 General

The UK offshore industry has been required to provide data on loss of containment incidents ever since the Piper Alpha disaster and the subsequent enquiry led by Lord Cullen [7] In his report, Cullen stated; *"I am convinced that learning from accidents and incidents is an important way of improving safety performance"*. One way of achieving this was to systematically record incidents that occurred on installations in UK waters. This was intended to provide a structured set of data that would identify trends and could be used as a source of information for risk analyses.

Thus, the UK HSE set up a reporting system to gather data on loss of containment incidents that occur on installations operating in UK waters alongside the population of equipment in which these incidents were occurring so that frequencies could be derived.

2.2 Incident Data

The latest publicly available information from the HCRD gives details of 4656 incidents until 31st December 2015. Provisional data for 2016 and 2017 has also been made available to a joint industry project for research purposes.

The data collected up until 2015 has 118 fields of information for each record. The most important of these are;

- Date of incident
- Installation Type
- Process System Type
- Equipment Type
- Quantity of Hydrocarbon Released
- Equipment Size
- Equivalent Hole Diameter
- System Pressure at the time of the release

Other information provides more details on the installation, the area where the release took place, the cause of the leak, how it was detected, information on ignition and any emergency action which followed as a result. While these are important and merit detailed analysis, they are not required to establish frequency distributions for various types of equipment.

Figure 1 shows the variation in the number of incidents reported on a year by year basis. This shows a downward trend indicating that appropriate process leak frequencies used in risk assessments may also have decreased; however, the population of process equipment should also be taken into consideration.

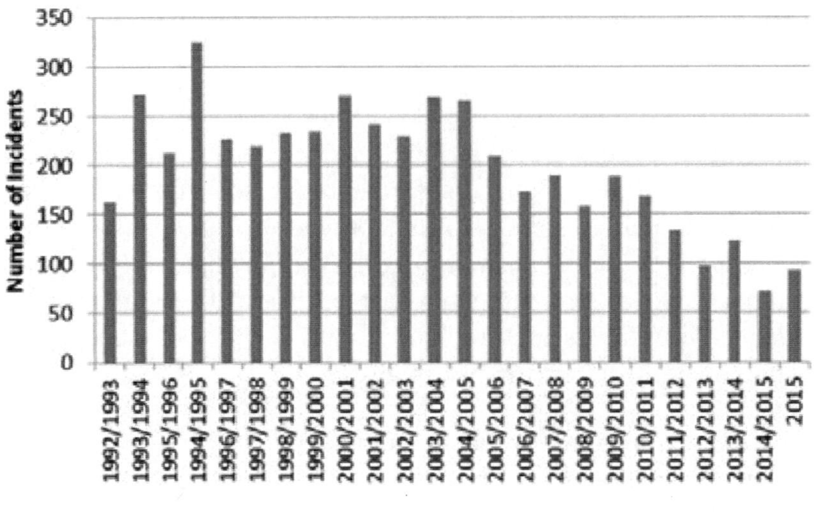

Figure 1: Number of Reported Incident per Year in the HCRD

2.3 Population Data

The successful operation of the HCRD depends not only on the correct submission of reportable hydrocarbon releases, but also on the retention of an accurate record of all duty holders (operators/owners) together with full details of their systems and equipment population data.

Between 1992 and 1994, the HSE collected systems and equipment population data for UKCS installations. This provided data on the total population of 120 categories of process equipment in the UKCS. In 2002 an on-line system was introduced to enable operators to submit and update their equipment population data.

3 UK HSE Dataset Update

3.1 Population Data

Some anomalies were observed in the population data set and in 2016 this prompted members of the industry body Oil and Gas UK (OGUK) to work in collaboration to analyze the available data and estimate appropriate values. This included an exercise to provide estimates of missing data through comparison with equivalent process systems and installations. The dataset produced by this exercise was made available on the HSE web site [3] and is one of the main inputs to the derivation of the IOGP correlations.

The UK industry has continued to work to improve the estimates of the equipment population. In 2020 a system will be implemented which requires operators to review the equipment counts for their installations when updating their Safety Case documents, a 5-yearly process. This process should ensure an improved estimate of the equipment count and enable refinements to the parameters used in the leak frequency correlations.

3.2 *Incident Data*

The incident data collected is extensive with approximate 100 data fields applied to close to 5000 records.

As a result of legislation introduced by the European Union in 2014 [8], the requirements for reporting changed. The new ROGI form [9] had a mixture of compulsory and voluntary information. However, some data considered necessary by risk analysts, such as the equipment type and the equivalent hole diameter fell into the voluntary category reducing the completeness and quality of the data.

4 Norwegian Database

4.1 *General*

Previously, risk assessment of offshore installations in Norwegian waters used the HCRD as the leak frequency data source. However, a recent study involving a number of operators and consulting organizations in Norway adopted an alternative approach for deriving frequency estimates; the Process Leak for Offshore Installations Frequency Assessment Model (PLOFAM2) [5]. Although, the model is mainly validated against available Norwegian Continental Shelf (NCS) leak data, it also takes account of data from the HCRD. Leak frequencies are based on the period from 2006 until 2017 while the whole period (2001-2017) is used for the relative leak rate distribution.

4.2 *Incident Data*

Information was gathered on 260 incidents recorded at installations located on the Norwegian Continental Shelf (NCS) in the period 2001 to 2017 inclusive.

4.3 *Population Data*

Data was obtained from equipment parts counts compiled for use in quantitative risk analyses for 79 installations which were in operation in the NCS at some point during the period 2001-2017. For a further 25 installations, where equipment counts were not available, estimates were made by comparing them with similar installations.

5 Derivation of Leak Frequencies

5.1 *Selection of Relevant Incidents*

Not all incidents reported are considered relevant to a QRA study. In addition, the population data needs to match with the incident data. Therefore, incidents relating to installations not included in the population data set are excluded as are systems such as water treatment, flaring/venting systems or utilities. Some holes were very small and those less than 1 mm in diameter were excluded along with incidents occurring in depressurized process systems.

Taking all of these criteria into account, the number of incidents from the HCRD data set was reduced to 1964 from the original 4563.

Similar but not identical criteria were applied to the NCS dataset resulting in 217 relevant incidents for use in the PLOFAM model.

Both studies based the frequencies on experience from 2006 onwards.

The hole size distribution is also important. It is assumed that the underlying distribution has been constant over the whole period. The hole size model thus uses the selected data over the whole period; 1992 Q3 to 2015 Q4 for the UK data and Q1 2001 to 2017 Q4 for the PLOFAM study.

5.2 Frequency Calculation

The frequency of leaks greater than or equal to 1 mm can generally be determined by dividing the estimated number of leaks in that period by the population to which it applies measured in equipment years. For cases where the equipment type was not recorded or there are no recorded incidents further statistical analysis was conducted.

5.3 Hole Size Distribution

The available data for a given type of equipment can be arranged in order of hole size and a plot of the frequency of exceeding that hole size produced. Typically, this produces a plot similar to the blue line shown in Figure 2.

5.4 Mathematical Correlation

The general equation used in the correlations reported by IOGP [6] is of the form

$$F = Cd^m + B \qquad \text{for } d = 1 \text{ mm to } D$$

$$F = 0 \qquad \text{for } d > D$$

Where F is the frequency of a hole greater than or equal to d

d is the hole size

D is the equipment size

and C, m and B are coefficients for each equipment type.

For some equipment types there is sufficient data to determine a variation in the value of these parameters as a function of the equipment size. For other types, constant values are assumed.

A least squares curve fitting process was used to determine the values of the C, m and B parameters which gave the best fit to the historic data. This gives a function represented by the black line in Figure 2.

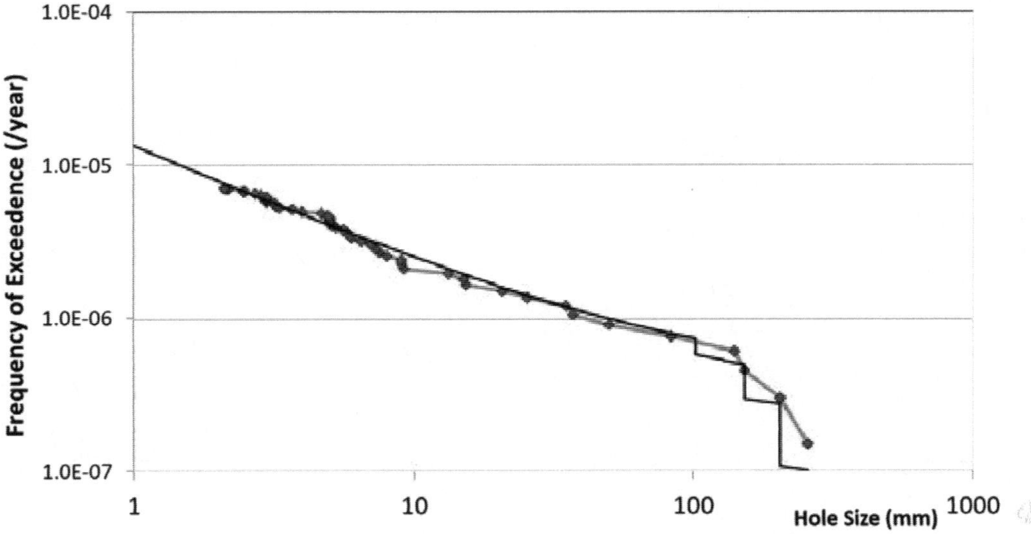

Figure 2: Mathematical Correlation Fitted to Historic Data

The IOGP Guidance [6] provides details of the parameters along with graphical and tabulated representations for 24 equipment types. As an example, Figure 3 shows the graph for flange joints which indicates the effect of equipment size.

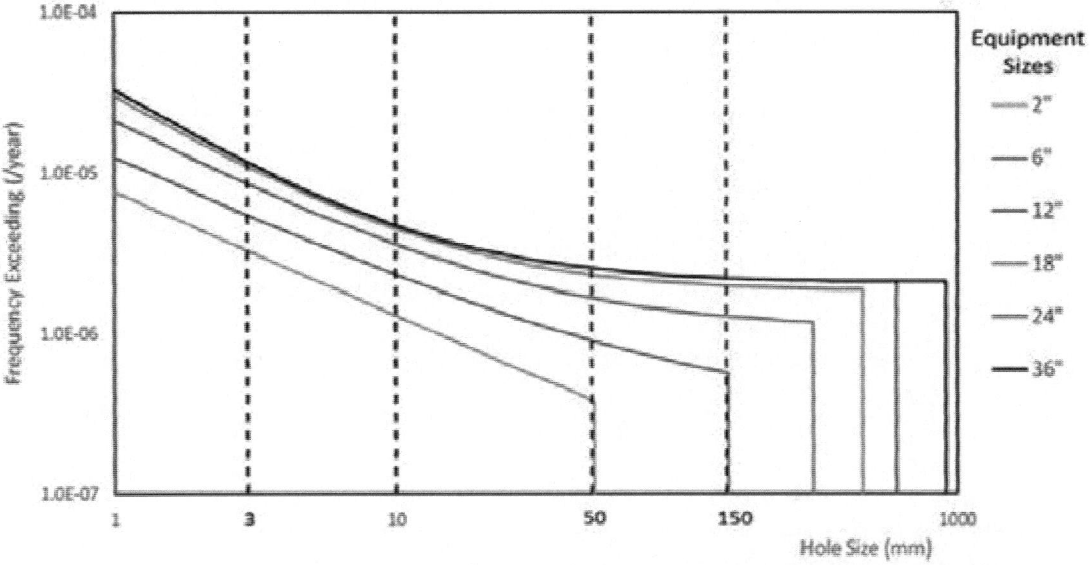

Figure 3: Frequency Exceedance Curves for Flanges Using the IOGP Correlation

The frequency of leaks within a certain hole size range, e.g. between 10 mm and 25 mm is evaluated by subtracting the value exceeding the upper range from the value exceeding the lower range.

5.5 PLOFAM Model

The PLOFAM study proposed a more complex formulation;

$$F(d, D) = [F_0(D) - \alpha F_D(D)]\, d^{m(D)} + \alpha F_D(D), \qquad 1\,mm < d \leq D$$

$$F(d, D) = 0, \qquad d > D$$

Where, $F_0(D) = F_{hist}(A_0 D^{M_0})$

$$F_D(D) = F_O(D)\,(A_D D^{M_D} + B_D)$$

and $m(D) = \frac{log(F_D - \alpha F_D) - log(F_0 - \alpha F_D)}{log(D)}$

where $F(d, D)$ is the hole size frequency distribution per equipment-year.

F_0 is the total leak frequency per equipment-year. The subscript 0 indicates the total leak frequency for an equipment and hence the "starting point" on the y-axis.
$F_0 = F(d = 1, D)$.

F_D is the total full bore hole frequency per equipment-year. The subscript D indicates the frequency for getting a hole equal to the equipment diameter D.
$F_D = F(d = D, D)$.

d is the hole diameter in millimetres

D is the equipment diameter in millimetres

m Slope parameter

F_{hist} is the average leak frequency (independent of equipment diameter) for the relevant equipment type per equipment-year

A_0 is a parameter in equation for total leak frequency, F_0

M_0 is a parameter in equation for total leak frequency, F_0

A_D is a parameter in equation for full bore hole frequency, F_D

M_D is a parameter in equation for full bore hole frequency, F_D

B_D is a parameter in equation for full bore hole frequency, F_D

α is a dimensionless parameter, independent of equipment diameter D, $\alpha \in [0, 1\rangle$.

As an example, Figure 4 shows the exceedance curves for a standard flange using the PLOFAM model.

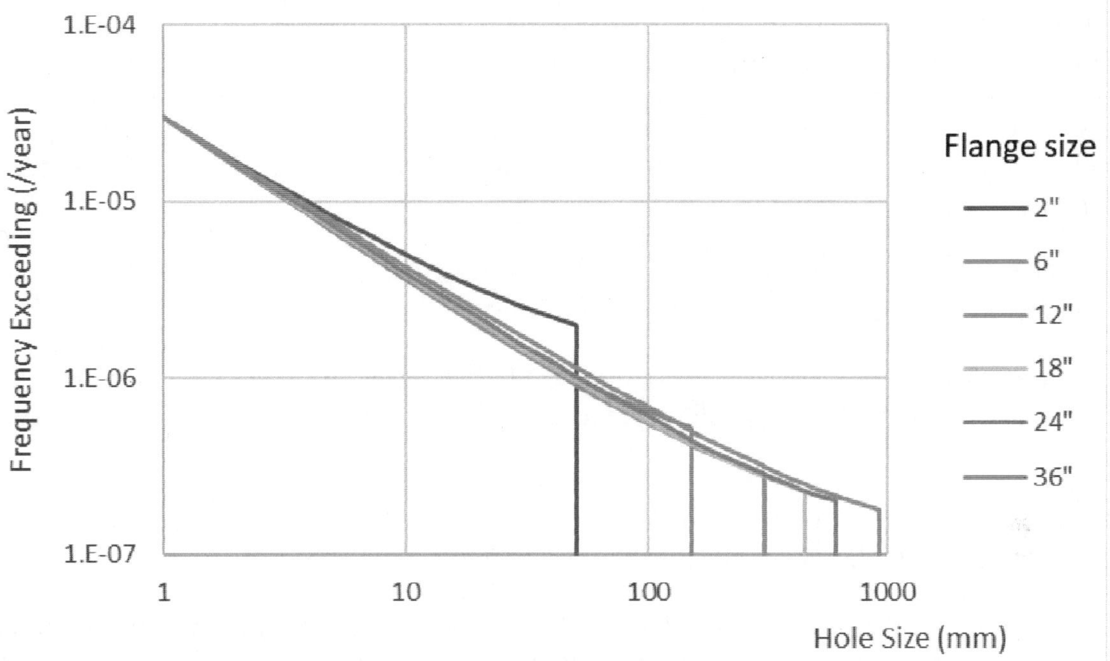

Figure 4: Frequency Exceedance Curves for Standard Flanges Using the PLOFAM model

6 Effect on Using the New Leak Frequency Models

Both the IOGP and PLOFAM models give estimates of significantly lower frequencies than previously used. In the case of the IOGP formulation the values are approximately 60% lower. Given that the risk is directly proportional to the leak frequency, the calculated risk will fall by a similar amount. As such, remedial measures which were previously judged to be required may now be considered unnecessary.

The main factors for the reduction in the IOGP estimates are;

- The reduction in the annual number of incidents reported
- The decision to base frequencies on the last 10 years of available data
- A reassessment of the incidents to be included in the analysis.
- A reassessment of the equipment population data.

The PLOFAM model produces estimates for leak frequencies which are lower than the IOGP correlations and this is particularly the case for larger hole sizes. This implies a bigger reduction relative to previous estimates compared with those obtained from the IOGP model.

7 Future Work – Towards Improved Accuracy and Derivation of More Value

The collection of data in UK and Norway should continue in order to gain more statistical significance for the various equipment types. It is also necessary to confirm or otherwise that the downward trend continues. Some observers believe that the current level of reporting marks a low point in the frequencies of events and that a combination of aging and the effects of the downturn in the oil industry on maintenance levels will lead to an increase in coming years.

A review of the population data in the UK is required to ensure a robust database is available to accurately assess the leak frequencies.

The quality of the incident data collected needs to be actively managed with improvements in the UK reporting form and encouragement to complete all data fields as accurately as practical.

These two initiatives will improve and maintain the quality of the data gathered in the UK in future years and thereby make the complete dataset more valuable.

More information can be obtained from the existing data. Relatively few of the data fields are used in the calculation of the leak frequency functions described in this paper. The data can be used to address some important questions being asked by the industry such as;

- Is there an increase in the frequency during start-up operations compared with normal operating conditions?
- Do leaks on normally manned installations occur more frequently during the period that a workforce is present?
- What proportion of leaks are due to design/maintenance issues compared with procedural/operational failures?
- What effect does the age of the equipment have on its likelihood of failure to contain the hydrocarbon?
- Does the trend of leak frequencies vary between types of equipment and can these changes be related to improved design or operation?

8 References

[1] HSE, Spreadsheet entitled "Offshore Hydrocarbon Releases 1992 – 2016" http://www.hse.gov.uk/offshore/statistics.htm (accessed Dec 31, 2019).

[2] HSE Spreadsheet entitled "Offshore Hydrocarbon Releases 2015 – 2016" available from http://www.hse.gov.uk/offshore/statistics.htm (accessed Dec 31, 2019).

[3] HSE Spreadsheet entitled "Offshore Hydrocarbon Population Data 2015 – 2016" available from http://www.hse.gov.uk/offshore/statistics.htm (accessed Dec 31, 2019).

[4] Bain, B, "Updated Leak Frequency Modelling Based on the UK Hydrocarbon Release Database, Hazards 27 Symposium, 2017.

[5] Lloyd's Register Consulting, Process Leak for Offshore Installations Frequency Assessment model – PLOFAM(2). Report No. 1007566/R1 Rev: Final, December 2018.

[6] IOGP, Risk Assessment Data Directory, "Process Release Frequencies", Report 434-01, September 2019.

[7] Cullen, The Hon. Lord W. Douglas 1990. The public inquiry into the Piper Alpha disaster. London: H.M. Stationery.

[8] European Union, 2014, "Commission Implementing Regulation (EU) No. 1112/2014.

[9] Offshore Safety Directive Regulator, Report of an Oil and Gas Incident (ROGI) Form, https://www.hse.gov.uk/osdr/assets/docs/rogi.pdf (accessed 31st December 2019).

Application of Management of Change (MoC) in Refinery

Seckin Gokce, Process Safety Chief Engineer & Gozde Turk Eski, Process Safety Superintendent, Tupras Izmit Refinery, Kocaeli/TURKEY

Many companies are forced to make new investments / innovations in various fields to make them stronger due to changing and developing conditions or to sustain current operational availability. Facilities need to consider the changes to be implemented in terms of safety, environment, health and reliability. If the planned changes are applied on site without appropriate hazard analysis the risk of a process safety accident could increase significantly. This paper covers management of change and risk assessment studies with the examples of refinery applications.

INTRODUCTION

In all chemical process plants especially in refineries, it is critical to achieve high operating plants operational and mechanical utility. In such continuously working systems, there is maintenance activities done within a schedule but there may be unplanned maintenance. Mostly maintenance activities are repair and replace in kind. Besides these in order to improve operational availability, operational safety or keep up with the technological improvements there are many changes done in the units. Change means here any kind of process, hardware or technology differentiation in the process plant related with the process. All these changes come up with a need but also they may cause new risks or create new needs. Any kind of change inside a process plant should be reviewed and discussed with all the related disciplines (operations, technology, maintenance, design) in order to define all the possible outcomes of the change and assure all hazards are adequately addressed to meet company and regulatory requirements. Poor change management mostly causes new Process Safety related incidents which may lead to deaths, huge asset damages, irreversible environmental damages and also reputational effects. In chemical industry around the world there are so many examples of poor change management (for example Bhopal, Flixbroug).

Management of Change, as a Process Safety Management System task, should be rigidly addressing the scope of change, process of change management and closure and approval of the change. All the people in a company can come up with the change request that is why it is important to clearly convey the message from top to bottom that all kind of changes must be following up the MOC process regarding to the company standards.

MANAGEMENT OF CHANGE APPLICATIONS

Equipment Changes

In the refinery, most typical equipment changes we come across are pump renewals, new pump implementation, instrument additions, pipeline changes and rarely new process equipment like drum or column additions. Any not-in kind changes/additions, including emergency repairs and replacement, are in the scope of MOC if they are in the same kind. For example, think about the renewal of a pump if there is any difference from the current pump like material, shut off pressure, temperature specification, mechanical seal plan etc. that is under the scope of MOC. As defined above this change should follow the MOC path. During this MOC process, all design parameters and current process conditions should be compared to decide the change is safe to continue or not. In current application, we mostly go over a checklist besides what if or HAZOP risk assessment in order to be sure that we covered all parts. These checklists are the attachment of MOC standard, which is specific to equipment type.

These kind of changes are the ones which can be detected by people and easier to do risk assessments. There are other important tasks. This includes training, maintenance records and project documents revisions. Frequently these tasks are poorly done.

When we first initiated the MOC standard the hardest part was making people finish the change process completely. They think that if the approval is taken and equipment is implemented on site change is completed. Actually no, change is continuing until you finish all the documentation and trainings. That is why there is a closure approval in MOC system in order to make both parts, user of the change (which is mostly operations) and doer of the change (which is mostly maintenance or project departments), efficiently handover the change.

Process Changes

Another type of change in the refinery is process changes, which covers the operating temperature/pressure changes, chemical changes, process flow changes or by-pass of equipment or systems etc. These changes are mostly requested and done by the same team, which are operations and process development team. That is why most of the time they try to by-pass the MOC process intentionally or unintentionally. This kind of changes might be seen as innocent test runs but industry experience shows that these are the ones that have the highest potential to cause the incidents.

Here the critical thing is all related disciplines, to be affected by the change, gathering around the table and discussing the change and potential outcomes of the change. Defining all related disciplines is important and needs a good knowledge about the process, company standards and experience. In some cases you may need to gather the team again and again, and ask for third parties like process designer company etc. Experience level of the risk assessment team is important for all MOCs but that is more critical for process changes compared to other types. It is important that during the hazards review any new risks are identified including discussion of potential failure modes that cold be introduced.

Service Changes

So many plants, so many products and so many pipelines… What you see when you glance over the refinery! The term operational availability sometimes requires to use the same pipeline or equipment for different purposes. That is the reality inside the refinery, we have a good tool to manage the need of this flexibility which is MOC. These changes also easily captured under the scope of MOC since it is unusual path or service to be used. Here the critical thing is risk assessment teams are more crowded since that is mostly affecting most of the plants and disciplines. There may be meeting scheduling problems but beware of that all the team should be together for a better brainstorming. Especially, in MOC related risk assessments brainstorming and welcoming all levels of questions during the risk assessments is important.

Organizational Changes

At first glance, organizational changes seems like an HR issue but it is more than that. Since as chemical industries we have been dealing with risky processes, competency in the job is critical. Especially for HSE critical roles, any kind of organizational changes should be assessed in terms of competency.

In order to have a systematic MOC approach for organizational changes the first step is having HSE critical roles and their competency level requirements already defined.

Temporary Changes

Any kind of temporary technical, facility or organizational changes sometimes assumed as a minor issue and not classified as an MOC. However, experience and critical chemical industry accidents show that there are so many examples of major accidents caused by these temporary changes. Since they are seen as minor, most of the time these are not very well assessed. Any changes should follow the same risk assessment procedure the only change in temporary changes is to follow up the change in order to re change it. According to our standard any changes that will last less than 6 months is considered as temporary changes.

Management of Change Steps

Any change that affects process chemicals, technology, equipment, procedures, or facilities, as well as any organizational, personnel, and policy changes that could impact these elements. For applying on facilities successfully, MOC system needs to be well organized to make good risk assessment with relevant steps. Facility has own MOC software and employees use it when they have any demanding changes which is related with scope of MOC.

Step 1: Determine if an MOC is required or not.

Facility has own standard about MOC and it gives definition and scope of changes. The process safety department decide it. If it is appropriate change demanding the MOC steps are started or rejected.

Step 2: Process Hazard Analysis (PHA) and Coordination

After determining of the desired change is required MOC, PHA coordinator start to pre-assessment. Firstly, the change package need to include such kind of detailed information; description of change, type of change (temporary/permanent), related site, drawings, technical calculations & data etc. It is very critical and important that all related documents are provided and studied well by the proposing the change. If all needed items are prepared properly, PHA coordinator decides the risk assessment tool or tools (HAZOP, What if, HAZID, Fault Tree and Event Tree Analysis, Scenario Modeling (Toxic Gas Dispersion, Fire and Explosion Analyses), Facility Siting Studies and organize meeting with all related departments to evaluate the risks against corporate risk criteria (usually a risk matrix) and define the necessary additional barriers to risk.

Process Safety Management System starts with managing the process related risks and from the first step of establishing a new process unit that is needed to evaluate the risks and define the necessary measurements. Process Hazard Analysis methodologies are the systematic tools that help to evaluate the risks in a process. For different needs there are several methodologies like DOW FETI, HAZOP, What if, HAZID, Fault Tree and Event Tree Analysis, Scenario Modeling, Facility Siting etc. Some of these methodologies are semi quantitative some are quantitative methodologies. In order to define and prevent major industrial accidents in our refinery we use DOW FETI, HAZOP, FTA and ETA and Scenario Modeling. Moreover for some kind of projects we may also apply What if and HAZID analysis instead of HAZOP depending on the complexity. Facility Siting is another tool for risk related studies that is used to evaluate the risks for buildings in order to protect people inside the buildings.

The workflow for a full unit PHA study starts with DOW FETI calculation that is used for risk ranking of al static equipment inside the unit. DOW FETI calculation is done in accordance with the equipment temperature, pressure, hold-up amounts, components and also reaction types if there is any. Then risk assessment process continues with HAZOP studies that are for scenario identification. In HAZOP, first step is chopping the unit into small parts to be evaluated that is called nodes. After that all nodes are evaluated using P&IDs and PFDs by keywords like temperature, pressure, flow, level etc. HAZOP helps to understand what kind of scenarios are possible inside the unit and these critical scenarios are taken through FTA and ETA studies. HAZOP is a semi quantitative risk assessment method where FTA and ETA are quantitative methods. FTA is applied for critical scenarios in order to find out the probability of top event then ETA is then applied in order to find out the probability of scenarios like fire, explosion, toxic dispersion etc. caused by this top event. FTA and ETA analysis all use the statistical PFD and failure rate values of all initiating events and barriers defined in HAZOP for the related scenario.

Scenario Modeling is the following step that is used for understanding the effect of the probable scenario. The most probable scenarios coming from ETA is modeled to see the effect zones. These effect zones are used to plan Emergency Response Studies for all these potential major accidents.

Facility Siting is another check point in a unit PHA study. All potential scenario modeling results are used to evaluate the risks for the buildings and also if there are any in the effect zones of fire, explosion or toxic release then there should be actions defined.

Also in this step, PHA coordinator and facility superintendent give the decision of application of Prestart-up Safety Review (PSSR) before the operate it.

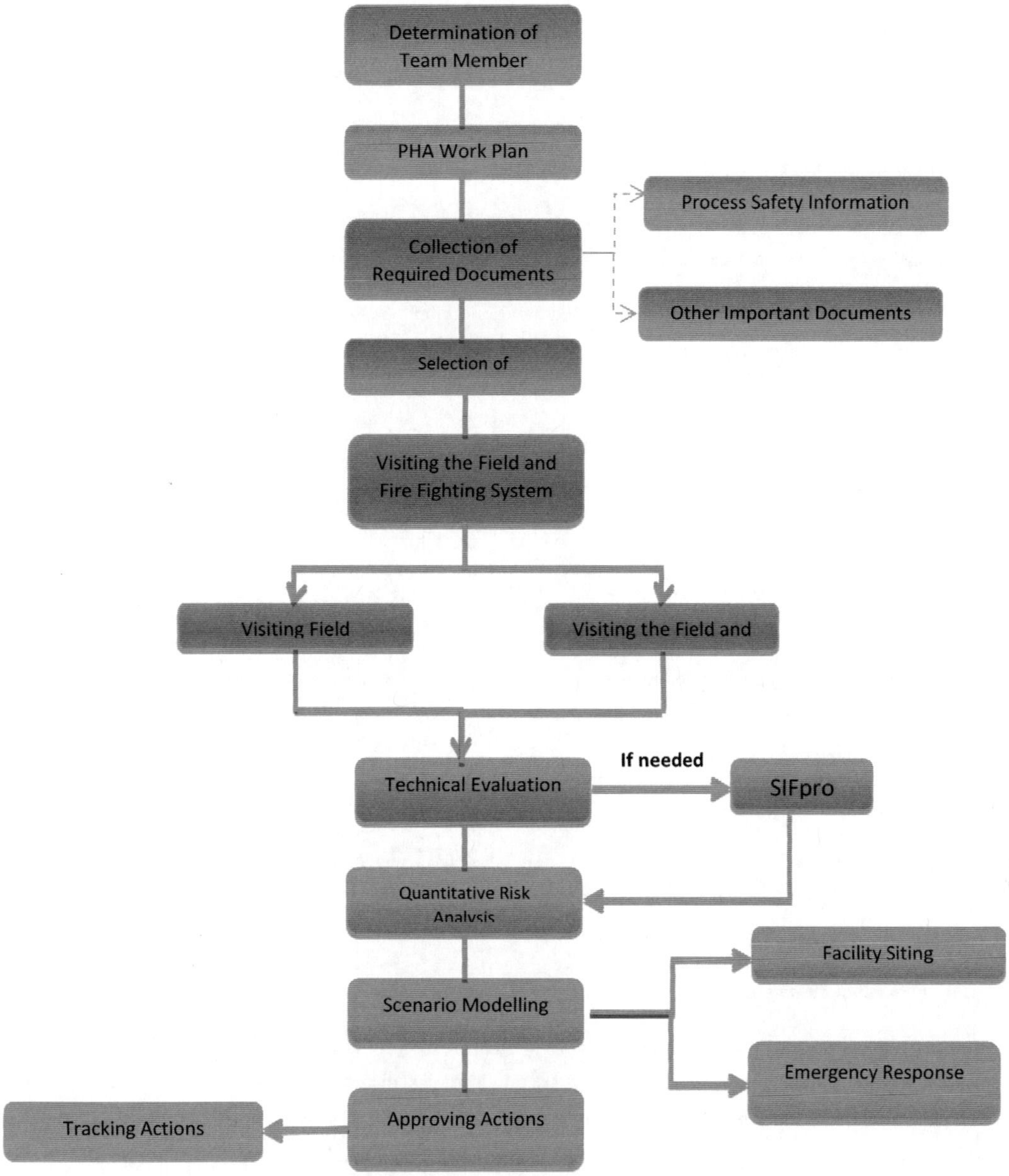

Figure 1: General Process Hazard Analysis Step

Emergency cases;

Sometimes, in facilities have emergency cases that change needing immediate implementation, the process should allow for administrative steps to be done quick. If such cases occur, facility's responsible person need to get verbal confirmation from site manager then should apply the desired change. In 48 hours, the proposer or implementer have to create change package on the MOC system then detailed risk assessment will be done and documented by PHA team.

Step 3: Technical Approval-PHA members

After the PHA studies, team analyze the request and evaluate the risk if it is acceptable risk level or not. If it is not, the team can make suggestions to reduce risk or rejected the request due to the safety issues. If the change request's details are proper and covered all safety and technical issues, the request is approved by team members.

Step 4: Technical Approval-Site Managers

The change request will be sent as a package that includes all risk assessments and technical drawings to the all related site managers for reviewing and approving. If the approval steps could be done then the change request will be started applying on site.

Step 4: Implementation of Change

Following the approval process, the implementer can start his desired change on the field. During the PHA studies if there is an additional suggestions/actions are advised due to the mitigation of risk, the proposal change have to cover it.

During that period the revision of the relevant documents should be done, employees and affected sites to be provided training about the changes by their site leaders. It's very important that to share new situation's risks and its barriers/procedures, which are new for related, sites. All steps of the risk assessment studies and updating of documents are applied and tracked by MOC system.

Step 5: PreStart-up Safety Review (PPSR)-If needed

PSSR ensure that all necessary checks have been made before commissioning a plant, and to ensure that all issues that can be checked for commissioning so far are safe. PSSR is the last checkpoint to inquire whether all equipment has been installed or maintained according to the standards and all necessary checks have been performed in accordance with the Process Safety Management System. PSSR team uses check lists during the evaluation. There are two possible results from the PSSR. First are necessary items that must be completed before start-up, and secondly, items that must be completed after start-up. For items to be completed after start-up, completion dates need to be describe, along with the person responsible for completion. The additional actions are tracked by MOC software too.

Step 6: Implementation of Action Items (Pre-Startup & Post-Startup Actions) and Tracking Actions

Actions items that were deemed "post-start-up" in the evaluation phase are completed at this step. While these action items were not required for cutover to the new state, they should be completed before the MOC is closed to ensure the sustainability and on-going safety of operations. Overdue actions are tracked on MOC software as KPI's.

Step 7: Closing Approval and Commissioning

In fact this step is final control step. All issues relevant with the desired changes are implemented on site, technical documents are updated and uploaded on system, training & instructions are provided related employees and additional recommendations are done. The implementer and site manager review the all steps and agree that MOC package satisfy all requirements. They both approve it close the MOC package.

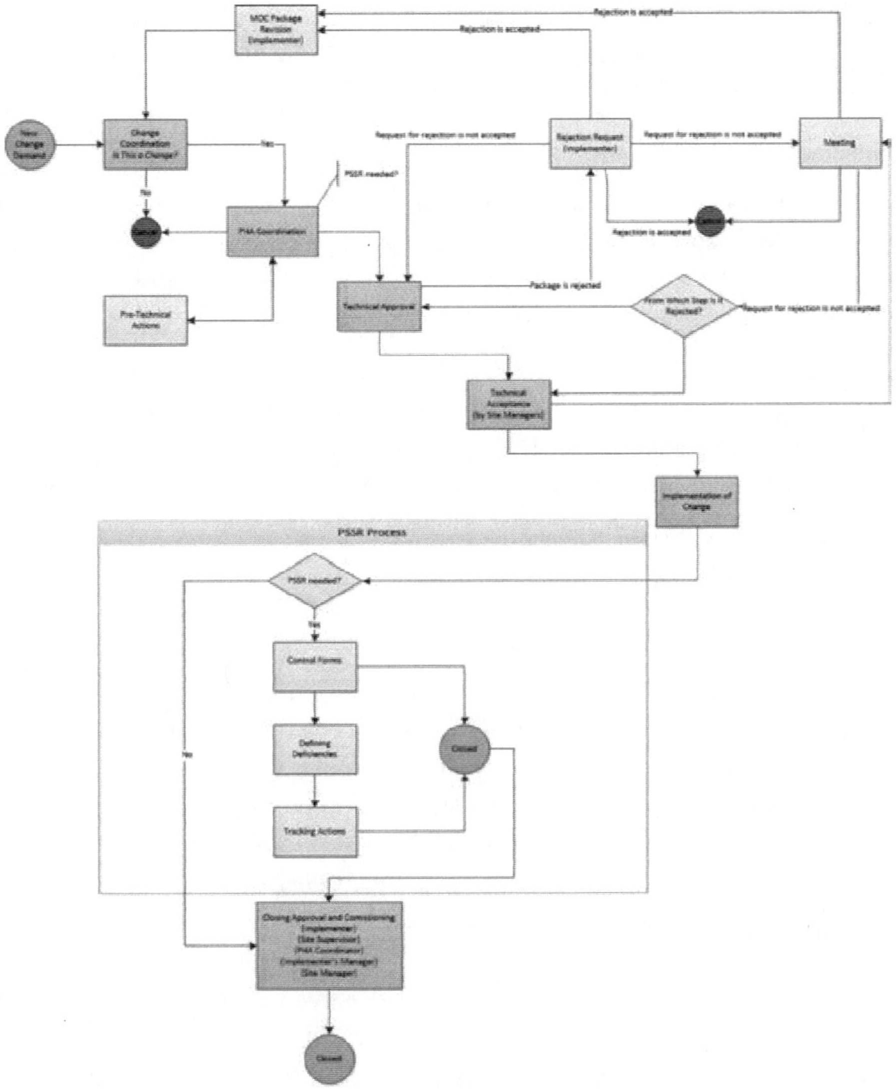

Figure 2: MOC and PSSR Steps

Auditing

Management Systems are audited in accordance with the Independent Auditing Standard for Internal Auditing Standards in order to ensure that the MOC is operated as specified in this standard, to find and remove the deficiencies in the system.

In order to identify, classify and resolve the deficiencies that occur in the Management of Change, the Change Coordinator informs the refinery management at the Annual Review of Management. In accordance with the decisions taken in this meeting, the problems that prevent the operation of the system are eliminated.

Example of MOC Studies

Case 1: Capacity Increasing Demand on Hydrocracker Unit

Case Description

Faculty has Hydrocracker Unit, which has been operated as 8800m3/day capacity. Due to the some profit issues, management would like to increase capacity. To define acceptable capacity value, process simulation team has started the studies if the desired change is possible or not.

What Happened?

Process team has created a new MOC package to discuss their findings. After the studies, they found it 9200 m3/day for new capacity value. PHA coordinator started the change organization with related departments. It was very important that the process hazard analysis-meeting schedule fit to everyone. Before the meeting, all data/reports/documents/graphs were prepared by process team and shared with the coordinator to send it all PHA team members.

During the meeting, all information was discussed by team members. In such cases, What If analyze form has used to describe all new potential risks, their results and existing barriers. All items were examined carefully and checked if the risk of item was ALARP level or not. If not, additional recommendation was given by team and it was tracked on MOC system. According to the result, team has signed What If analyze form, and then it was uploaded on system which was available for all employees. Then, Operation Manager and Refinery Manager also gave technical approval to final form.

Since the unit is very critical, its all parameters need to be checked and evaluated carefully.

For example;

- Checking unit vendor's report – operational parameters/design parameters
- Evaluation of new speed values and related erosion effects in critical lines that was determined by Reliability Department according to simulation results.
- Evaluation of relief valve capacity considering with blocked outlet scenarios were done by Process Department.
- Evaluations of pumps and compressors capacities were done by Mechanical Department.
- Evaluations of control valves capacities were done by Instrument Department.
- Evaluation of furnace load and static equipment's capacities were done by Process Equipment and Development Department.
- Evaluation of process parameters (reactor temperature profiles, residence time etc. effect on system.) were done by Process Department.
- Evaluation of supports due to new speed and load change in lines were done by Mechanical Department and Reliability Department.
- Reviewing of refinery flue gas/recycle gas flow effect on DEA system due to increasing of those flows were done by Process Department.
- Evaluation of the sufficiency of sulfur, hydrogen and contaminated water stripping units according to the simulation results were done by Process Department.
- Evaluation of consumption of washing water and corrosion inhibitors according to the simulation results were done by Process Department.
- Evaluation and suitability of process instruments (set value, critical alarm, range wide etc.) according to the simulation results were done by Project Instrument and Process Department.
- Evaluation of electrical motors, reviewing power consumption were done by Project Electrical Department.
- Checking and evaluating the steel structures from the relevant documents according to the new situation were done by Civil Engineering Department.
- Evaluation of fan capacity according to the simulation results was done by Process Department.
- Evaluation of catalyst activity at new situation was done by Process Department.

What Happened If MOC hadn't been Done?

Hydrocracker facilities are operated under very high temperature and pressure. Other danger is toxicity due to H2S content. Any type of change that would like to implement on facility's any process item need to be evaluated. A project that is intended to be applied to the site without a MOC coordination may result in larger industrial accidents, leading to a loss of life and property.

In that case, the desired would have been implemented to site without any considering safe design & operation parameters/risk assessment and then possibly facility had been faced serious process safety incidents.

Case 2 : Tank Service Change-Waste Water Treatment Unit (WWTU)

Case Description

WWTU unit had decided to make tank service change due to some operations causes. In the previously, storage tank had been used to hydrogen peroxide (H2O2) tank then services was changed to iron 3 chloride (FeCl3) storage tank. The unit made this change on its own.

What Happened?

After a short time, during the routine field check, operator had saw the corrosion in tank's transmitter connection cables which was used as FeCl3 storage tank.

Picture 1: Deformation of connection cables

The unit had bypassed the MOC system so such type of near misses occurred. WWTU had stopped chemical operations and inform the Process Safety Department. MOC package was created to make risk assessment study. Same change was made on other WWTU too. After this near misses, other WWTU checked their equipments and connection cables. They realized that tank feed pump had corrosion too. Same mistake, similar near misseses…

What is the Correct Approach?

Sometimes, the units want to make quick changes and apply it on site without any risk assessments. It is very important to be able to work safely in the field. But how? Operational disciplines are key parameters for every employee. If WWTU has such kind of intentions, firstly they had to create MOC package to make comprehensive risk assessment study.

What Would Happen If MOC Package Created?

PHA coordinator organized meeting with all related disciplines. The team considered desired change conditions if it was fit or not. For example, material of tank, connection cables, instruments materials and services, impact of process, pump materials etc. were evaluated according to new service.

Also, one of other critical point is unit's chemical interaction matrix. In the meeting, this matrix was checked by members. If the MOC was done, it was certain that the tank service change could not be done due to chemical interaction matrix. As it is seen below, this two chemicals has redox reaction.

$$2FeCl_3 + 3H_2O_2(l) = 6HCl(a) + Fe_2O_3 + 1.5O_2(g) \quad \Delta H(20C) = -520.0kJ$$

WWTU	Sodyum Karbonat	Fosforik Asit	Karbon giderici	KOI giderici	H202	Sülfürik Asit	Kostik	Cationic Polielektrolit (Sıvı)	FERROFOS 8411	Sodyum Hipoklorit	Hidroklorik asit	FECl3	Ascorbik Asit
Sodyum Karbonat		Heat+CO2+ Pressure			Heat	Heat+CO2 + Pressure					Heat+CO2+ Pressure+HCl		Heat+CO2+ Pressure
Fosforik Asit					Heat		Heat			Heat+Cl2		Heat	
Karbon giderici													
KOI giderici													
H202						Heat+O2+ Pressure	Heat			Heat+O2+ Pressure	Heat+O2 Pressure	Heat+O2+ Pressure	Heat
Sülfürik Asit							Heat			Heat+Cl2		Heat	

Figure 3 : Part of WWTU Chemical Interaction Matrix

Also, there was tank service change checklist, it's also examining chemical compatibility.

What is the Key Lessons?

In fact, even it was a near misses this event has many messages to employees. Applying and dominating the standards, importance of communications and using MOC to make comprehensive risk assessment study are main messages to everyone. Sometimes the change is not as simple as it seems. Every desired changes should be evaluated carefully to work and implement safely.

References

1-CCPS, Guidelines for the Management of Change for Process Safety

2-CCPS, Guidelines for Process Safety Metrics, 2010, 4.5.1 Choose Metrics Appropriate for the Operation, 48p

Explosion mitigation using aqueous solutions of flame inhibitors

Kees van Wingerden
Gexcon AS
Fantoftvegen 38, 5072 Bergen, Norway
kees@gexcon.com

Dirk Roosendans
Total SA
Tour Michelet, La Défense, A0819, Paris, France
dirk.roosendans@total.com

Pol Hoorelbeke
Total SA
Tour Michelet, La Défense, A0819, Paris, France
pol.hoorelbeke@total.com

Prepared for Presentation at
American Institute of Chemical Engineers
2020 Spring Meeting and 16th Global Congress on Process Safety
Houston, TX
March 30 – April 1, 2020

Explosion mitigation using aqueous solutions of flame inhibitors

Kees van Wingerden
Gexcon AS
Fantoftvegen 38, 5072 Bergen, Norway
kees@gexcon.com

Dirk Roosendans
Total SA
Tour Michelet, La Défense, A0819, Paris, France
dirk.roosendans@total.com

Pol Hoorelbeke
Total SA
Tour Michelet, La Défense, A0819, Paris, France
pol.hoorelbeke@total.com

Keywords: explosion mitigation, flame inhibition, water mist, superheated water, water deluge

Abstract

The efficiency of mists of aqueous solutions of flame inhibitors as an explosion mitigation means was investigated on small-scale showing that by increasing the amount of flame inhibitor (potassium carbonate) in water, an increase of the degree of mitigation (reduction of flame velocity and explosion overpressure) is observed up to full quenching.

To generate water-mists directly at a larger-scale in sufficiently high concentrations dispersion of superheated aqueous solutions of potassium carbonate was chosen. Without inhibitors dispersion of superheated water resulted in a clear explosion mitigation in a 50 m^3 obstructed vented volume. Introducing superheated aqueous solutions of potassium carbonate however resulted in a reduction of the mitigation effect.

As an alternative an industrial water deluge system was applied. Again, both pure water and inhibitor-solutions were applied. In this case the explosion itself generates fine droplets due to water droplet break-up. At high water loads explosion peak pressures were reduced considerably by the water deluge but a clear effect of adding inhibitor to the water could not be differentiated.

The paper discusses the possible reasons for the difference of behaviour on small and large scale.

1 Introduction

Vapour cloud explosions are among those events causing the highest losses in the petrochemical industry. Hence, preventive measures are given highest attention: correct design of the facilities (incl. material choice), its maintenance and inspection and choice of equipment approved for use in potentially explosive atmospheres. Research in the past has especially been devoted to predicting the consequences of these events [1, 2, 3]. Less attention has been given to mitigation of vapour cloud explosions, i.e. limitation of consequences.

Considering research that has been performed water deluge has been shown to be effective as a means of explosion mitigation. The water deluge system shall be activated before ignition. The positive impact of water deluge has been documented extensively [4, 5]. Only droplets smaller than 10-20 μm will affect flame propagation due to evaporation in the flame (in methane-air mixtures). Larger droplets need to break-up before having an impact. Flow accelerations occurring during explosions in congested areas are able to break-up droplets larger than typically 200 μm [5]. Hydrodynamic forces acting on droplets in an accelerating flow will cause droplets to break up if these droplets are not able to adapt to the flow accelerations.

In [6] it was suggested to use solid flame inhibitors dispersed into the vapour cloud upon its release as an alternative mitigation technology. Relatively low concentrations (50-100 g/m^3) of solid flame inhibitors such as sodium bicarbonate, sodium chloride, potassium carbonate and potassium bicarbonate are able to reduce flame speeds and overpressures in hydrocarbon-air mixture explosions considerably [7]. A system to disperse the solid flame inhibitors in practice has been developed and its properties to disperse the inhibitor [8] and to suppress explosions [9] has been tested on full scale.

In [10] it was suggested to investigate the possibility of combining the two: applying aqueous solutions of flame inhibitors to mitigate vapour cloud explosion. Experimental investigations were performed on small-scale using water mist-potassium carbonate solutions. Whereas water mist alone already had an impact and caused a reduction of flame speeds and overpressure introducing potassium carbonate in the water resulted in an additional reduction.

The current paper briefly describes the results of the small-scale experiments performed in a 135 l channel but especially addresses the results of tests performed with aqueous solutions of flame inhibitors at a larger-scale. Two sets of large-scale experiments were performed: experiments using superheated aqueous solutions and experiments where aqueous solutions were dispersed using a water deluge system. The main aim of these experiments was to verify the effects seen on small-scale on a larger scale and the possibility of using this mitigation technique on industrial scale. Both sets of large scale experiments were performed in a congested 50m^3 chamber provided with explosion venting. In all experiments potassium carbonate was used as a flame inhibitor.

2 Description of the test set-ups

The small-scale experiments were performed in a horizontal 135 l test channel with a length of 1.5 m and a cross section of 0.3 m x 0.3 m. The large-scale experiments were performed in a 50 m^3 vessel with a length of 8 m and a cross section of 2.5 m x 2.5 m. Both facilities are described below.

2.1 Small-scale 135 l channel

The small-scale vessel is only open at one of the short ends. Ignition is effected at the other closed short end of the channel. One of the long walls is transparent allowing for high speed filming. Baffles can be inserted in the channel at three different locations.

- at the ignition end (31.0 cm from ignition end)
- in the middle (67.5 cm from ignition end)
- at the open end (148.5 cm from ignition end)

Each baffle consists of five 1.5 cm strips evenly distributed over the cross section and blocking 25 % in total. During the tests none, 1 or 2 baffles were used.

As a flammable gas propane was used which was added to a gas recirculation system, gradually, and mixed with the air until a homogeneous set concentration is reached. During filling the open end is closed with a thin plastic sheet. To inject aqueous solutions a total of 10 fogging nozzles (type BETE® PJ6) are used (5 in the roof, 5 in the back wall). The nozzles generate droplets with 90% < 25 μm at a pressure of 70 bar g which give a flow rate of 0.114 l/min per nozzle (assuming pure water).

Figure 1 shows the 135-l channel.

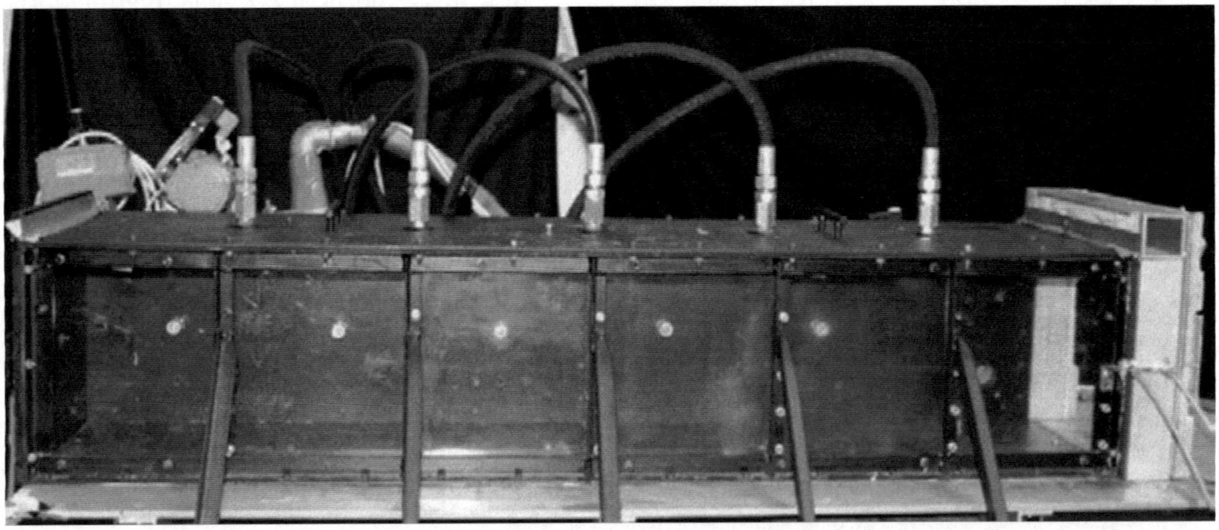

Figure 1 135-l channel looking towards transparent long side wall. Five (5) of the water fog nozzles are directly connected to the water hoses seen on top of the channel. The recirculation system is visible on the left side behind the channel.

The gas concentration is measured using a Servomex XA4200C1 gas analyser. The explosion pressure is measured at 3 locations in the channel. A high-speed video camera (3000 fps) was sued to record flame propagation in the channel. The water/aqueous solution mist concentration was estimated on the basis of the total water/aqueous solution injected and collected at several locations in the channel and the injection time. In the majority of the tests the injection of the mist lasted 10 s resulting in average water concentrations of 30 – 80 g/m^3 (4 – 11 mol %). Concentrations were generally higher close to the open end of the channel.

2.2 Large-scale 50 m^3 vessel

The two series of large-scale tests were performed in a congested chamber 8.0 m long, 2.5 m high and 2.5 m wide, thus having a volume of 50.0 m^3. The chamber has a coarse steel grating held by beams forming a mezzanine deck approximately 1.25 m above the floor of the vessel. During the current tests one or both of the short ends were used as vent openings. As Figure 2 shows the chambers contains several objects (obstructions) which cause explosion flames to accelerate due to turbulence generated at these obstructions.

Figure 2 50 m^3 explosion chamber used to investigate effect of aqueous solutions of flame inhibitors

Also during the large-scale experiments propane was used as a flammable gas. As on small-scale the propane-air mixtures were prepared by adding the propane to a gas recirculation system, gradually, and mixed with the air until a homogeneous set concentration was reached. During filling the open ends of the chamber were closed with a thin plastic sheet. The gas concentration of the mixtures was monitored using an infrared gas analyser (type Servomex 4900).

To generate mist-air mixtures as obtained during the small-scale experiments would imply the use of a very large number of nozzles (3700) all to be provided with hoses and pressurised water/aqueous solution. This is not a feasible solution for industrial applications. To generate water-mists directly at a larger-scale in sufficiently high concentrations a dispersion of superheated water/aqueous solution seemed to be the only viable solution. To investigate the possibility of using superheated water in the 50 m^3 vessel three 4 litre suppression bottles supplied by IEP Technologies UK were applied. Each of the bottles were modified and provided with a 1500W heating jacket and a controller from IsoHeat, a thermocouple and a pressure transducer. The pressure inside the bottles was controlled by adjusting the temperature of the heating jackets. The pressure inside the bottles was also monitored during the dispersion of the water/aqueous solution. The temperature of the water/aqueous solution was monitored using thermocouples. The bottles were either filled with 2 or 3.5 litres of water or aqueous solution. Each bottle was connected to a nozzle and mounted along the centre line of the roof of the explosion chamber, regularly divided over its length.

Ignition was effected 6 s after the moment of dispersion to limit the effect of turbulence caused by the injection. The ignition source was an exploding wire located close to one of the short walls of the chamber (1 m from the end in the centre).

As a second part of the experimental work, the superheated water/inhibitor-solution dispersion system was substituted by a regular deluge system consisting of 7-11 nozzles of type MV12 125° with K-factor of 18.7. The nozzles were mounted regularly divided against the roof of the chamber. The droplet size delivered by these nozzles was not measured but on the basis of results presented for other MV nozzles it is expected to be between 200 and 600 μm with an median value of about 400 μm [11].

For the experiments using water deluge the configuration of the 50m^3 explosion chamber was changed from two open ends to one open end and one closed end. Ignition was effected close to the closed end to provoke higher flow velocities promoting droplet break-up. The total water/aqueous solution application rate was varied between 9.5 $l/min/m^2$ to 22 $l/min/m^2$.

3 Results

3.1 Small-scale tests

The parameters varied during the small-scale experiments in the 135-l channel were especially the propane-air concentration, the degree of congestion and the composition of the aqueous solution. All small-scale experiments were performed using potassium carbonate (K_2CO_3).

Generally speaking, the experimental results show that water mist in itself already causes a minor reduction of both flame speeds and overpressures (for the water concentrations used of 30 -80 g/m^3). Introducing potassium carbonate into the water the degree of mitigation is observed to increase up to full quenching of the explosion flame (for potassium carbonate concentrations of 60-70 g/l in water. This was seen for all obstacle configurations investigated including a channel without obstacles. An example of typical results obtained is summarised in Figure 3.

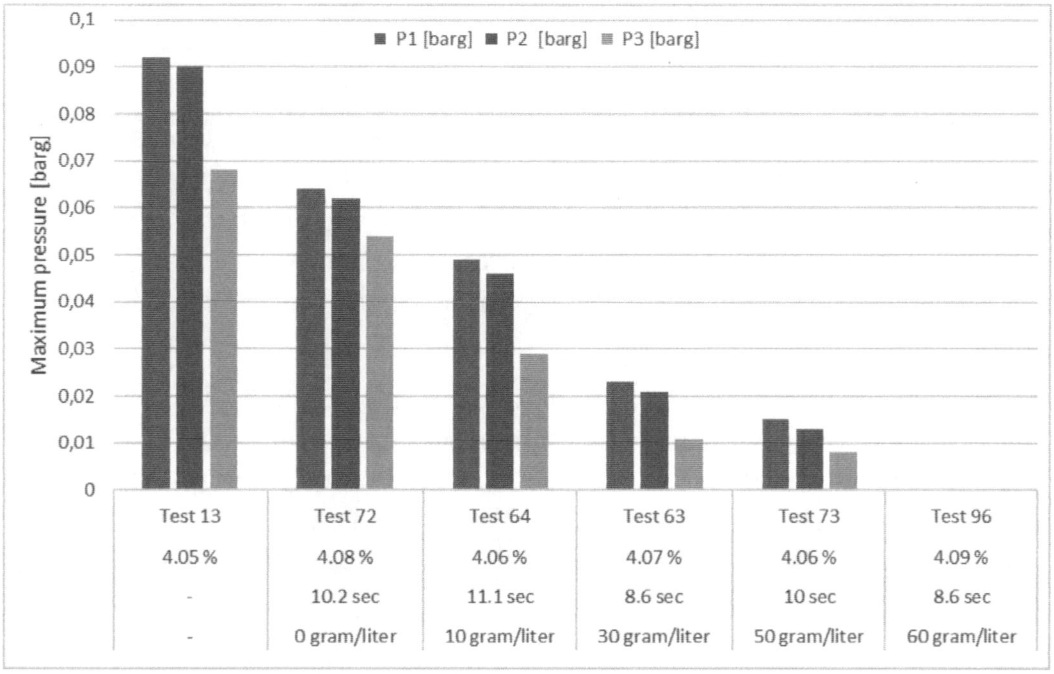

Figure 3 Effect of aqueous solutions of potassium carbonate on explosion overpressures in 4.05-4.09 % propane-air mixtures in a 0.135 m^3 channel with two baffles (at 31.0 cm and 148.5 cm from ignition end). Overpressures are measured at 3 locations (P1-P3) for increasing amounts of the flame inhibitor dissolved in water (increasing from 0 to 60 g/litre). The water mist concentration is determined by the injection duration varying from 8.6 to 11.1 s.

The overall results show that water droplets of a size of 90% < 25 μm in a concentration of 30-80 g/m^3 appear able to reduce the overpressures in the channel by 44-83% depending on the obstacle configuration. Adding potassium carbonate to the water causes this reduction of overpressures to increase. The more potassium carbonate is added to the water the stronger the mitigation of the propane explosions is up to full quenching (i.e. no ignition) when the potassium carbonate concentration in the water is 60 g/l or more.

3.2 Large-scale tests: Superheated aqueous solutions

The principle of superheated water is using stored thermal energy. The bottles are heated up to temperatures of 180 °C or higher resulting in a saturated water pressure in the bottles of 10 bar or higher. The thermodynamic condition of the water is now such that upon a sudden removal of the confinement (opening of the bottle) the water flashes immediately resulting in a water vapor cloud. This cloud cools off rapidly upon contact with the surrounding air and condensates giving rise to a cloud of very fine droplets. A part of the superheated water does not flash resulting in larger droplets. As a result, discharge of a superheated water system as described above results in a bimodal droplet diameter distribution. For the system used in the current investigation the droplet distribution was not measured but in [12] a similar system was reported to have a bimodal droplet distribution with modes located at 10 μm and 120 μm. The relative composition by weight was reported to be 25 % of 20 μm droplets and 75 % of 180 mm droplets.

In the current tests the distribution may be different due to generally higher temperatures used (on average 206 °C). It is expected that this thermodynamical condition leads to a higher fraction of fine droplets.

Nevertheless, assuming that approximately 25 % is present as small droplets one can assume that after 6 s delay time this is the only fraction left in the gas-air mixture. The terminal velocity of droplets of 180 μm is about 0.7 m/s and since the initial jet is pointing downwards this seems a valid assumption. If 25 % of the initial dispersed water is present in the mixture at the moment of ignition this implies a water fraction of 4 mol % if 6 l is dispersed and a water fraction of 7 mol % if 10.5 l is dispersed.

Initially reference tests were performed to determine the explosion overpressures in the obstructed 50 m^3 vessel without any mitigation except from the vent openings. Next tests were performed with superheated water without the introduction of inhibitor into the water. Results obtained for 4.15 % propane-air mixtures are summarised in Figure 4 showing a clear reduction of overpressures when increasing the amount of superheated water released from 0 to 10.5 l. The effect of the water can also be recognised from the delay the maximum overpressure is generated.

In [12] the design efficiency D_E of the superheated water mitigation system was expressed as:

$$D_E = \frac{Measured\ value\ of\ C_E}{Theoretical\ maximum\ value\ of\ C_E\ for\ the\ mitigation\ method\ used}$$

Where $C_E = (P_u - P_S)/P_u$

With P_u = the maximum pressure without the superheated water system activated

P_s = the maximum pressure with superheated water system activated

Using this as a way of presenting the results of the experiments performed the effect of the superheated water system using water without adding any inhibitor would be as shown in Figure 5 presenting the results for both lean (3.5 vol % propane-air), optimal (4.1 vol % propane-air) and rich propane-air mixtures (5.0%vol. propane-air). The theoretical maximum value of C_E has been assumed to be 1 implying full quenching of the flame as seen for the small-scale experiments. The Figure shows that the mitigation is similar for lean and rich mixtures showing an increase of the design efficiency when the amount of water released increases. For rich mixtures the effect of the water is however marginal: hardly a reduction of the overpressure is seen even for a 7 mol % water fraction.

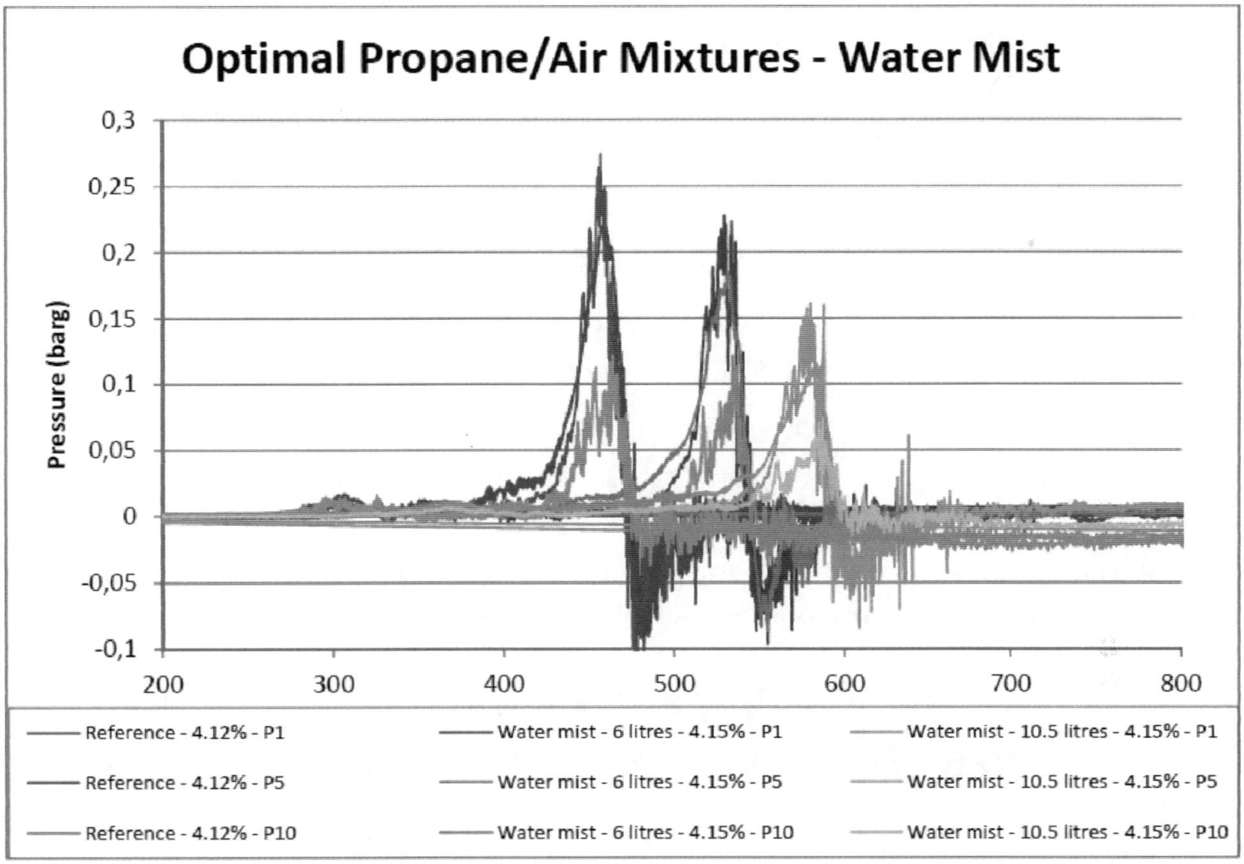

Figure 4 Effect of superheated water on overpressure generated by optimal propane-air explosions in a vented 50 m^3 obstructed chamber. The amount of water released was increased from 0 to 10.5 l.

As a comparison: the small-scale experiments (for optimal propane-air mixtures) resulted in a design efficiency of 0.32 considering pure water mists (for a water fraction of 4-11 mol %) which is very comparable.

Introducing potassium carbonate into the water in relative amounts up to exceeding those used in the small-scale experiments did not result in a considerable additional reduction of the overpressures seen. In fact, pressures seem to increase slightly the more potassium carbonate was added. An example of results is shown in Figure 6. The tests were performed injecting 10.5 l water, a water fraction of 7 mol % (assuming that 25 % of the initial water injected remains air bound as small droplets (< 20 µm)). The potassium carbonate concentration in the water was gradually increased from 30-210 g/l (the latter is far more than used in the small-scale experiments where full quenching was observed for concentrations of 60 g/l). Figure 6 shows that flame propagation seems to be affected in the initial phase causing the moment of occurrence of the maximum overpressure to be delayed for increased amounts of potassium carbonate. This could indicate a reduction of combustion rates at least during the initial phases. The maximum pressure is however increasing. The reason for this behavior is not clear. [13]

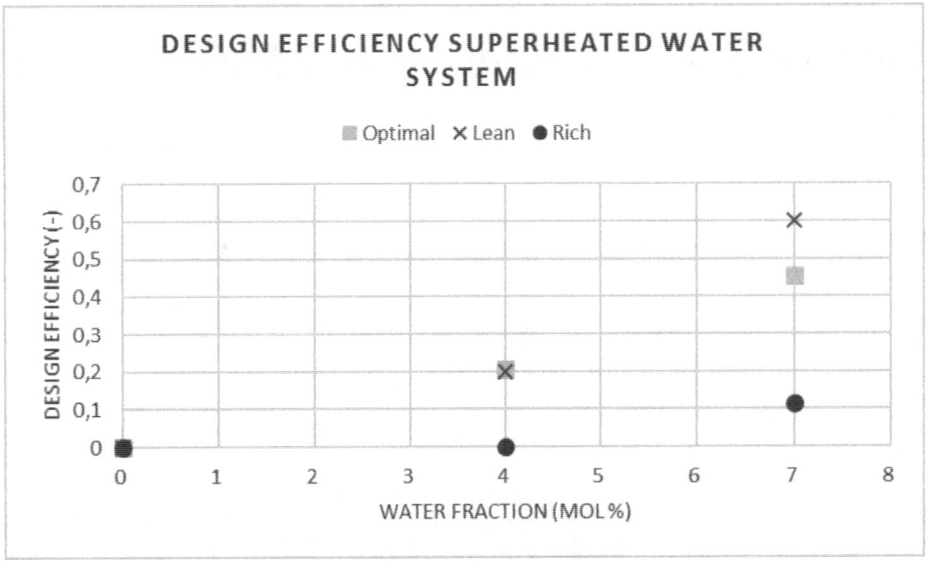

Figure 5 Design efficiency of superheated water system for lean, rich and optimal propane-air mixtures as a function of the water fraction.

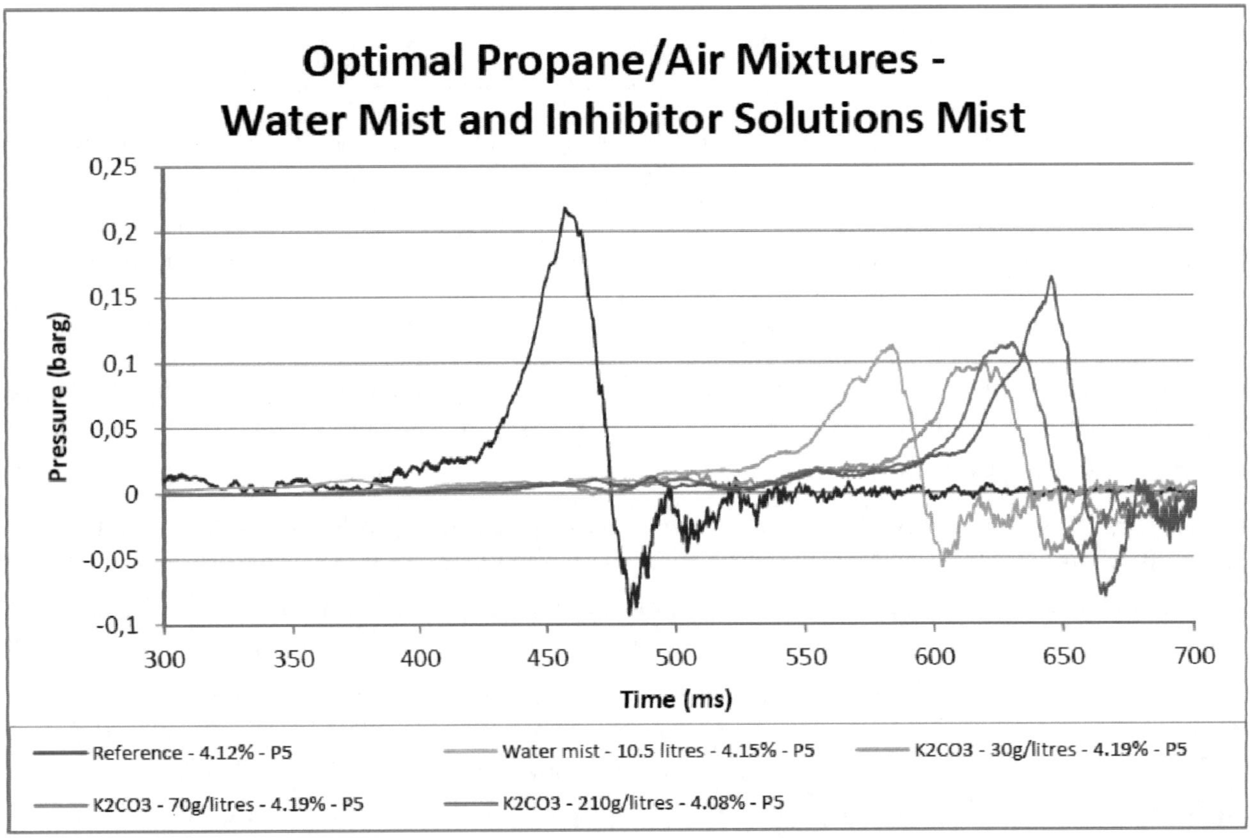

Figure 6 Effect of superheated aqueous solutions of potassium carbonate on overpressure generated by optimal propane-air explosions in a vented 50 m^3 obstructed chamber. The amount of aqueous solution released was 10.5 l. The potassium carbonate concentration was increased from 0 to 210 g/l water.

points in the direction of the higher viscosity of the solution as a possible explanation. A higher viscosity would however result in a higher concentration of fine droplets.

A better explanation is to be found in [14] where it was concluded that dissolved salts tend to displace superheated conditions to higher temperatures. The higher the potassium carbonate concentration the higher the temperatures necessary for obtaining superheat conditions and hence the lower the efficiency of the mitigation system. Using superheated aqueous solutions of flame inhibitors as a technology to generate fine mists to mitigate explosions is therefore not a feasible technology.

3.3 Large-scale tests: water deluge system

Since the superheated water/aqueous solution system did not result in the mitigation seen on small-scale it was decided to apply an industrial water deluge system instead. In this case the explosion itself generates fine droplets due to water droplet break-up. This implies that the aqueous solution mitigation system, if it works, only can be applied in congested areas where flame accelerations, due to the interaction between combustion generated turbulence and the combustion, occur. Droplet break-up occurs when a droplet is exposed to a flow resulting in hydrodynamic forces stronger than the forces keeping the droplet together, viz. surface tension. This can be expressed by the Weber number:

$$We = \rho v^2 d/\sigma$$

where

We = Weber number

ρ = density of the gas mixture stream (kg/m^3)

v = velocity of the gas mixture stream relative to the velocity of the droplet (m/s)

d = diameter of droplet (m)

σ = surface tension of liquid (N/m)

The critical Weber number where droplet start to break up is We = 10-12 [5].

The experiments were performed in the 50 m^3 vessel with a similar obstruction layout. The only difference was that one of the open ends was closed resulting in stronger flame acceleration and thereby higher pressures. Ignition was effected close to the closed end of the vessel. involved both reference tests, tests with water and tests with aqueous solutions. Figure 7 shows an example of results obtained. The maximum pressure in case of no mitigation was 1.7 bar. Using a relatively high water application rate of 21.9 l/min/m^2 results in a maximum pressure of 1.2 bar. Using the same application rate for an aqueous solution of potassium carbonate of 100 g/l results in a maximum pressure of 1.05 bar, implying some effect of the inhibitor but not as much as one would expect from the small-scale experiments. The application of water deluge as a mitigation system implies that the explosion is allowed to develop initially up to the moment droplet break-up occurs and where mitigation sets in. Turbulence generated by the water deluge system causes the flame to propagate faster during the initial phase than the non-mitigated test. The maximum pressure obtained is therefore a combination of the effect of turbulence during the initial phases and the degree of mitigation afterwards. Figure 8 shows the effect of varying the water application rate on the maximum explosion overpressure. A lower water application rate implies less turbulence generation initially but also lower water fractions after droplet break-up, i.e. less mitigation.

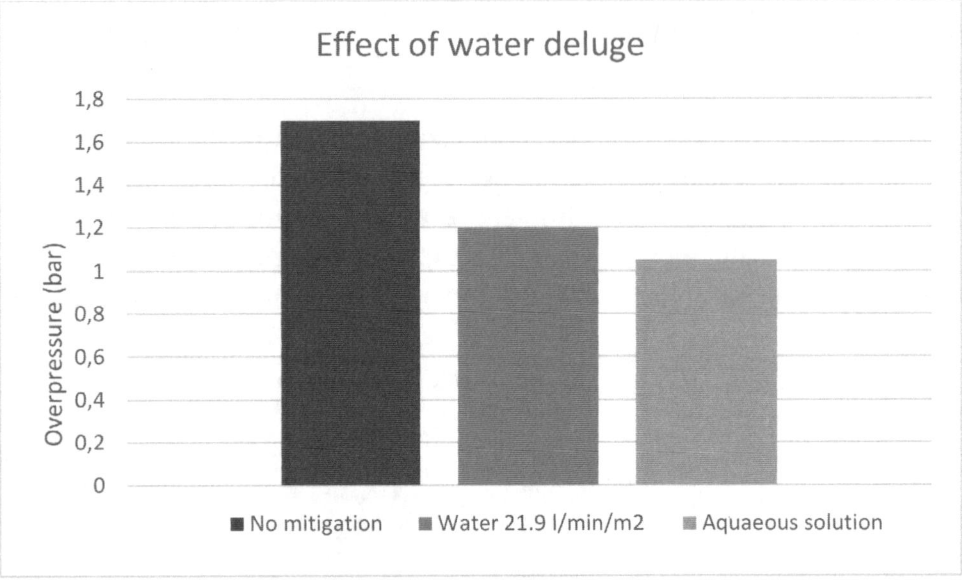

Figure 7 Maximum overpressure obtained for 4.1 vol % propane/air explosions with a water loading of approximately 21.9 l/min/m² with and without a potassium carbonate solution of 100 g/l, compared to the dry reference test.

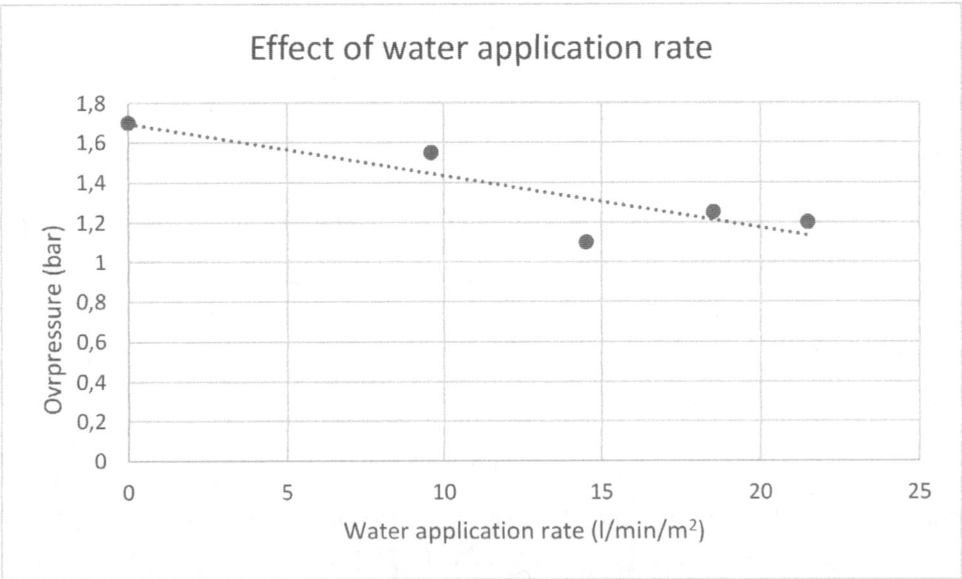

Figure 8 Maximum overpressure obtained for 4.1 vol % propane/air explosions varying the water application rate.

In an attempt to distinguish the effect of initial turbulence from the mitigating effect of the aqueous solution tests were performed with low application rates reducing the initial turbulence whereas the mitigating effect was still expected to be considerable. This, however, appeared to be disappointing as seen in Figure 9.

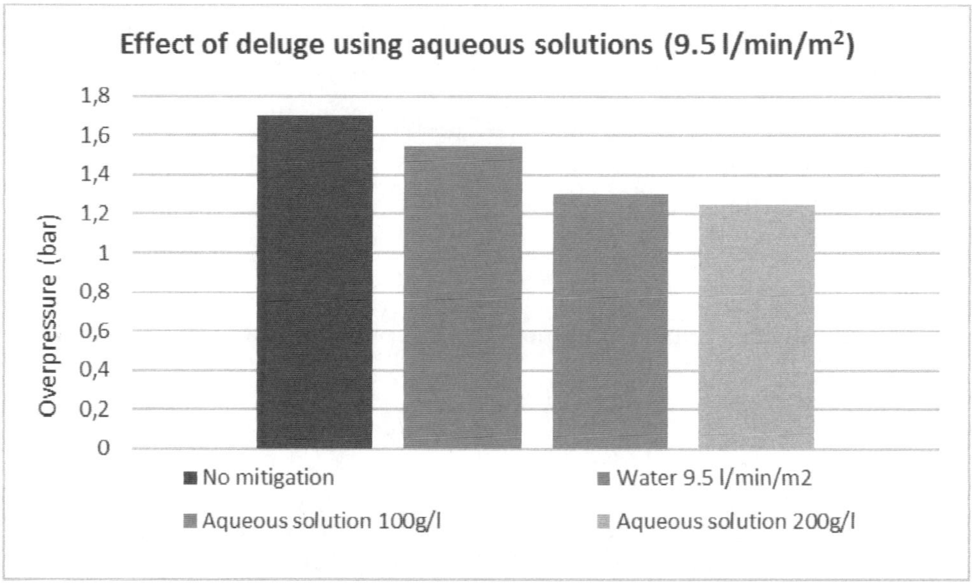

Figure 9 Maximum overpressure obtained for 4.1 vol % propane/air explosions for a low application rate (9.5 l/min/m²) with water and aqueous solutions of 100 g/l and 200 g/l of potassium carbonate.

Although aqueous solutions do result in some improved mitigation compared to pure water the effects are limited. A possible reason may be that some salts (such as NaCl) cause an increase of the surface tension of water making it more difficult for droplets to break up [15] by flow as the Weber number suggests. The effect of an increased water surface tension and increased viscosity due to the introduction of potassium carbonate into the water has hardly any influence on the droplet characteristics of the water nozzle used [16].

4 Conclusions

Whereas direct injection of mists of aqueous solutions of water and potassium had considerable improved mitigation effects on explosions in a small-scale congested channel compared to injection of water mist only a similar effect could not be reproduced on a larger scale.

Generation of water mist using a superheated water injection system resulted in a considerable reduction of explosions overpressures in a congested 50 m^3 chamber when using pure water. This effect was however not improving any further when applying aqueous solutions of water and potassium carbonate. An explanation for this is that the adding of potassium carbonate causes superheated conditions to be displaced to higher temperatures. The higher the potassium carbonate concentration the higher the temperatures necessary for obtaining superheat conditions and hence the lower the efficiency of the mitigation system. Using superheated aqueous solutions of flame inhibitors as a technology to generate fine mists to mitigate explosions is therefore not a feasible technology.

Using a water deluge system instead, aqueous solutions were shown to have a small improved mitigation effect on explosions compared to that of pure water only. A possible reason may be that introduction of potassium carbonate in the water causes an increase of the surface tension of water making it more difficult for droplets to break up in an accelerating flow.

As such there is a challenge to apply aqueous solutions as a measure for mitigation of large-scale vapour cloud explosions since generation of sufficiently fine mists of aqueous solutions on an industrial scale is not possible with current technologies.

5 Acknowledgments

The authors acknowledge the financial support obtained from Total SA for performing these experiments.

6 References

[1] Van den Berg, A. C., The multi-energy method: A framework for vapour cloud explosion blast prediction, Journal of Hazardous Materials, 12 (1), (1985), 1-10

[2] Tang, M.J. and Baker, Q. A., A new set of blast curves from vapor cloud explosion, Process Safety Progress, 18 (4), (1999), 235-240

[3] Puttock, J.S., Developments in the Congestion Assessment Method for the prediction of vapour-cloud explosions, Loss Prevention and Safety Promotion in the Process Industries, Proceedings of the 10th International Symposium (2001), 1107-1133

[4] Catlin, C.A, Gregory, C.A.J., Johnson, D.M. and Walker, D.G., Explosion mitigation in offshore modules by general area deluge. Trans. I.Chem.E. 71B, (1993), 101-111

[5] van Wingerden, K., Mitigation of gas explosions using water deluge. Process Safety Progress, 19 (3), (2000), 173-178

[6] Hoorelbeke, L., Numerical and Experimental Study of Vapor Cloud Explosions, PhD Thesis. Univ. of Brussels, 2011

[7] van Wingerden, K. and Hoorelbeke, P., On the potential of mitigating vapour cloud explosions using flame inhibitors, American Institute of Chemical Engineers, Seventh Global Congress on Process Safety (GCPS), Chicago, Illinois, USA, 2011

[8] Roosendans, D., Hoorelbeke, P., and van Wingerden, K., Mitigation of vapour cloud explosions by chemical inhibition, IChemE Symposium Series 158, (2012), 313-326

[9] Davis, S.G., Merilo, E., Engel, D., Ziemba, A., Pinto, M. and van Wingerden, K. Large scale detonation testing: New findings in the prediction of DDTs at large scales, Journal of Loss Prevention in the Process Industries 48, (2017), 345-357

[10] Roosendans, D., van Wingerden, K., Holme, M.N. and Hoorelbeke, P., Experimental investigation of explosion mitigating properties of aqueous potassium carbonate solutions, Journal of Loss Prevention in the Process Industries, 46, (2017), 209-226

[11] Wighus, R., Droplet size, velocity and area distribution – Deluge nozzles, SINTEF NBL report no. A12114, 2012

[12] Tam, V.H.Y., O'Connell, M., Pedersen, G. and Renwick, P., Testing of the Micromist device: an active soft barrier for explosion control, Journal of Loss Prevention in the Process Industries, 16(1), (2003), 81-88

[13] Roosendans, D., Mitigation of Vapor Cloud Explosions by Chemical Inhibition using Alkali Metal Compounds, PhD Thesis Vrije Universiteit Brussel, 2018

[14] Thiéry, R., Loock, S. and Mercury, L., Explosive properties of superheated aqueous solutions in volcanic and hydrothermal systems, Metastable Systems under Pressure (NATO Science for Peace and Security Series A: Chemistry and Biology), 2010th Edition (ed. Rzoska, S. et al), 293-310

[15] Okura, H.I., Chena, Y., Wilkins, D.M. and Rokea, S., The Jones-Ray effect reinterpreted: Surface tension minima of low ionic strength electrolyte solutions are caused by electric field induced water-water correlations, Chemical Physics Letters, 684, (2017), 433-442

[16] Kooij, S., Sijs, R., Denn, M. M., Villermaux, E. and Bonn, D., What Determines the Drop Size in Sprays? Physical Review X 8, 031019 (2018)

585375 Contain the Tiger: Accounting for All Energy from Reactive Chemicals in Process Design

Abstract:

Chemical manufacturing processes rely on intended chemical reactions to create desired products. During this process, energy from both intended and unintended sources can be generated, and needs to be safely accommodated through process design and safety systems. This paper will identify test methods for characterizing the energy produced by chemical reactions and review process design strategies including inherently safer design options. Several types of chemistries will be reviewed using methods contained in the Center for Chemical Process Safety (CCPS) book Guidelines for Chemical Reactivity Evaluation and Application to Process Design.

A case study will be reviewed to illustrate the concept and application of these critical process design elements for reactive chemical systems, focusing on the proper use of kinetic parameters, thermal stability data, and heat transfer requirements to develop an appropriate Basis of Safety for the reactive system design.

The following learning objectives will be achieved:

1. Introduce a practical approach for Reactive Energy Hazard Evaluation.
2. Overview a disciplined approach for reviewing test data to identify potential energetic scenarios.
3. Understand critical elements of process design for reactive chemical systems.

Chemical Process Safety and Fire Protection – the Role Fire Protection Plays in Your Process Hazard Analysis

Ron Melucci, Technical Manager – Fire Protection
Janet Persechino, Technical Manager – Chemical Process Safety
Engineering Planning and Management, Inc.
959 Concord Street
Framingham, MA 01701
rbm@epm-inc.com
jap@epm-inc.com

Abstract

In any chemical process the fundamental principles of process safety should be applied to decrease the likelihood of a release. Similarly, the fundamental principles of fire protection should be applied for mitigation of the consequences of a release. Focusing on process safety guidance from the Occupational Safety and Health Administration (OSHA), the Environmental Protection Agency (EPA) and other relevant industry practices such as the Center for Chemical Process Safety (CCPS) Risk Based Process Safety and FM Global Data Sheets will decrease the likelihood of releases and catastrophic losses. Fire protection practices outlined in the National Fire Protection Association (NFPA) Fire Code and the International Fire Code are also essential for loss mitigation. These process safety and fire protection methodologies work in tandem and should not be considered as part of the overall risk assessment process.

Chemical plants require more specialized fire protection features and systems than most industrial facilities. The extreme hazards associated with the chemical processing industry (CPI) can quickly lead to a cascading sequence of events that can result in a catastrophic loss. An unwanted release of flammable liquids, if not properly controlled, can overtax sprinkler systems with several minutes and compromise the integrity of vessels, piping, safety interlocks and structural steel elements. This can lead to a collapse of process equipment/structures, loss of process control, devasting off-site consequences and significant business interruptions.

The CPI faces unique challenges when it comes to the automatic sprinkler protection placement and installation and other fire mitigation techniques. This paper will outline some of the different protection schemes and how to decide which protection option is best suited for a particular application.

The overall conclusion is that process safety must be applied for loss prevention and this should be combined with active and passive fire protection features to mitigate the consequences of a release should lapses in process safety occur. These topics should not be considered separately but viewed as working in partnership. It is important that these concepts are implemented and reviewed by experts in the fire protection and process safety fields to ensure adequate protection schemes are employed in a proper manner and the information is correctly covered and credited in the Process Hazard Analysis (PHA).

Introduction

In any chemical process the fundamental principles of process safety should be applied to decrease the likelihood of a release. Similarly, the fundamental principles of fire protection should be applied for mitigation of consequences. Focusing on process safety guidance from the Occupational Safety and Health Administration (OSHA), Environmental Protection Agency (EPA) and other relevant industry practices geared specifically for chemical process safety management will decrease the likelihood of releases and catastrophic loss. Fire protection practices are outlined in the National Fire Protection Association (NFPA) Fire Code NFPA 1 and the International Fire Code (IFC). These process safety and fire protection methodologies work in tandem to achieve loss prevention and mitigation.

The applicable codes (mentioned above) require Process Hazard Analyses (PHAs) and Fire Hazard Analyses (FHAs) for most new chemical plants and those undergoing changes. The codes state that the analyses are to be performed by someone with expertise in chemical process safety and fire protection. Additional building and system safeguards are often implemented as a result of these structured and systematic analyses. This can include improved exothermic reactor vessel cooling systems, isolation valve installation, pipe rack protection, etc.

FM Global Datasheet 7-14, *Fire Protection for Chemical Plants,* indicates that over 60% of the surveyed chemical plant losses between 2003 and 2014 were attributed to equipment design problems and operator error. These items are evaluated when the correct hazard analysis methodology is employed. A PHA will include an evaluation of adequate system safeguards and human reliability aspects. This type of comprehensive review should decrease the chance of having a loss due to a design or human element issue.

Chemical plants require more specialized fire protection features and systems than most industrial facilities. The extreme hazards associated with the chemical processing industry (CPI) can quickly lead to a cascading sequence of events that can result in a catastrophic loss. An unwanted release of flammable liquid, if not properly controlled, can result in a fire that can overtax sprinkler systems within several minutes. Such uncontrolled fires can compromise the integrity of vessels, piping, process safety systems or structural steel elements. This can lead to a collapse of process equipment/structures and a loss of process control which can result in:

- ✓ Fatalities
- ✓ Serious injuries
- ✓ Large fires/explosions
- ✓ Significant property damage
- ✓ Environmental issues
- ✓ Extensive Downtime
- ✓ Loss of market share
- ✓ Damage to company reputation
- ✓ Difficulty in obtaining insurance coverage
- ✓ Fines/penalties

A sprinkler system may control a chemical plant fire by cooling and wetting surrounding areas but may not extinguish fires involving flammable gases, liquefied flammable gases and low flash point flammable liquids. Additional consideration may therefore be needed for special fire protection systems that protect critical equipment from fire damage, particularly where loss of equipment/components can result in catastrophic losses as described above.

Fire Protection and the PHA

The fire protection systems should be designed in conjunction with, or in consideration of, the PHA. The PHA will include identification of critical equipment that is needed to prevent catastrophic loss and will estimate the likelihood and severity of unwanted releases.

Fire protection system design for chemical plants with flammable liquid hazards requires careful consideration of the fire code requirements as well as other requirements such as NFPA 30, *Flammable and Combustible Liquids Code,* and the guidelines in FM Global Datasheet 7-14, *Fire Protection for Chemical Plants.* Installation guidelines such as NFPA 15, *Standard for Water Spray Fixed Systems for Fire Protection* and NFPA 13, *Standard for Installation of Sprinkler Systems,* are also applicable.

To meet the fire protection objectives and thus loss mitigation, both passive and active (fire) protection features and systems should be implemented. Common and effective features include:

Table 1 Accidental Release Consequence Mitigation Methods

PASSIVE PROTECTION	ACTIVE PROTECTION
Spatial or Fire Barrier Separation	Area-type Automatic Sprinkler or Deluge Protection
Fire Proofing of Structural Steel/Equipment Supports	Fixed Directional Deluge Water Spray Protection
Containment and Emergency Drainage	Process Safety Features such as Safety Shut Off Valves

Active fire protection should cover all areas where a flammable liquid release may flow. Containment systems may include emergency drainage that drains to a secondary containment area to help remove the released liquid and discharged water to an impoundment area that is located far from equipment and buildings.

Fire Protection Installation Challenges for the CPI

Installation of sprinklers close to the ceiling may not be effective in many CPI applications. Sprinklers are often installed far below ceilings, or below open metal grating, which can significantly delay sprinkler activation. There are two common reasons for this:

- Stratification - There may be insufficient fire plume buoyancy and temperature at high ceiling levels. It is imperative that sprinklers activate in the first several minutes of a fire and even this type of small delay in sprinkler activation, especially given the high hazard materials involved in the CPI, can mean complete loss of fire control which can result in a catastrophic event

- There may be no ceiling jet formation to cause sprinkler activation as in the case of a fire occurring below open metal grating. Improper sprinkler system design can lead to complete loss of fire control and a catastrophic loss scenario

Due to the installation challenges, sprinklers with heat collectors installed far below ceilings and below metal grating, have become standard practice. For this type of design, sprinkler activation is only achieved if the sprinkler(s) are located within the fire plume geometry (i.e., close to the vertical axis of the fire plume), to intercept the buoyant hot gases needed to actuate the sprinkler fusible element that allows the application of water.

The industry guidelines for such sprinkler design are not specific, prescriptive requirements. Therefore, a performance-based analysis is required of the fire protection engineer. NFPA 13 *Standard for the Installation of Sprinkler Systems*, does not specifically address the spacing requirements for sprinklers installed under open metal grating. However, section 8.5.5.3.4 of the 2016 edition of NFPA 13 contains the following verbiage: "*Sprinklers installed under open gratings shall be of the intermediate level/rack storage type or otherwise shielded from the discharge of overhead sprinklers*." The annex material elaborates that shields (not heat collectors), should be installed on sprinklers under open grating. The shields are intended to protect sprinklers from discharge of overhead sprinklers, but the intent is often misunderstood as heat collection.

NFPA 13 Section 8.5.4.1.4 states: "*Heat collectors shall not be used as a means to assist the activation of a sprinkler*." Studies have shown that "heat collectors" can actually delay the activation of the sprinkler by forming a dead air pocket. For sprinklers installed far below solid ceilings or below metal grates, sprinkler activation is only achieved as stated above, if the sprinklers are within the fire plume geometry.

Sprinklers are sometimes installed far below the ceiling or below metal grate platforms for protection of critical equipment. This may include directional sprinklers or nozzles to provide exposure protection to the vessel or equipment, in lieu of fire extinguishment or control. This can often be the preferred fire protection approach to mitigate fire damage. The exposure protection objective described in NFPA 15, *Standard for Water Spray Fixed Systems for Fire Protection*, notes that the protection objective is to keep the vessels and other critical equipment sufficiently wet and cool to protect these assets from fire exposure.

Conclusion

The CPI faces unique challenges when it comes to automatic sprinkler protection. This type of protection can be used for different purposes in the CPI. An assessment of the risk, plant layout and overall loss mitigation goals will help decide which protection option is suited to the application. Due to the subjectivity and lack of definitive industry guidance for every situation, a qualified fire protection engineer who is very familiar with these requirements and guidelines, should make the fire protection design decisions described herein.

The overall conclusion is that process safety must be applied for loss prevention to decrease the likelihood of a loss occurrence and this should be combined with active and passive fire protection for mitigation of consequences. These topics should not be considered separately, but viewed as working in partnership as proactive and reactive systems. It is important that these subjects are administered and reviewed by experts in fire protection and process safety fields to ensure adequate protection schemes are employed.

583240 Experimental Study on the Characterisation of Flammability of Methane Enriched in Hydrogen in the Framework of Risk Assessment of Injection of Hydrogen in Natural Gas Network (concept "POWER TO GAS")

Abstract:

In the context of planning a decarburization of the European Union energy sector, in order to reach a reduction of carbon emissions up to 95 % by 2050, several studies on the compatibility of the existing gas networks with blends of natural gas enriched with hydrogen up to 20% of hydrogen have been tested in industrial scale projects.

The explosion process of multi-component gas mixture is extremely complex and may cause serious disaster effects. The safety issue concerning explosion of multi-component gas mixture is urgent to be investigated on account of its wide range of applications. In current work, series of experiments to characterize severals parameters (Lower Explosive Limit - Upper Explosive Limit - Lower Oxygen Concentration - Maximum Pression - Gas explosion constant - Minimum Ignition Current - Maximum Experimental Safe Gap - Ignition Sensitivity) were performed in a 20 L spherical explosion vessel at initial conditions of 1 atm and 293 K, involving methane-hydrogen/air mixtures.

In relationship to the European Union ATEX Directive 2014/34/EU and protective systems intended for the use in potentially explosive atmospheres, gases and dusts are classified according to their ignition properties. The explosion group for gases and vapors is determined by the maximum experimental safe gap (MESG) and/or minimum ignition current (MIC) ratio values, both defined in the ISO/IEC 80079 (Explosive Atmosphere—Part 20-1: Material characteristics for gas and vapour classification—Test methods and data) standard.

This paper presents the experimental results produced by Ineris highlighting the influence of the addition of hydrogen on the flammability and explosive properties of the natural gas-hydrogen mixture (0-20% Vol.).

The presence of hydrogen has a negligible influence on certain properties such as:

o the lower explosive limit,

o the oxygen limit concentration,

o the maximum explosion pressure,

o the self-ignition temperature;

The following properties are more sensitive to hydrogen content:

o the upper limit of explosiveness increases linearly with the hydrogen content in methane;

o Industrial methane, like natural gas, is classified in Group IIA on the grounds that it contains no more than 25% Vol. hydrogen. A mixture of methane with other compounds of group IIA of any proportion is classified in Group IIA;

o the maximum pressure-mounting speed remains constant and equal to that of methane up to about 15 % hydrogen, but increases rapidly for higher hydrogen levels;

o the minimum ignition energy decreases rapidly as the hydrogen content increases, suggesting an increase in the probability of inflammation with the hydrogen content (which, however, has yet to be confirmed by specific tests). However, for low hydrogen levels (10% max), the probability of ignition, although greater than for natural gas alone, would remain in the same orders of magnitude.

587043 Maintenance Error Management for Improving Mechanical Integrity

Abstract:

Weak management systems lead to Human Errors that lead to accidents. The paper covers significance of maintenance errors (ME) in process industry, what and why of ME, principles of ME, extent of ME, and how to deal with ME. It proposes to have a structured program in place and what must be considered while formulating the program at a company. Limited deployment of the program at some intense maintenance environment at process facilities has shown promising results.

Maintenance errors have been among the principal causes of several major accidents in a wide range of technologies, for instance Piper Alpha disaster. For process industry, maintenance, particularly maintenance of safety critical equipment, has always been addressed as important means to prevent major accidents. Maintenance is indissolubly linked with safety in a number of ways - errors in planning, executing or checking maintenance could cause system faults; or inadequate or erroneous maintenance may also mean that existing failure or degradation is not discovered and corrected – and thereby contributes to production shut-downs, work accidents and/or major incidents. Maintenance errors not only endanger lives and assets, they are also extremely bad for business.

Process plant maintenance is a particularly challenging task due to several factors such as frequent removal and replacement of a large numbers of components, cramped, arid, poorly lit spaces and different groups work on the same item of equipment. Turnaround maintenance poses further heightened risk.

The remedy is to have a structured Maintenance Error Management Program. A wide variety of error management (EM) measures are resorted to targeting person and team, workplace and task, and organizational factors. Different types of error require different counter-measures; engineering solutions, collective mindfulness and organizational culture that is just, willing to report its unsafe acts and able to learn from them. It is not enough to deploy some tools but need to be watched, adjusted and managed on regular basis.

Aviation industry has recognized this and has mature programs in place. These could serve as examples for chemical process industry. For instance Maintenance Error Management System by Civil Aviation Authority, UK and Boeing's human factors tool called the Maintenance Error Decision Aid (MEDA) for investigating contributing factors to maintenance errors.

Human error in process plant maintenance in the past has not been given the attention it deserves. It is a pressing Human Factors issue. Practitioners of Mechanical Integrity element of PSM need to factor its impact and develop actions to mitigate its effect. This paper would help them present the case to management for recognizing human errors in process plant maintenance as a major risk factor that ought to be controlled by having a structured program in place.

A Machine Learning-based Approach to the Identification of Key Material Failure Modes of the Process Facilities

Rustam Abubakirov
Safety and Risk Engineering Group,
Dept. of Chemical and Materials Engineering, Nazarbayev University
Nur-Sultan city, Kazakhstan
rustam.abubakirov@nu.edu.kz

Ming Yang
Safety and Security Science Section,
Faculty of Technology, Policy, and Management, Delft University of Technology
Delft, The Netherlands
m.yang-1@tudelft.nl

Prepared for Presentation at
American Institute of Chemical Engineers
2020 Spring Meeting and 16th Global Congress on Process Safety
August 17 – August 21, 2020

A Machine Learning-based Approach to the Identification of Key Material Failure Modes of the Process Facilities

Rustam Abubakirov
Safety and Risk Engineering Group,
Dept. of Chemical and Materials Engineering, Nazarbayev University
Nur-Sultan city, Kazakhstan
rustam.abubakirov@nu.edu.kz

Ming Yang
Safety and Security Science Section,
Faculty of Technology, Policy, and Management, Delft University of Technology

Keywords: Asset Integrity, Reliability, Damage Mechanisms, Process Safety.

Abstract

Identification of credible failure modes and associated damage mechanisms is a critical part of asset integrity management program for the process units in the oil and gas industry. The Failure Mode and Effect Analysis (FMEA) has been widely used for these purposes. At the stages of units' design and early operation, subject matter expert's opinions (SMEs) with qualitative or semi-quantitative approaches for the damage risk assessment are often used. However, the quality of the assessment can be heavily dependent on the experts' knowledge and information available for the studied system. The present study aims to improve the reliability of the existing credible failure mode identification by using machine learning techniques of regression and feature engineering in the framework. The developed model assesses the contribution of individual damage mechanisms to the overall component's risk level given the available operational, design, and inspection metrics. By means of feature engineering, the limited input information is expanded for damage factor/applicable mechanism analysis by API RP 581 and API RP 571, used as a reference. Meanwhile, the several regression techniques will be used for risk analysis of individual damage mechanisms, in terms of their relative probabilities of occurrence and consequences. Based on which, the damage mechanisms are prioritized. The performance of the model will be tested on the example of several process units, including sour gas treatment, thermal cracking, separators, and several simulated flowlines.

1. Introduction

Material failures and their degradation or corrosion present a significant challenge to the present-day process industries. Wood et al. (2013) note that corrosion-related phenomena were the primary contributing factor of 20% of major accidents across the EU refineries since 1984. The range of the affected equipment included almost all process units, including trays, drums, towers, with the piping being the primary group (71%). With this regard, early identification of credible damage mechanisms, their characterization, and development of mitigation strategy is the critical task for maintaining the components' integrity.

There are several practiced methodologies for damage mode identification and its cause and effect characterization with FMEA (Failure Mode and Effect Analysis) among the widely-used ones. FMEA implies a systematic quantitative framework for probabilistic risk assessment to describe the state of all credible failing events of the considered component. Applied in asset integrity management (AIM), the framework deals with the characterization of all potential damage mechanisms throughout the component's lifecycle, during both normal and transient operations where the equipment is subjected to the interaction with the aggressive environment/ conditions. The framework is generally reliant on the subject matter experts' (SME) opinion in estimations of its target metrics as risks or risk priority numbers. The holistic quantitative methodologies as API RP 581 (2016) might be applied for these tasks as well. In API RP 581, Probability of Failure (PoF) assessment is based on the concept of the statistically derived generic failure frequency for different types of components and damage factors quantifying the severity of material degradation. The framework provides extensive calculation protocols covering a wide range of damage mechanisms inherent to the petrochemical industries. However, the massive data inputs and its pre-processing time is often not entirely available for quantitative assessment; thus, making qualitative and semi-quantitative approaches more commonly used for the risk assessments (Bai and Bai, 2014). This explainins active SME involvement in the analysis. The involvement of SMEs is not something undesirable in risk assessment. Nevertheless, it can be subjective and dependent on the surveyor's experience and the information available for the assessment.

There have been several attempts to introduce alternative ways of quantitative judgment into the risk assessments. The works might complement the existing standard frameworks or might offer its own probabilistic or reliability-based theory for the risk assessment. For example, Rachman and Ratnayake (2019) proposed a machine learning-based approach for the component screening assessment as per API RP 581. Paltrinieri et al. (2019) have developed a neural network for the dynamic risk assessment in drilling operations. Similar applications of machine learning are available in adjacent fields as process control and optimization (Min et al., 2019; Wu et al., 2020). In general, the main advantages of these approaches lie in their efficiency in solving high-dimensional problems as the discussed cases are and availability of computational environments and tools for it. Another important competitive aspect of machine learning approaches, which is less discussed in the area of chemical process safety, is their usefulness in cases with missing information inputs. There is a significant advancement in missing information handling in the area of bioinformatics (Lopes and Ribeiro, 2012; Thomas et al., 2020). Recently Martinez-Luengo et al., (2019) illustrated missing values imputation through employing neural networks in wind turbine fatigue life assessments. In the process component integrity assessments, the issue of missing information is not less critical: some commonly reliant metrics as thicknesses or corrosion rates are scarcely available in real operations.

The aim of the study is the development of a flexible and transparent approach supporting quantitative identification of critical failure modes in the cases with incomplete information available for conducting conventional quantitative assessments. The emphasis of missing parameters handling simulations will be put on cases with inspection metrics missing for the considered component. The approach will use the methodologies on damage mechanisms evaluation of API RP 571 (2011) and API RP 581 (2016) to develop a model allowing risk assessment of each damage mechanism through machine learning techniques. Performance and applicability of two regression algorithms of Deep Neural Networks and Random Forest will be illustrated throughout the study. To deal with the cases of incomplete input information the approaches of data imputation and integrated model development with the support of missing features will be introduced. Performance of the proposed method will be evaluated through the case study on several chemical process units, including sour gas treatment, thermal cracking, separators, and several simulated flowlines operating in a corrosive environment.

2. The proposed methodology and its application

2.1 The basic principles

The main challenge for operators to perform a valid quantitative risk assessment is their limited access to some types of data prescribed in the standard frameworks. Moreover, some critical metrics in the assessments, such as measured thickness or actual corrosion rates, are scarcely available due to their relatively high costs. It is rarely prioritized to obtain these data during the component's early operational history. To assess the component's integrity state quantitatively in these cases, we may adopt the power of statistical and machine learning approaches. Figure 1 presents a schematic flowchart of the proposed approach.

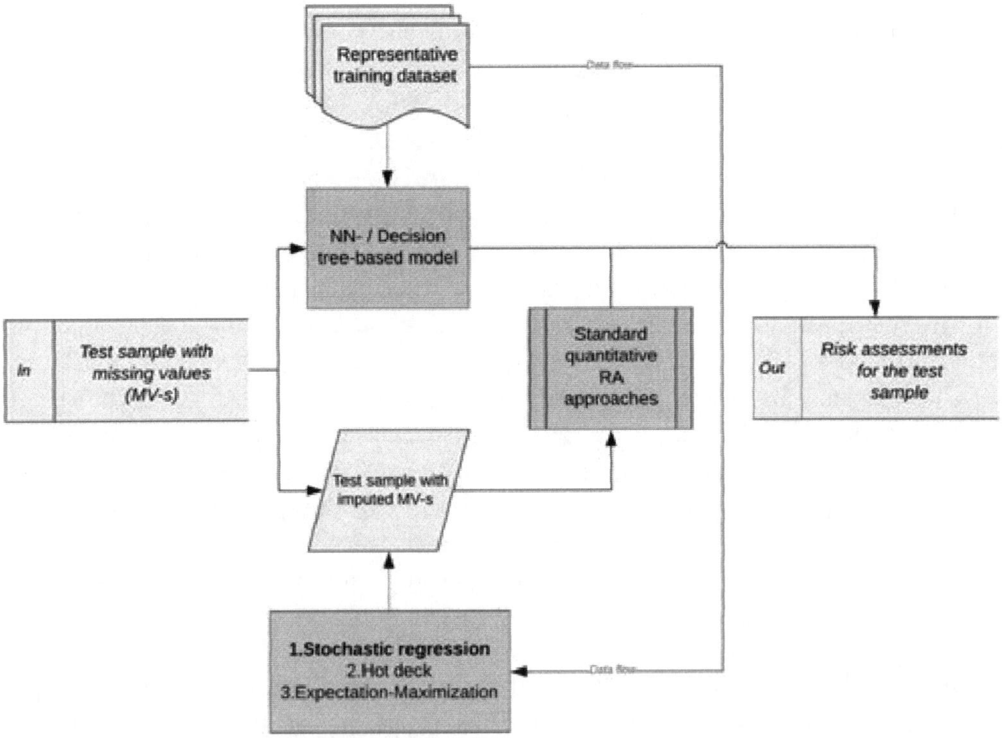

Figure 1. Flowchart of the proposed approach

In the identification of critical damage mechanisms and failure modes, we may encounter some sample components with possibly missing input data on the design, operational, or other inspection parameters. In these cases, these input data need to be pre-processed with all missing values imputed. This step will help to prepare the sample ready for a standard risk assessment. The multiple imputation algorithms like expectation–maximization, hot deck, or stochastic regression can be used for that purpose. Alternatively, the models like Neural Networks or Decision Tree-based algorithms with built-in support of missing values can be powerful as well (Lopes and Ribeiro, 2012). The latter ones require less time to implement them but at the cost of higher required computational power. However, the successful implementation of both approaches is dependent on the quality of the training datasets used. An ideal training dataset is believed to cover a full range of credible damage mechanisms inherent to the potentially assessed components (i.e., being representative); meanwhile they should be consistent in the data entries over the relationships that they carry.

2.2 Data inputs and their pre-processing

A training dataset comprised of 100,000 samples with labeled 62 features and 13 target values has been created for model training and validation. The features represent the standard design, operating, and inspection metrics used in API RP 581(2016) procedures with 13 target variables, 12 of which were the damage factors, and 1 being the consequences of the failure (CoF) all calculated by the framework above. The damage factors have been accounted for damage mechanisms as general thinning, both internal and external, stress corrosion cracking, mechanical fatigue, high-temperature hydrogen attack, and brittle fracture. The detailed list of the considered mechanisms is presented in Table 1.

Table 1. Damage mechanisms considered within the present study

#	Governing damage mechanism	Individual damage mechanism	Nomenclature
1	Internal thinning	General internal thinning	Int Thin
2	External thinning	General external thinning	Ext Thin
3		Corrosion under insulation	
4	Stress corrosion cracking	Amine cracking	SCC – AC
5		Caustic cracking	SCC – CC
6		Chloride stress corrosion cracking	SCC – ClSCC
7		Hydrogen induced cracking	SCC – HIC
8		Sulfide stress cracking	SCC – SSC
9	High temperature hydrogen attack	High-temperature hydrogen attack	HTHA
10	Brittle fracture	Brittle fracture	BF
11		885° embrittlement	885F
12	Mechanical fatigue	Mechanical fatigue	MF

The samples' feature distribution was based on several operational service classes (e.g., sour hydrocarbons or lean amine) with corresponding fluid, material, control and corrosion parameters interrelatedly distributed within regular operational ranges. The majority of the remaining features were not apparently dependent on any other one. They were set to be randomly distributed within their characteristic ranges. The entries in the training dataset do not need to be real-world operational data. Even on the contrary, the authors believe that the sufficient presentation of instances with deviations from the normal operational conditions is crucial while training. It would allow the model to learn from a full spectrum of potential failure conditions and their effects. This property is called representativeness of a dataset (Schaafsma and Vihinen, 2018).

Data consistency and bias are other two important aspects to consider before the model training stage. Apart from its general usability, consistency accounts for the absence of contradicting entries within the dataset. Especially it is relevant to the integrated datasets comprised of smaller datasets coming from differing origins, i.e., simulations to the real-world entries, etc. For example, in completely dry and clean hydrocarbons, internal thinning is not expected; while even a small fraction of contradicting samples in the dataset (noise) could significantly disturb the learning. Bias in the data is something more challenging to trace. It is related to the representativeness and is the state when the variables capturing the desired phenomenon are misrepresented in the dataset due to the lack of actual observations or their prejudice. The case in the area of AIM can be misrepresentation in training of rarely captured metrics as for corrosion under insulation or, in contrary, including only heavily corroded well-observed samples to the process.

Once the quality of data is ensured, it needs to be pre-processed for the use of ML algorithms. Feature selection has been performed because of relatively high dimensionality of the data. The procedure aims to eliminate redundant and irrelevant features compromising the accuracy and speed of model learning (Maldonado and Weber, 2009). Several methods can be performed, which include knowledge-based selection, where domain knowledge is used for well-understood cases like RBI. The alternatives can be the filter methods checking for excessive correlation between the features or the wrapper methods utilizing the ML algorithms themselves to set the best list of features (Rachman and Ratnayake, 2019). API RP 581 clearly defines the set of required metrics required for the assessments with API RP 571 providing detailed information on applicable damage mechanisms. Thus, the subsets of features for each DF and CoF were manually defined for the reference model (no missing values). Even though a few features had both Pearson and Spearman correlation above 0.95, suggesting the repetition of information. These features, along with the ones lacking any variability within itself, were excluded from the dataset. These and further procedures have been performed in the Google Colaboratory environment with Python 3.7 and standard packages as Pandas, NumPy, Scikit-learn and Keras.

2.3 Machine learning

Two types of regression algorithms have been used for predicting the target variables (damage factors and CoF-s) based on the feature set provided for each sample component. These were Random Forest Regressor and Deep Neural Network. Generally, the choice of the algorithm is primarily dependent on the data to which it is applied, its type, quality, and the available computational power. Therefore, no universal algorithm can be applicable and effective for a wide range of cases, even though some general performance trends might be drawn. In some studies, such as Rachman and Ratnayake (2019) and Paltrinieri et al. (2019), they have discussed the application of machine learning to process risk estimation. According to the works, Neural Networks and the ensemble algorithms as Random Forest were expected to perform effectively on presumably similar in structure highly non-linear data. Olson et al. (2018) in its widely referred comparative study of the most common algorithms also notes about the powerfulness of the ensemble algorithms. However, the main accounted advantage of the chosen methods was their ability to handle missing inputs. The standard algorithms and especially the networks are not able to handle the missing values on their own; it generally requires extensive modifications. Nevertheless, given the successful model performance on the complete dataset, the model's re-training on a dataset with reduced dimensionality, which fits a specific problem, makes it possible to deal with missing values. Application of such integrated model is believed to require significantly less computational time compared to imputation techniques with the prediction efficiency retained.

Deep Neural Networks (DNN) uses multiple layers to progressively extract the features in increasing complexity from the input information (Kelleher et al., 2015). Within its structure, the first layer will represent the input set of unnormalized features, the last layer defined as the output with the layer in between being a set of model-derived layers. The overall length of the model chain there defines its deepness, referring to the term "deep" in its name (Goodfellow et al., 2016). Within the present study, a standard Neural Network Sequential functional in Keras open source library has been used for the DNN model.

Random Forest Regressor (RFR) is an ensemble algorithm which comprises a set of decision trees built independently and with a different structure. Building several decision trees and taking their weighted vote, rather than relying on a single decision tree, generally prevents the overfitting, which is inherent to the decision tree-based algorithms (Pal, 2005). Overfitting is an undesired condition when the obtained model fits extensively to the data at which it was trained, including noise in data. Random Forest algorithm can also be used to calculate feature importance distribution based on Gini impurity reduction. For the practical application, to some extent, it allows tracking the structure of the model and its decision criteria, which is challenging otherwise with DNN. In the present study, a standard Random Forest Regressor estimator in sklearn.ensemble library has been used.

There are several metrics used for the model performance assessment in regression problems. The list includes but not limited to the Mean Absolute Error (MAE), Mean Squared Error (MSE), Mean Absolute Percentage Error (MAPE), with each having its application specifications dependent on the type of model, data structure, the fraction of outliers, etc. For illustration purposes, MAPE was used due to its intuitive simplicity and independence from the scale of the assessed parameters. MAPE is given by:

$$MAPE = \frac{1}{n}\sum_{t=1}^{n}\left|\frac{A_t - P_t}{A_t}\right| \qquad (1)$$

Where A_t is the actual value; P_t is the predicted value of the target variable.

In addition to the standard metrics, the results with the following discussion will be illustrated through the examples of four chemical process units: sour gas treatment, thermal cracking, and two separators for which the identification of critical failure modes has been performed. Another two representative simulated flowlines have been picked up from the testing dataset and added to the analysis to assess the overfitting potential of the resultant model. A short specification of the components is presented in Table 2.

Table 2. General information on the process components used for model benchmarking

Sample #	**1**	**2**	**3**	**4**	**5**	**6**
Component type	Vessel	Vessel	Vessel	Vessel	Pipe (simulated)	Pipe (simulated)
Process medium	Sour HC gas, amine	High temperature sweet HC gas	Sour HC liquids, saline water	Sour HC liquids, saline water	Sour HC liquids, lean amine	Sour HC liquids, saline water

3. Results and discussions

Benchmarking of the model has been performed on the six process components in Table 2. Figures 2-3 illustrate the results. Figure 2 presents the probabilities of the failure caused by the governing damage mechanisms. They are predicted by the DNN-based model based on different data availability scenarios and compared to the reference predictions as per API RP 581. Concerning inspection metrics, three cases have been considered:

(1) All data is available. The model's accuracy is tested.
(2) Measured remaining thickness with missing data.
(3) Measured remaining thickness and both internal and external corrosion rates with missing data.

Figure 3 presents a hierarchical risk value chart for each damage mechanism - failure modes, based on which the prioritization and identification of critical failure modes could be performed. The data used to plot the figure, including all the damage factors (DF), were taken from the same predicted outcomes as those in Figure 2 was constructed; thus, not accounting for the natural randomness in the model. The consequence of failure (CoF) for each corresponding case was predicted by the same DNN-based model, input data for which is based on Level 1 CoF Analysis as per API RP 581.

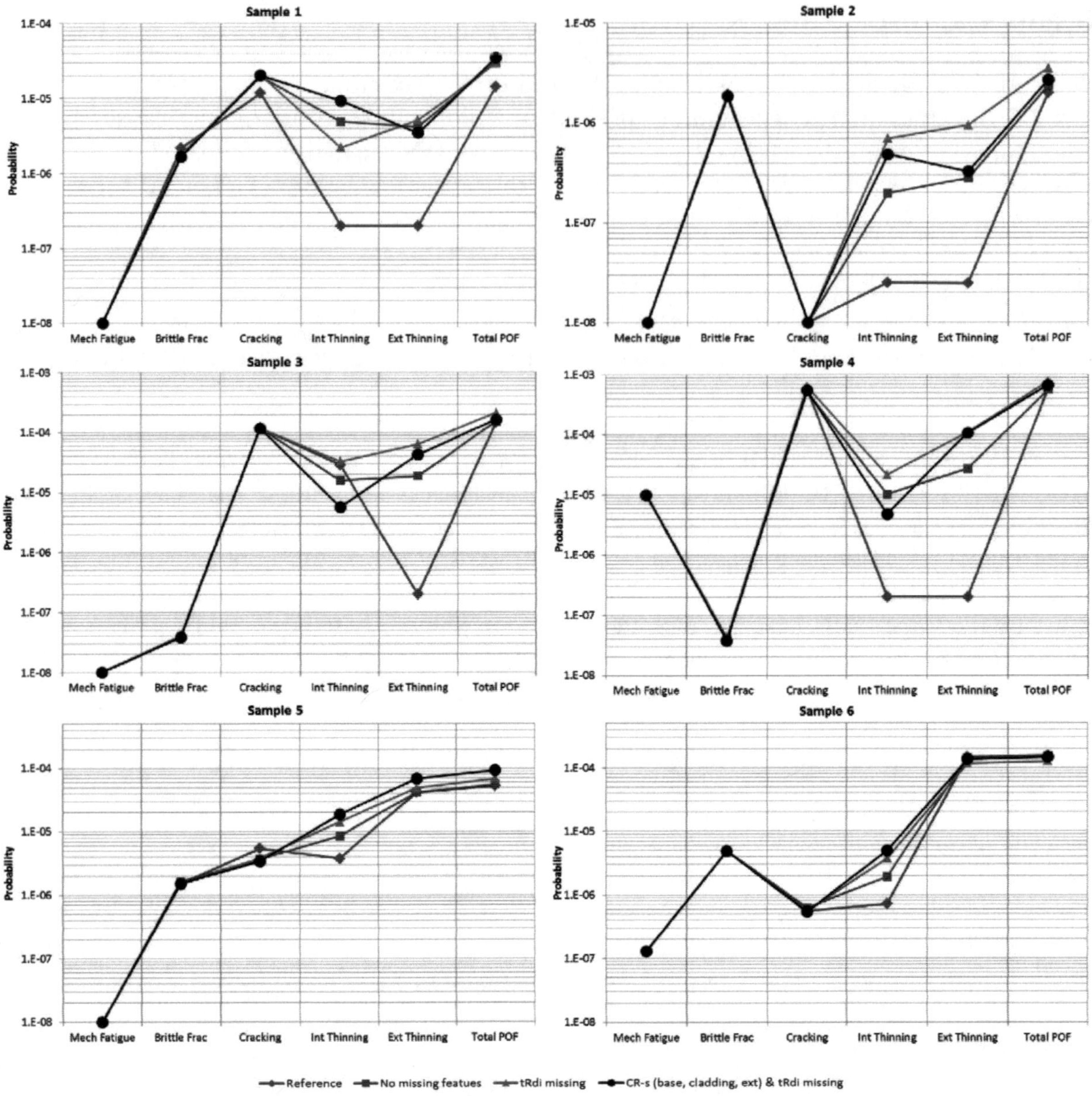

Note: Red line indicates the reference predictions as per API RP 581.

Figure 2. PoF-s by the major damage type groups at different data availability scenarios obtained by DNN model (blue lines).

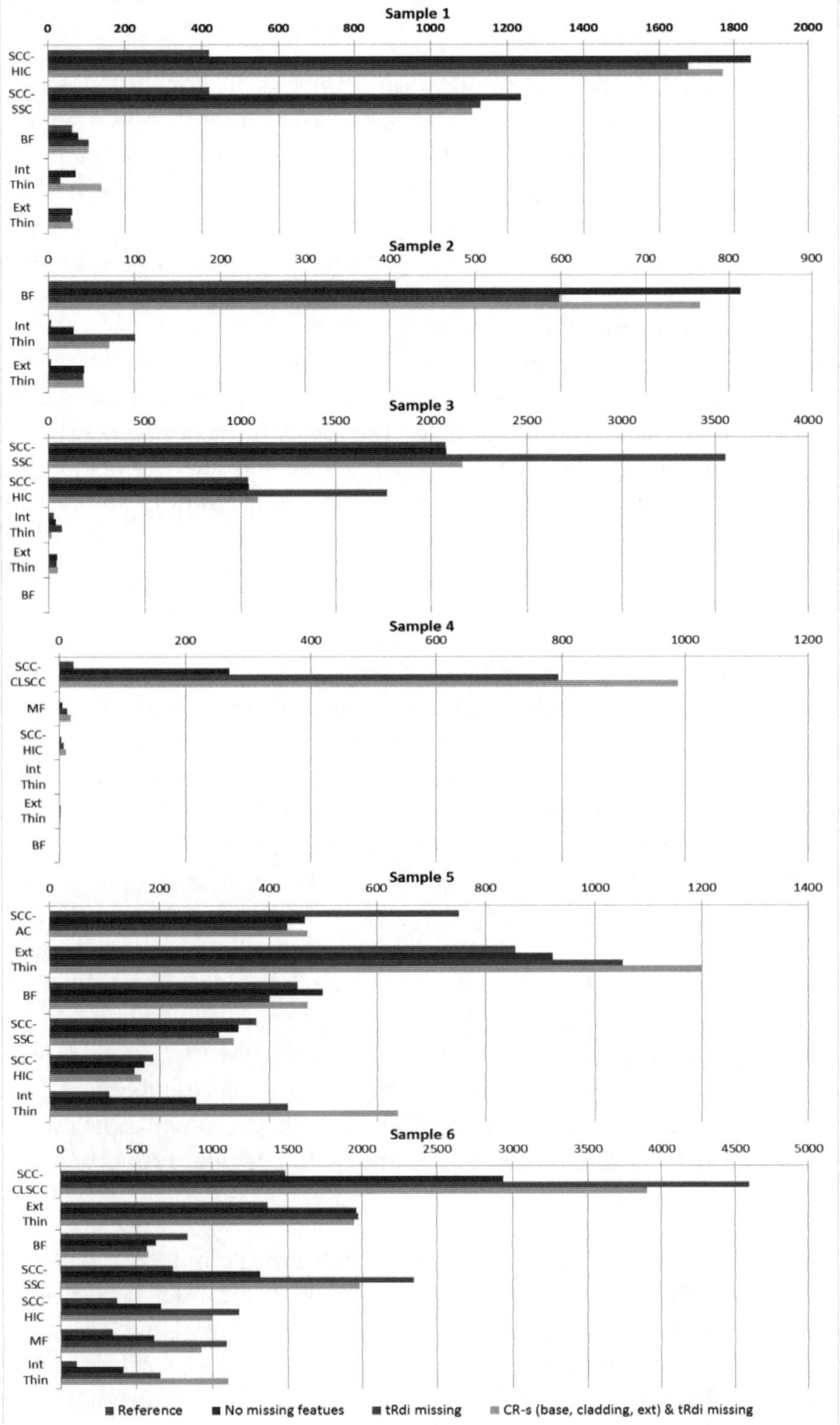

Note: Predictions produced by DNN model at different data availability scenarios are marked by blue, reference calculations as per API RP 581 by red.

Figure 3. Estimated risk values ($/year) for each individual failure mode inherent to the corresponding process component.

3.1 Integrated models

As indicated by Figures 2-3, the central question of interest is concerned about the quality of the obtained predictions and overall reliability of the proposed approach as a whole. The problem statement required identification of critical failure modes and their risk ranking; and, as Figure 3 illustrates, the hierarchical order of the failure mode risks was successfully predicted even in the cases with limited input data. However, the significant discrepancies were inherent to the magnitudes of the obtained predictions. In some cases, as for Samples 2, 3 and 4 from Figure 2, the difference between the reference values and baseline predictions with no missing values for PoF values accounted for two orders of magnitude, i.e., between 2E-7 and 2E-5. There is ample space for discussion on these two numbers. In the domain of statistics, various assessment metrics, e.g., MAE, MSE or MAPE, being applied would eventually indicate tremendous discrepancies. On the other hand, looking from the probabilistic perspective, the difference between "practically impossible" and "barely possible" as the two numbers imply would not seem as significant. To the authors' point of view, the interpretation of the results may vary depending on the user's objectives of the performed analysis. However, the authors also want to emphasize the importance of the signs of two phenomena causing the obtained discrepancies: data bias and model overfitting.

The primary assumption of the machine learning theory lies in the premise that the distribution of the test samples would be identical to the distribution of training ones (Mitchell, 1997). Once either of the sample sets is not comparable to each other and does not represent the overall actual population, the condition is called bias. The phenomenon can be classified into several types depending on the quality of the samples or prejudice of their measurements, and the distribution of the results in Figures 2-3 shows the signs of sample bias in the training data. For example, according to the review of the training samples, the absolute majority (93%) of them had a moderate or severe level of internal thinning (DF >10); meanwhile only 2% of the entries having the damage factor of around 1; thus, explaining the general conservatism of the predictions. A similar trend was inherent to the CoF data, where high-consequence entries were dominant as well. Despite the seeming simplicity of the phenomenon, managing bias in the data might not be as easy: eventually, the properties distribution is dependent on the user's expectations on the test samples. Ideally, the training dataset represents evenly-distributed data with the full set of relevant features. However, even with the ideal data model, precision is critically dependent on the used ML model itself, one of the critical properties of which is bias–variance tradeoff.

Underfitting and overfitting are the two opposing states in the model's bias-variance balance. Overfitting occurs when the model includes excessive features than was needed and fits the training data replicating its structure. Overfitting is undesired in machine learning as it not only complicates the model wasting resources, but also leading to incorrect predictions (Hawkins, 2004). From Figure 2, the significant difference in the accuracy of predictions was observed between Samples 1-4 and Samples 5-6, i.e., between the component sample groups being slightly different in the property's distribution. The model performance was significantly better in the predictions on simulated samples, which were firmly related to the model's training samples. This indicates overfitting. Considering the conservative values of predictions, the DNN model, in analogy with the discussed bias phenomenon, possibly prioritized the widely represented biased-high corrosion rates related features over the actual thickness measurements when it was available in the reference model. Later, when the thickness and corrosion rates were not available, the extended set of fluid property and material specifications features were used to support the conservative trend. The general ways to combat the overfitting are improving procedures of the model cross-validation by providing more test data to it and tightening the feature selection process.

Figure 4 presents the model's feature selection process and also overfitting. It represents the feature importance distribution charts for Internal Thinning DF predictions. Random Forest models were developed at different data availability scenarios. As it can be observed for the reference model, the features which were actually used in the Thinning DF occupy only the upper 0.8 fractions of the overall domain. Meanwhile, features like "MatS" – material sulfur content with the rest 47 features having no straight relationship to the thinning estimations occupied the remaining domain; thus, complicating and worsening the model. This is the sign of overfitting, which could be minimized by proper manual feature selection. However, such flexibility allows the model to perform prediction even when some of the features were missing. When "tRdi" (remaining measured thickness) was not present, the model placed "tOv" (design thickness) in the first place with the corrosion rates and time features re-distributed as well. In the third case, when no inspection metrics were available, the model placed fluid chemistry and material properties features like "pH" or "MCrC" (material chromium content) to the higher orders of importance trying to relate them to the missing corrosion rates. In this case, the missing features of measured thickness and corrosion rates could be called latent or 'hidden', as the information on their values might be extracted from a more extensive set of known parameters with some degree of certainty. Thus, the proper feature selection is highly important and it is always a balance between the model's simplicity and its flexibility. A great number of irrelevant features lead to the model's complication, overfitting, and unnecessary computational resources. On the other hand, some indefinite set of features might carry additional information which might be unrepresented otherwise in the models.

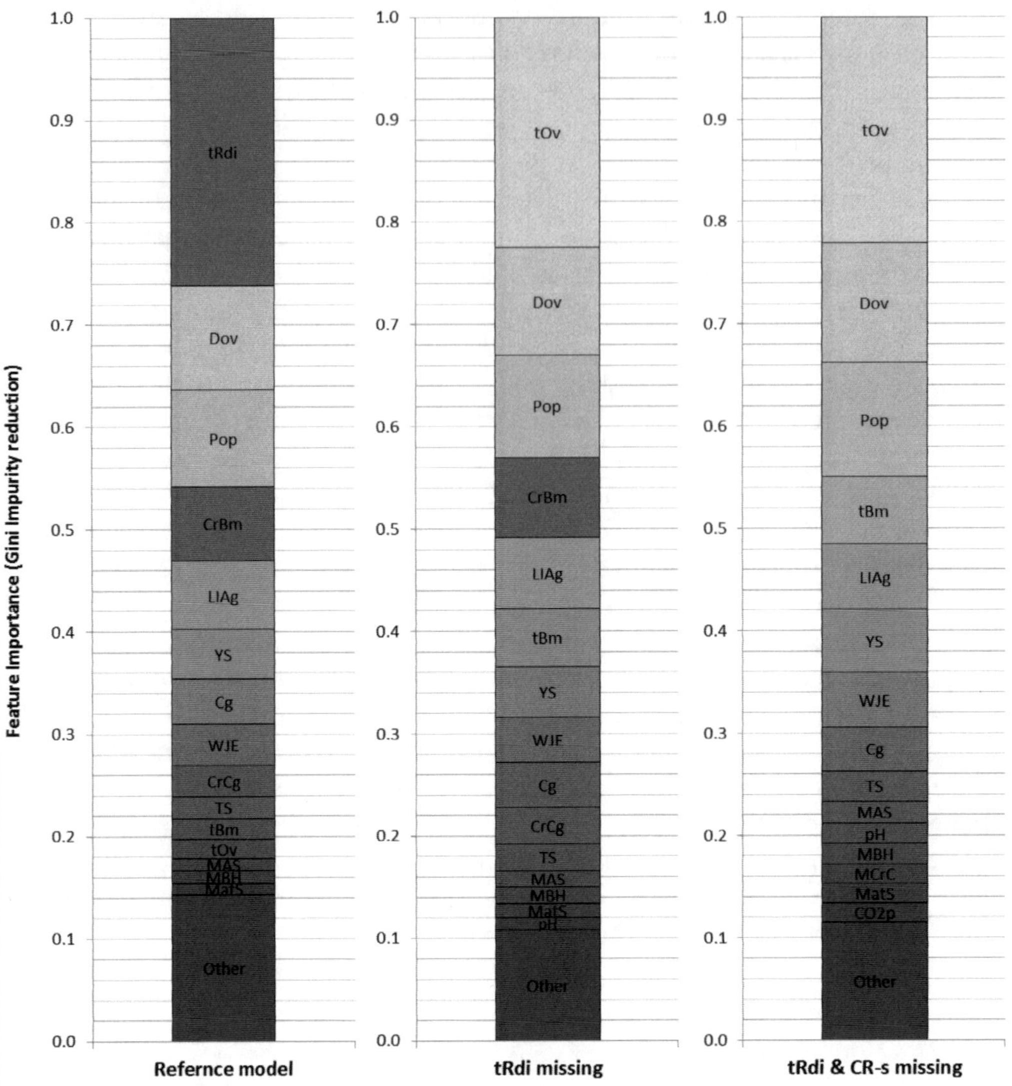

Figure 4. Feature importance distributions for Internal Thinning DF prediction model produced by Random Forest algorithm at different data availability scenarios.

3.2 Imputation models

Another proposed option for performing the risk assessment with limited data concerned imputation of the missing features employing stochastic regression. Unlike the integrated models, which predicted the required risk metrics directly, the approach implies a prediction of the missing metrics first with them to be used in the standard risk assessment calculations next. Theoretically, prediction of simpler metrics, e.g., corrosion rates or thickness instead of the damage factor, allows better control over the input set of features; thus, reducing the chances of overfitting or bias. Similar to the integrated models, two DNN models have been trained to predict Internal and External Thinning DFs at different data availability scenarios. To account for natural randomness of the models, the predictions have been made repeatedly for multiple times, allowing obtaining the whole distribution of the predicted metrics. Models' assessment has been carried on the test dataset to diminish the data bias effect with MAPE chosen as the assessment metrics. Figure 5 presents the obtained results in comparison with the integrated models set to predict the DFs directly.

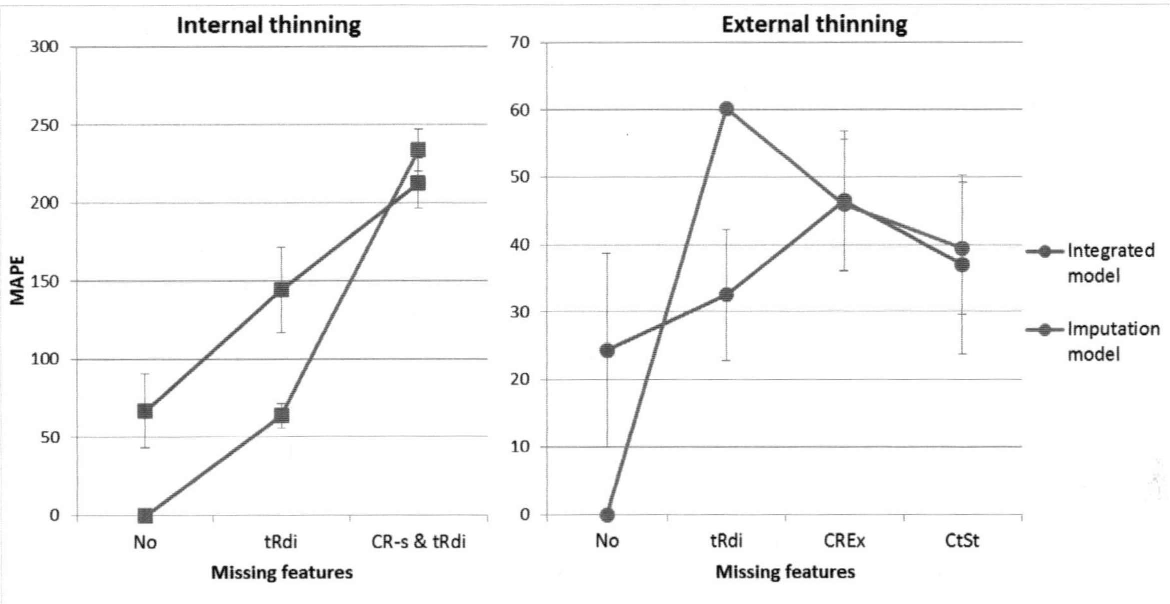

Note: the error bars on the chart indicate standard deviation of the results with dot indicating their mean.

Figure 5. Imputation-based and integrated model performance on the Internal and External DF predictions with denoted features missing in the analysis.

Despite the seemed competitiveness of the imputation approach, the results of the simulations did not reveal any significant advantages. In cases with the measured thickness, corrosion rates or coating integrity state missing, the approaches' performance was approximately equal with imputation results being generally less variable. Meanwhile, there was no practical meaning to assess imputation with no missing features. In general, the obtained results might indicate that the main complexity associated with predictions of the safety metrics with limited data lies in recovering of thickness, corrosion rate information or any other metrics missing. Thus, regardless of the applied approach of the chosen, the models' performance is critically dependent on the training data quality and the set of model features chosen to predict the optimal parameters.

4. Conclusions

Identification of credible damage mechanisms and associated failure modes is a critical task in asset integrity management for the chemical process components. FMEA is a framework widely used for that purpose. It relies on SMEs-led quantitative estimations of the risk associated with individual failure modes. Quantitative frameworks for the risk assessments as API RP 581 might be applied as well. The major challenges for their direct application are the amount and accessibility of data it requires. Some of the data entries, especially inspection metrics as measured thickness or actual corrosion rates, might be difficult to obtain to the operators. The present paper demonstrated a structured approach to deal with the limited data inputs by incorporating existent knowledge on the missing parameters into the calculations using machine learning techniques; thus, providing a quantitative base for decision-making in the analysis with missing parameters present. The methodology has been presented through a case study on several process components for which the failure mode risk assessments have been conducted. Two general ways of doing it have been illustrated: through data imputation and integrated models, with Deep Neural Network and Random Forest algorithm applied.

Results of the study shew that the approach's performance in risk metrics prediction is critically dependent on models' training data quality and bias-variance characteristics of the used model. Concerning the data quality, its representativeness, consistency, and bias level are essential. The data bias phenomenon, which significantly compromised the model's prediction accuracy for the real cases, has been traced throughout the results. Therefore, ensuring that the training samples' feature distribution is statistically close enough to the real population of samples is critical. The proper model tuning across bias-variance properties is not of less significance as well. From one side, the model variance allows it to optimize its feature set in cases with limited inputs; from the other, it might be prone to unnecessarily complicate the model leading to the loss of accuracy. Finding an optimal model for the risk assessment with limited input data includes a balance between the properties mentioned above in ratios that need to be further investigated.

Acknowledgements

This study was supported by Young Researchers Alliance and Nazarbayev University Social Development Fund through the Fostering Research and Innovation Potential (FRIP) program.

References

1. API RP 571. *API Recommended Practice 571, Damage mechanisms affecting fixed equipment in the refining industry.* USA: American Petroleum Institute, 2011.
2. API RP 581. *API Recommended Practice 581, Risk-based Inspection Methodology, Third Edition.* USA: American Petroleum Institute, 2016.
3. Bai, Yong, and Qiang Bai. *Subsea Pipeline Integrity and Risk Management.* New York: Gulf Professional Publishing, 2014.
4. Goodfellow, I. J., Y. Bengio, and A. Courville. *Deep Learning*. Citeseer, Cambridge, MA, USA: The MIT Press, 2016.
5. Hawkins, Douglas M. "The Problem of Overfitting." *J. Chem. Inf. Comput. Sci.*44, 2004: 1-12.

6. Kelleher, JD, Mac Namee B, and D'Arcy A. *Fundamentals of machine learning for predictive data analytics: algorithms, worked examples, and case studies*. Cambridge, 2015.
7. Lopes, Noel, and Bernardete Ribeiro. "Handling Missing Values via a Neural Selective Input Model." *Neural Network World*, 2012: 357-370.
8. Maldonado, Sebastián and Richard Weber. "A wrapper method for feature selection using Support Vector Machines." *Information Sciences*, 2009: 2208-2217.
9. Min, Qingfei, Yangguang Lu, Zhiyong Liu, Chao Su, and Bo Wang. "Machine Learning based Digital Twin Framework for Production Optimization in Petrochemical Industry." *International Journal of Information Management*, 2019: 502-519.
10. Mitchell, Thomas. *Machine Learning*. McGraw-Hill Education, 1997.
11. Olson, Randal S., William La Cava, Zairah Mustahsan, Akshay Varik, and Jason H. Moore. "Data-driven advice for applying machine learning to bioinformatics problems." Pacific Symposium on Biocomputing, 2018. 192-203.
12. Pal, M. "Random Forest Classifier for Remote Sensing Classification." *International Journal for Remote Sensing*, 2005: 217-222.
13. Paltrinieri, Nicola, and Louise Comfo, Genserik Reniers. "Learning about risk: Machine learning for risk assessment." *Safety Science*, 2019: 475–486.
14. Rachman, Andika, and R.M. Chandima Ratnayake. "Machine learning approach for risk-based inspection screening assessment." *Reliability Engineering and System Safety*, 2019: 518–532.
15. Schaafsma, Gerard C. P., and Mauno Vihinen. "Representativeness of Variation Benchmark Datasets." *BMC Bioinformatics*, 2018.
16. Thomas, Rajat M., Willem Bruin, Paul Zhutovsky, and Guidovan Wingen. "Dealing with missing data, small sample sizes, and heterogeneity in machine learning studies of brain disorders." In *Machine Learning. Methods and Applications to Brain Disorders*, edited by Sandra Vieira Andrea Mechelli, 249-266. Academic Press, 2020.
17. Wood, Maureen Heraty, Ana Lisa Vetere Arellano, and Lorenzo Van Wijk. *Corrosion-Related Accidents in Petroleum Refineries*. Luxembourg: European Commission, Institute for the Protection and Security of the Citizen, 2013.

Ecolab Process Safety Qualification Process: Executive Summary

Ecolab, a global petro/chemical manufacturer, had recently seen their Process Safety performance begin to plateau, and not see the improvements of past years. They were in danger of not hitting their targets across their global plant sites for Process Safety Event Reduction.

They had also recently developed an Engineering Career Ladder within Human Resources. With help from the Global Process Safety Team, and Global Engineering, the career ladder created consistent job titles and roles for all engineering roles with criteria that employees must achieve to move through the ladder. This created a path for engineers in the company to follow when developing their own careers.

With these two key data points in mind, the Process Safety Training Program, and senior leadership, recommended developing a consistent *Process Safety Qualification Process* to drive performance improvement. There were two goals for developing /deploying this process:

1. Ensure engineering employees at all global plant sites have the knowledge and expertise to complete key Process Safety tasks at the plant level. These tasks are crucial to Process Safety Event Reduction, and driving a strong consistent program across all global plant sites.
2. Provide tools, support, and resources for engineering employees to develop their careers, and complete the requirements listed in the Engineering Career Ladder

Outline of Qualification Process:

1. Employee completes self-assessment
2. Employee completes training and testing
a. Passes tests in all 12 core elements
4. Employee shows skill application with examples of their work and coaching
a. Submits project to Review Panel to show skill application
6. Review Panel confirms qualification through consensus

30% Training, Testing, Evaluation
70% Examples of experience and skill application

Core Elements of Qualification:

Compliance with Global Standards	RBPS/PSM Program Knowledge	Managing/Using Process Safety Information	Regulatory Compliance/ RAGAGEP
Chemical Compatibility/ Red List	Management of Change	HIRA/ PHA Tools	Plant Design
Measurement and Metrics	Root Cause Analysis	Critical Equipment	Process Safety Culture

The process is currently in the pilot phase, and will be ready for global deployment by the end of 2019. This sets the template for future evolution of the global program, and will be incorporated into other disciplines within the organization. This project was selected by the 5th CCPS Global Summit on Process Safety, Singapore, 2019. Nathan Thompson was also selected as co-leader of the *CCPS Early Career Professionals* project, now launching in North America.

Nathan Thompson: Global Process Safety Training Program Manager, St Paul, USA

585087 Right Sizing Combustible Particulate Management for R&D Processes

Abstract:

Combustible dust hazards are routinely recognized in a manufacturing setting but are commonly overlooked in laboratories. Most companies use research and development (R&D) departments to create new products but fail to recognize all new potential hazards associated with their development. Part of the struggle is the recognition of a combustible dust hazard and completing the right size review for an R&D setting.

This session will guide participants through educating non-safety personnel on combustible dust hazards, how 3M is screening combustible dust processes to drive the appropriate level of hazard analysis, and how to right size the R&D Dust Hazard Analysis (DHA) based on equipment in an R&D setting.

584646 Right Sizing Combustible Particulate Management While Meeting NFPA 652 Requirements

Abstract:

Working in the manufacturing industry at a Fortune 500 company with multiple commodity classes and size of manufacturing locations presented an interesting challenge when trying to control combustible particulate solid hazards and implement requirements from NFPA 652. Each location dealt with different combustible solid particulates, different processing methods and unique equipment. To handle this challenge, it was important to develop a combustible solids particulate management program that would meet the needs of every facility.

The global Combustible Solid Particulates management program that 3M has developed and is in the process of implementing, involves a three-step process which includes creating an inventory of solid particulates, analyzing the hazards of each process utilizing combustible solid particulates, and managing the hazards identified.

586586 Engineering Design for Ergonomics - the Missing Link

Abstract:

Process safety addresses many issues related to the safety of employees in the workplace and human performance. Human factors is included in the safety culture element, however ergonomics is often neglected. It has been said that human factors addresses worker safety from the neck up (cognitive issues) and ergonomics addresses safety from the neck down (anthropometric aspects). Admittedly, this is a gross analogy, however it is useful for discussion purposes.

During the design phase, it is rare that ergonomics is included as a priority in the decision process. How many times have we seen valves controlling processes placed eight feet off the ground where operations personnel are required to place chains to articulate them. This is often the cause of injuries we see surface in PHAs. These hazards are often disregarded as unavoidable because it was part of the original design, however if ergonomics would have been considered in the original design, then the valve would have been properly placed where operations personnel could easily access the valve without causing injury.

This paper will discuss the importance of incorporating ergonomics principles into the design of process equipment with the goal of reducing the potential for employee injury.

585298 New Wine in New Cups: Measuring Operator Performance for Safety-II Perspective

Abstract:

Standard operating procedures (SOPs) play a critical role in achieving safety and productivity of day-to-day operations in the chemical industries. Investigations of previous incidents also indicate that a majority of adverse events during routine and non-routine operations are attributed to issues associated with SOPs. For example, a recent investigation of fatalities that occurred at DuPont La Porte facility points out the absence of formal procedure required to remove plugging (e.g., hydrate) from a waste gas piping system and inadequate emergency procedures for hazard notification and evacuation as major causes of the incident [1]. To ensure the safety of a complex system such as chemical plants, there have been two viewpoints towards human operators' use of SOPs: *Safety-I* and *Safety-II*. First, Safety-I perspective defines safety as the absence of undesired events [2]. Thus, it looks for things that went wrong (e.g., slip, mistake, violation) and seeks to minimize the deviations from prescribed tasks, which is assumed to be a cause of adverse incidents . Second, Safety-II views safety as the presence of desirable and successful outcomes. Therefore, this standpoint highlights things that went right (e.g., adaptive behavior, workaround), and considers performance variability of human operators from SOPs to be inevitable and even necessary to maintain safe operations of the system [3]. In favor of Safety-I approach, conventional measures of SOP performance have long been established and utilized to indicate the degree of safety in relations to procedural systems. Nonetheless, no measures of human operators' SOP performance that adopt or support Safety-II framework exist to date. To address such a gap, this paper identifies limitations of traditional and dominant measures of SOP performance and then proposes a novel idea that harmonizes the two views towards safety (i.e., Safety-I and Safety-II). The new measurement concept embraces not only the conventional measures regarding the implementation of SOPs (i.e., following or not following a procedure), but also incorporates human adaptive behaviors. To instantiate the new measures of SOP performance, case studies using real-world examples in the chemical industries are presented. Following this, implications for the proposed measure of SOP performance based on Safety-II viewpoint and future research proposals to support the benefits of the measure are provided.

References

1. U.S. Chemical Safety Board, *Investigation Report: Toxic Chemical Release at the DuPont La Porte Chemical Facility*. 2019.
2. Hollnagel, E., *Safety-I and Safety–II: The Past and Future of Safety Management*. 2014, Surrey, UK: Ashgate Publishing.
3. Sujan, M.A., H. Huang, and J. Braithwaite, *Learning from incidents in health care: Critique from a Safety-II perspective*. Safety Science, 2017. **99**: p. 115-121.

586223 Incorporating Abnormal Modes of Operation into Pressure Relief System Design and Documentation

Abstract:

Pressure relief systems are commonly referred to as the last line of defense in oil refineries and chemical facilities; however, the design of pressure relief systems does not typically consider abnormal modes of operation such as startup, shutdown, and other temporary operations. Past studies have shown that a majority of incidents occur during abnormal modes of operation including several significant incidents that helped shape the state of process safety as it is today. This paper will review potential sources of overpressure during various abnormal modes of operation and analyze how these overpressure scenarios can be incorporated into pressure relief system design and documentation.

20AIChE Spring Meeting
16TH GLOBAL CONGRESS ON PROCESS SAFETY
HOUSTON • 2020

Facilitating Barrier Integrity Management During PHA Revalidation

Edward M. Marszal
Kenexis
3366 Riverside Drive, Columbus, OH 43221, USA
edward.marszal@kenexis.com

Austin Bryan
Kenexis
3366 Riverside Drive, Columbus, OH 43221, USA
austin.bryan@kenexis.com

Prepared for Presentation at
American Institute of Chemical Engineers
2020 Spring Meeting and 16th Global Congress on Process Safety
Houston, TX
March 30 – April 1, 2020

Facilitating Barrier Integrity Management During PHA Revalidation

Edward M. Marszal
Kenexis
3366 Riverside Drive, Columbus, OH 43221, USA
edward.marszal@kenexis.com

Austin Bryan
Kenexis
3366 Riverside Drive, Columbus, OH 43221, USA
austin.bryan@kenexis.com

Keywords: bowtie, LOPA, integrity management

Abstract

Currently, risk in process industries is managed through analysis of process hazard and identification of barriers to prevent or mitigate hazards (typically through a PHA), a design team which takes the specification for barriers to perform detailed design and ultimately construction of barriers, and operations which monitors and tests barriers. However, current practices leave many potential gaps in the workflow which may allow for inadequate risk mitigation and intolerable risk within a facility. These can include incomplete or inadequate specification of barriers, incorrect design, or inability to validate designs once in operation. The use of numerous software applications with differing database structures leads to teams and individuals operating in isolation, often with little visibility on how information given to them was generated and with no impetus to follow up and ensure that the final design fulfills their recommendations / specifications in actual operation. Using enterprise software tools with a standardized database structure that allow personnel and applications that are used in the different phases of the plant lifecycle to communicate and synchronize data, and can organize and present data in a more easily understood visual representation, such as bow-tie diagrams. Such tools can allow for both feed forward and feedback of information, ensuring that pertinent information is available to people when needed, and validation of the barriers through the lifecycle is streamlined.

1 Introduction

Process Hazards Analysis (PHA) considers the risk of accident scenarios based on the consequences of realization of the scenario, the likelihood at which the scenarios can occur, and the safeguards (a.k.a., barriers) that prevent or mitigate the scenarios. Risk analysis teams rely on barriers to reduce risk to tolerable levels, but during the analysis process can only set performance targets. These performance targets and conceptual specifications are passed down to design teams as recommended specifications with the expectation that that they will be faithfully implemented as the hazard analysis team intended. Constraints on budget, timing, equipment availability, and plant logistics can prevent recommendations from being executed, or in some cases modifications to the recommendations are made to make implementation practicable. Furthermore, it is also expected that once those well-designed barriers are put into place, they will continue to operate in conformance with their performance targets throughout their entire lifecycle. This requires that they are tested, and preventive and corrective maintenance is performed as specified. Even when recommendations are implemented, if maintenance and testing expectations are not properly communicated or implemented, barrier performance may not match the design targets and can prevent tolerable levels of risk from being achieved.

Without some sort of system of follow-up, plants might operate for long periods of time with intolerable levels of risk due to improperly implemented barriers. In order to ensure that barriers achieve the desired risk reduction throughout their entire lifecycle, a feedback mechanism must be used. Initially, feedback to the PHA team from the design team can be performed, prior to plant construction, to ensure that all recommended barriers are in place as specified. If gaps between specifications and actual design are identified they can then be addressed prior to start up. During plant operation, continued achievement of barrier performance targets, considering actual testing, inspection, maintenance, and bypass, can be confirmed by having operational phase tools. This includes management of bypass systems, event documentation systems, and computerized maintenance management systems, that communicate and synchronize with analysis phase tools to allow the analysis phase team an ongoing view of the status of barriers, in real time, throughout the lifetime of the facility. This synchronization is then particularly important during PHA Revalidation where the automated tools and bowtie visualization allow for easy verification that design assumptions match actual performance, and if they do not, allow for modifications to barrier design or process operating parameters to allow risk to be moved back into the tolerable region.

Use of enterprise tools with a standardized database structure begins with the hazards analysis documentation tool being able to automatically exports lists of barriers to designers along with their performance requirements, automatically, without having to enter data twice. The design tools should be able to link backward and view the hazard scenarios that the barriers are intended to protect against to improve designer decision making because they have a better idea of the purpose of the barriers. After completion of design, the design phase performance calculations can be fed back to the analysis phase tool to allow comparison of analysis phase expectations with design phase specifications. The design phase information should then be fed forward into operation and maintenance phase tools such as computerized maintenance management software (CMMS) for scheduling and tracking of testing and maintenance, management of change software for the tracking of bypass, and event management systems to record barrier activation events. These tools should be able to feedback, throughout the lifecycle to design phase tools, and subsequently the analysis phase tools. In this way, all tools can have a complete view and barrier requirements and actual performance throughout the entire lifecycle. All along the way, bow tie diagrams provide an easily understood visualization mechanism, across all tools, for visualization and communication of barrier status and thus achievement of tolerable risk.

Defining Barriers and Setting Their Performance Targets

PHA is initially performed during the analysis phase of the safety lifecycle of a facility. In this phase, the design of the process equipment is analyzed by a multidisciplinary team and performance targets for barriers defined. PHA provides discussion prompts, usually deviations from design intent, that lead to hazard scenarios that are fully developed by the PHA team. These scenarios will include a cause, such as an equipment failure or improper operator interaction with the process. This cause, and potentially other causes can then result in a loss of containment of hazardous materials or energy from the process. The loss of containment can then result in one or more consequences, which represent harm being caused to assets, environment or people. The team will also discuss the safeguards or barriers that are in place to prevent or mitigate the consequences. Barriers are either preventive or mitigative. In general, preventive barriers prevent the loss of containment from occurring, preventing any related consequences. Mitigative barriers do not prevent the loss of containment. Instead, they function by reducing the magnitude of the consequences that result from the loss of containment. The PHA team then considers the scenario's likelihood and consequence in order to determine if the overall risk is tolerable or if barrier addition or improvement is required in order to make risk tolerable.

At this time, the work of a PHA team is generally performed using computer-based worksheets of information. Rows of data present scenarios, causes, consequences, and safeguards. In order to better allow teams and other stakeholders of the PHA to visualize the information contained in a PHA scenario bow-tie diagrams are being used more frequently. Furthermore, the development of Unified Hazard Assessment[1] is allowing for the interchangeability of data in HAZOP, LOPA, Hazard Register, and Bow-tie diagram formats – all automatically in a single tool through a unified data structure that serves all those purposes. Consider a typical hazard scenario, as shown in the figure below. The LOPA includes threats (green), preventative barriers (purple), the top level event (blue), mitigative barriers (yellow), and consequences (red). The rows of a LOPA worksheet can be easily represented in the associated bowtie, providing an easier to understand picture of process risk.

---

[1] *Applying Data Science to Hazard Analysis*, Edward M. Marszal, InTech, January/February 2020, ISA, Research Triangle Park; NC.

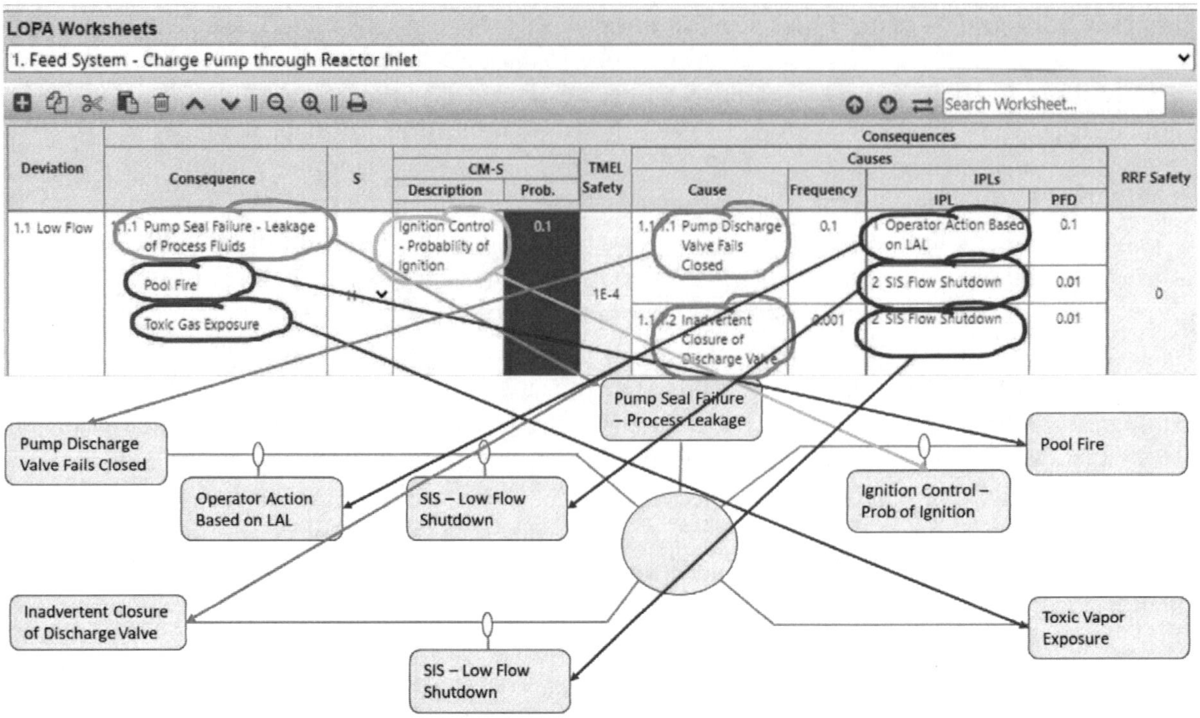

Figure 1 – Converting a LOPA Scenario into a Bowtie Diagram

A shortcoming of many PHA studies in this phase is that they do not develop the barriers in sufficient detail for persons responsible for the barrier design to effectively perform their task. This lack of data can manifest itself in poor choices by the design team that will ultimately prevent the expectations of the analysis phase team from being realized. This failure, at a minimum, could result in costly re-work late in the lifecycle of the plant and at worst could result in failure of tolerable risk goals to be achieved by the facility. Consider the following LOPA data sheet. While it appears to have a comprehensive listing of the information that is required by the design team, in fact, there are a few parameters that are required by the design team that are left out.

IPLs				
IPL	IPL Tag	PFD	IPL Type	Selected SIL
1 Operator Action Based on LAL-102	LAL-102	0.1	ALARM	
2 Charge Pump (P-101) Discharge Low-Low Flow Stops Pump	FZL-P101	0.01	SIF	SIL 2
2 Charge Pump (P-101) Discharge Low-Low Flow Stops Pump	FZL-P101	0.01	SIF	SIL 2

Figure 2 – Typical LOPA Safeguard Listing

Furthermore, some of these parameters are only appropriately specified by the analysis team and if left to the design team could be ignored or inappropriate choices made. Some of the common parameters that are often left out of analysis phase documentation include:

- Process Safety Time
- Shut off Valve Leak Tightness Requirements

The analysis phase is where the expertise to properly define these parameters resides. Furthermore, if the analysis team does not define them, these parameters are frequently ignored, potentially resulting in designs that will not allow a safe state to be achieved upon activation of the barrier. While the figure above shows the listing of barrier attributes in a worksheet format, there is no reason that the same information cannot be presented in an enhanced bow-tie diagram format as shown in Figure 3.

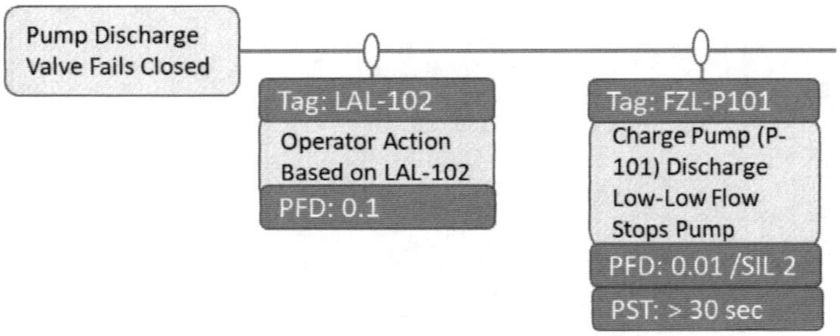

Figure 3 – Barrier Details View in Bowtie Diagram

It is even reasonable to expect that a comprehensive tool could allow the user to click on the barrier in the bow-tie diagram and have the data sheet(s) for the associated barrier be presented for a more comprehensive view. This is presented in Figure 4.

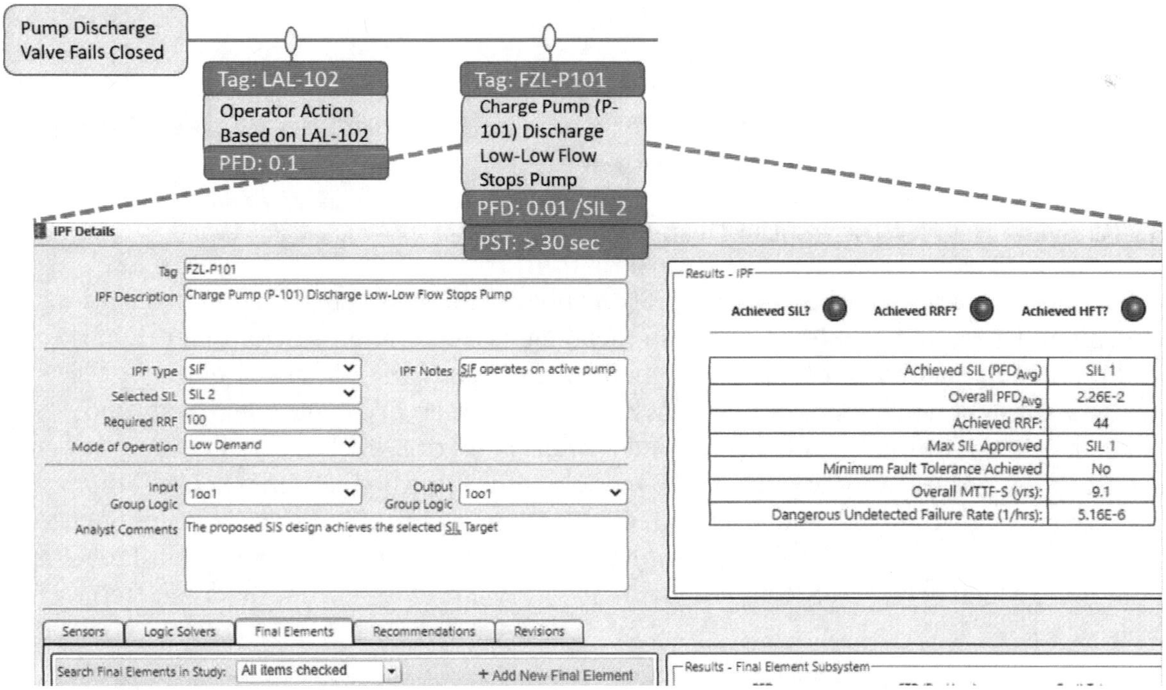

Figure 4 – Bowtie Linked to Barrier Specifications

Using advanced tools, and implementing administrative controls and training, will ensure a more comprehensive specification of barriers by the analysis team. This information is automatically moved, copied, or synchronized with the design phase tool. The synchronization needs to apply regardless of the type of barrier being employed, whether this is a SIF, a relief valve, or critical alarm. Automatic synchronization allows for complete and comprehensive transfer of requirements from the analysis phase to the design phase with a minimum of effort and no chance of error transposing information from one tool to another. The use of standard and open data structures for different portions of the plant safety lifecycle will greatly improve the workflow of barrier design.

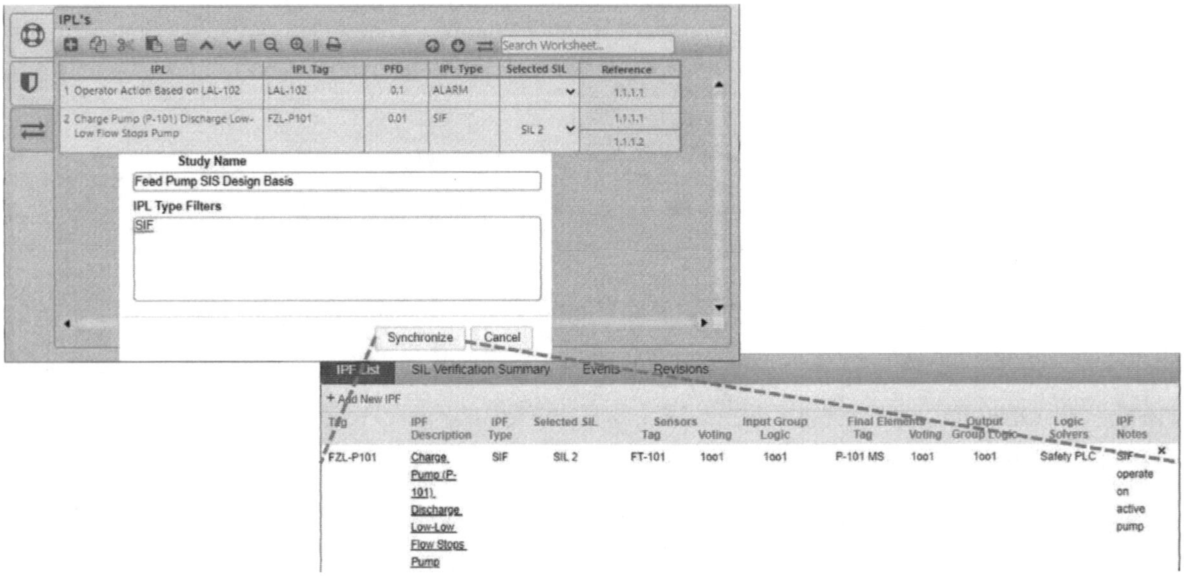

Figure 5 – Synchronization of Barrier Specifications Between Analysis Phase and Design Phase Tools

Ensuring Barrier Integrity During the Design Phase

Once barrier requirements are defined in the analysis phase and then transferred into design tools, conceptual design and detailed design can then begin. A common barrier that might be designed for a process facility is a safety instrumented function (SIF). Analysis phase requirements that might be defined for a SIF include a description of the measurements and actions taken by the SIF, process safety time, and a safety integrity level (SIL). Designers of SIF will make many design decisions about the SIF including the specific instruments used, the amount of redundancy employed, and the frequency and type of testing of the SIF components. All these factors are considered during the design process which culminates in a quantitative calculation of the actual achieved SIL in comparison with the performance targets selected by the analysis phase team.

As designers perform their tasks, there is great benefit in their understanding the hazard scenario when making decisions about the appropriateness of various options. For instance, a SIF designer may choose to implement a "double block" arrangement of shutoff valves (i.e., two valves piped in series) to lower the probability of system failure and allow the achievement of higher SIL targets. While this design choice might seem appropriate to the designer, there is a possibility that trapping of chemicals that are prone to self-accelerating decomposition (e.g., cumene hydroperoxide) in between the two closed valves could introduce a new serious unintended hazard.

For designers to make the best choices, the ability to directly view the hazard scenario for the equipment they are designing is often essential. While designers are working with options for their equipment, it should be possible to present the hazard scenario(s) that employ those barriers. Optimally, a single click in the design phase tool should be able to present a list of related PHA scenarios and their bow-tie diagrams. This allows the design team to better understand the scenarios developed by the analysis phase team, the rationale behind their performance target selection, and the ramifications of the decisions that they are making regarding the options for achievement of the selected targets. This concept is presented in the figure below.

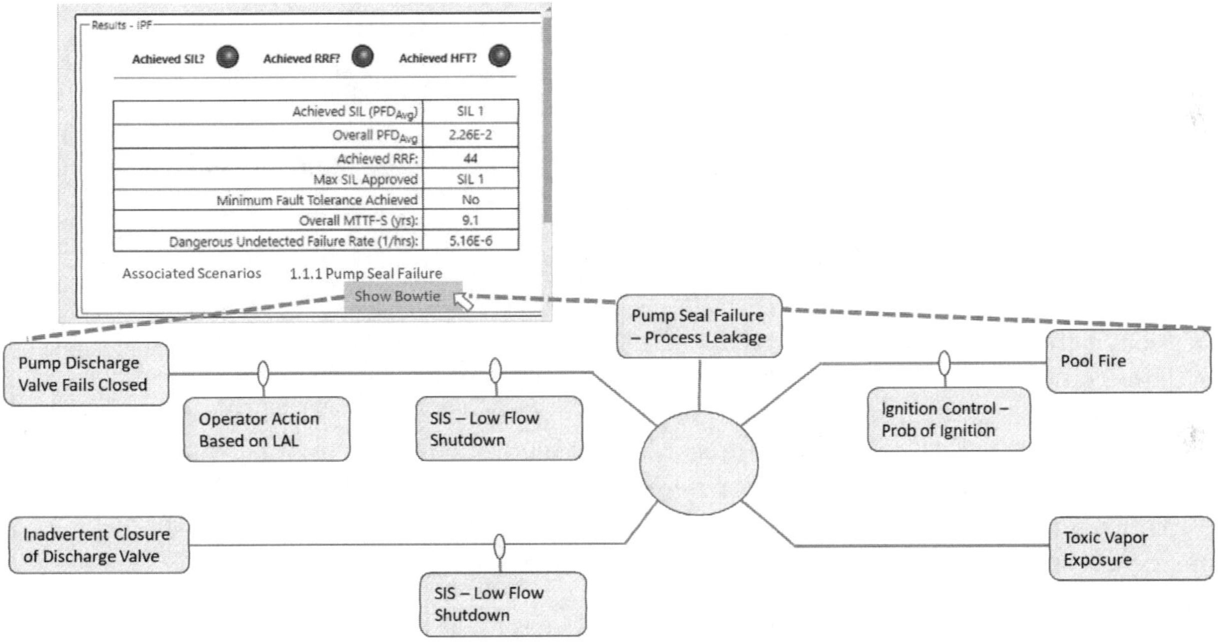

Figure 6 – Accessing Hazard Scenario from Design Phase Tool

After the design team has completed its work, or at least the first pass at its work, data developed during the design needs to be synchronized with other phases in the safety lifecycle of the facility. This requires both feed forward and feedback. The feed forward portion requires that design data be transferred to the operational phase. Operational phase data includes testing frequencies and procedures along with procedures for response to failures of components and subsystems of barriers. The feedback includes parameters that are achieved by the barriers as developed by the design team so that a comparison against targets can be made and confirmed. At this stage, it is critical for the analysis phase team to view the design team's output vis-à-vis their expectations that were developed in the analysis phase to ensure that the barriers, as designed, achieve the risk reduction the analysis team identified and do not create any additional hazard scenarios. Feedback of the design to the analysis tools (Figure 7) allows for rapid review of designs by the analysis team and identification of any gaps which will need to be addressed prior to installation and start up. Again, the bowtie diagram provides an easily understood visual for both the analysis team and management.

IPL	IPL Tag	PFD Target	PFD Actual	IPL Type	Selected SIL
Operator Action Based on LAL-102	LAL-102	0.1		ALARM	
Charge Pump (P-101) Discharge Low-Low Flow Stops Pump	FZL-P101	0.01	0.022	SIF	SIL 2

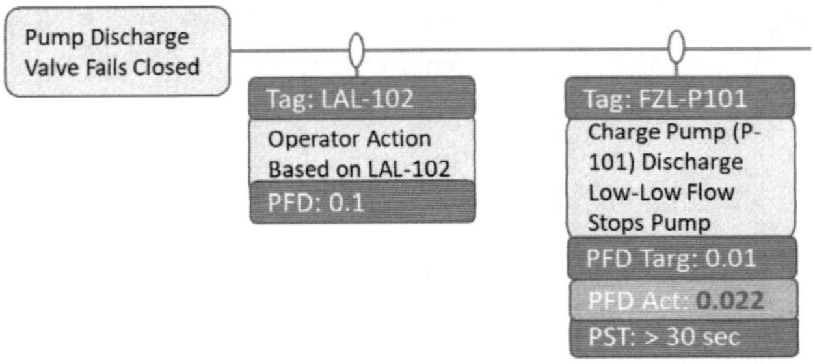

Figure 7 – Display of Design Result Feedback to Analysis Phase Tools – Worksheet and Bowtie

Ensuring Barrier Integrity During the Operational Phase

Interactions between the plant personnel and the barriers, including proof testing and inspection, are required during the operational phase. These activities are required to occur at a frequency and in accordance with the procedures developed during the design phase. The frequency and comprehensiveness of testing has a significant impact on the achievement of performance targets, as delayed or incomplete testing may impact the capability of the barrier to achieve its performance targets. Typically, during the design phase, a certain frequency and level of comprehensiveness of testing is determined to be required.

It is ideal if the design phase tool allows for automatic population of the operation phase tool, commonly a computerized maintenance management system (CMMS) (Figure 8). This population will need to include what devices need to be tested and inspected, at what frequency this testing should be performed and what procedures should be utilized for the testing. Seamless integration of software ensures the analysis and design teams' requirements are faithfully implemented and eliminate the potential for transcription errors.

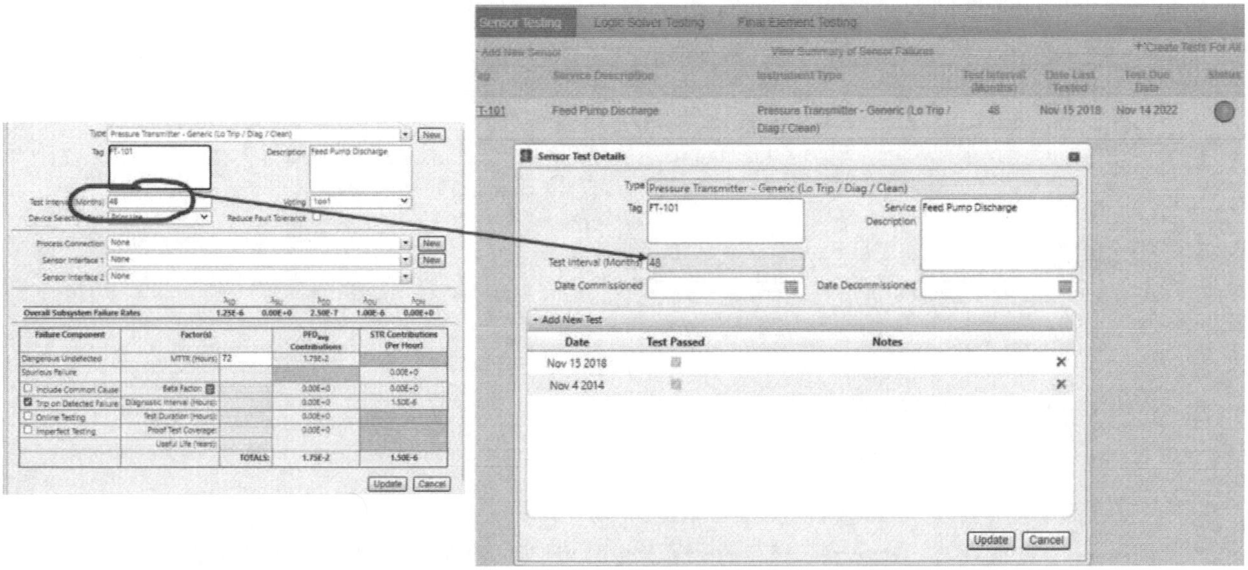

Figure 8 – Automatic Population of CMMS Database from Design Tool

Again, it is useful for the operational phase tool to be able to display information from the design phase, especially if a failure occurs during testing and repairs must be made, in some cases where replacement in kind might not be available. The data associated with the failures of barrier components is essential, as it will be compared with design phase assumptions during revalidation. As such, tracking them in the operational phase tool and feeding back summary information to the design phase tools becomes critical.

In addition, barriers might be fully or partially bypassed while a process plant is still in operation to allow maintenance and testing without adversely affecting the plant. These bypasses must be tracked and authorized carefully, often using a management of change software tool. During the bypass of a barrier, several items of information are critical to the team applying the bypass. First off, during the bypass authorization process, an analysis of the hazard scenarios impacted by the bypass of the barrier is essential. This analysis needs to be documented in the bypass authorization tool, and is greatly facilitated by having the barriers listed in the bypass authorization tool's data base and allowing selection of the barrier in the tool to provide access to a display of the hazard scenario in a bow tie diagram (Figure 9). Providing the bowtie diagram of the relevant scenarios allows for better understanding of the risk being mitigated and an informed decision on the duration of the bypass and potential additional mitigation that may need to be implemented while the bypass is in place.

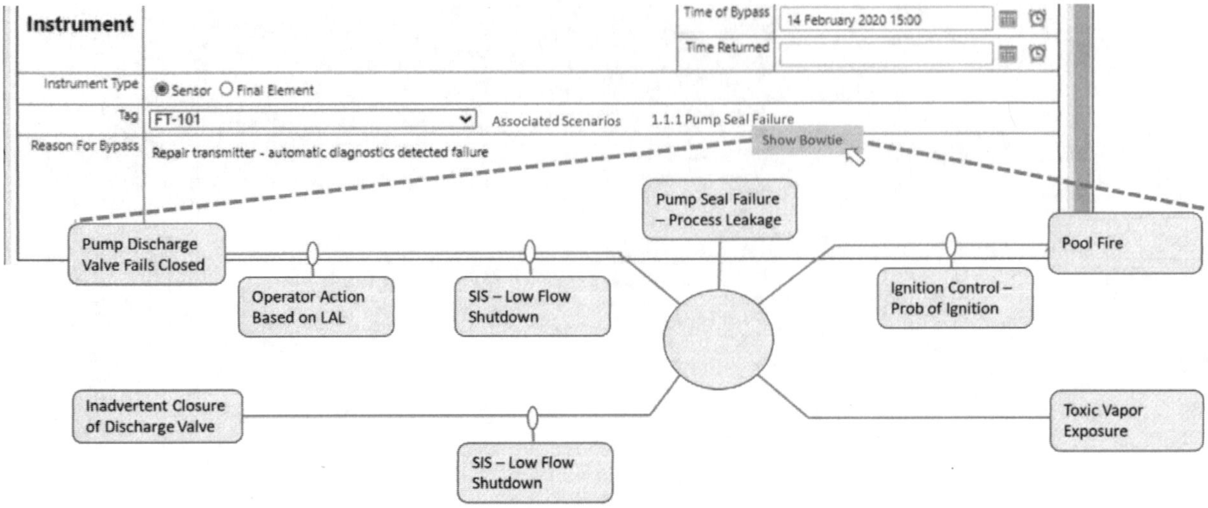

Figure 9 – Display Bowtie of Hazard Scenario from Bypass Authorization Tool

Also, during bypass of a barrier it may be important to have compensating measures, also known as an alternate protection plan, that will serve as a replacement for the bypassed barrier if the level of risk is determined to be beyond tolerable levels while the barrier is bypassed. Alternate protection plan information can easily be developed for each barrier during either the analysis phase or design phase of the plant safety lifecycle. Once developed, this information can be automatically included in a bypass authorization worksheet by selecting the barrier, and automatically communicated to operation staff through emergency SOP databases that can be linked to the design tools.

Instrument		Time of Bypass	14 February 2020 15:00
		Time Returned	
Instrument Type	◉ Sensor ○ Final Element		
Tag	FT-101		
Reason For Bypass	Repair transmitter - automatic diagnostics detected failure		

	Type	Description	Additional Action Required	
			Alternate Protection Plan	Bypass Risk Assessment
○	Type 1	Bypass an instrument for repair or maintenance; instrument is part of fault tolerance system where SIF will still activate upon process demand; repair completed in less than MTTR	No	No
○	Type 2	Bypass an instrument for repair or maintenance; instrument is part of fault tolerance system where SIF will still activate upon process demand; repair requires more than MTTR	No*	Yes
◉	Type 3	Bypass an instrument for repair or maintenance; instrument is NOT part of fault tolerance system; repair completed in less than MTTR	Yes	No
○	Type 4	Bypass an instrument for repair or maintenance; instrument is NOT part of fault tolerance system; repair requires more than MTTR	Yes	Yes
○	Type 5	Bypass instrument for any reason other than instrument repair or maintenance	Per Bypass Risk Assessment *	Yes

* May be required if the Bypass Risk Assessment indicates that it is necessary

Alternate Protection Plan

Item	Value / Description
What process variable or variables must be monitored?	Local Flow Indicator at Pump P-101 Discharge
What are the manual trigger points for the monitored variables?	Flow of less than 10% of linearized scale
Who is responsible for performing the process variable monitoring?	Dedicated outside operator
Who is responsible for performing the manual shutdown action?	Dedicated outside operator, same as monitoring variable
What specific actions must be taken to manually shutdown?	Set Pump P-101 Hand / Off / Auto switch to the OFF position
Can a manual shutdown be performed within the process safety time?	Yes
Is there sufficient independence between the normal operating staff and the alternate protection?	Yes, a dedicated separate operator is used
A Bypass Risk Assessment has been performed and is acceptable (if required)	No bypass risk assessment is required.

Figure 10 – Automatic Completion of Alternate Protection Plan from Design Phase Tool Data

The operational phase of the plant lifecycle is where actual performance metrics about the plant evidence themselves. Tracking actual performance and comparing it against design phase and analysis phase assumptions is critical to ensuring achievement of tolerable risk. Critical events that must be tracked include the occurrence of initiating events and activation of barriers. The key to tracking is to associate each initiating event occurrence and barrier activation with an alarm (audible or journal) that will be documented in the plant control systems alarm log. Then, on a periodic basis the alarm log can be parsed to identify the occurrence of the alarms that indicate initiating events or barrier activation. This information can then be fed back to the analysis phase for revalidation.

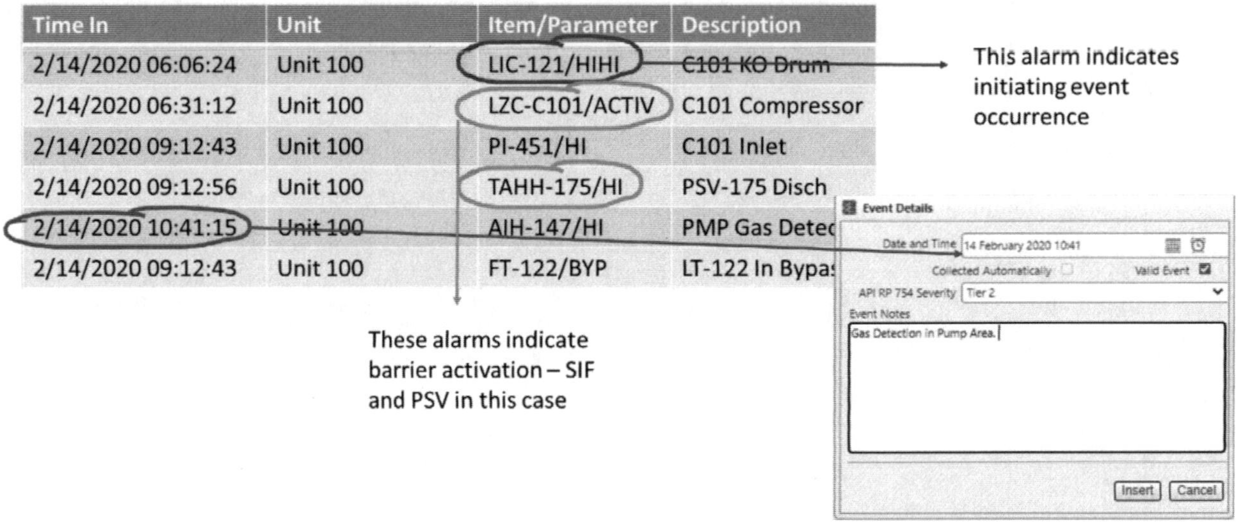

Figure 11 – Alarm Log Parsing for Barrier Activation and Other HSE Relevant Events

Revalidation Workflow for Barrier Consideration

PHA, the analysis phase of the process safety lifecycle, typically needs to be revalidated on a regular basis to ensure that design assumptions were appropriate, actual performance matches expectations, and any changes to the plant operating conditions or equipment modifications were appropriately handled. While revalidation is often considered a simple task of relooking at the PHA document it is the starting point for revalidating all facets of the plant process safety lifecycle including not only the analysis phase, but also the design and operational phases. Some of the items that need to be fed back and considered during revalidation include the following.

- Initiating event frequency
- Barrier activation rate
- Barrier failure rate (by component)

These actual numbers must be compared against design assumptions to determine if there are gaps between the risk of the plant as designed, as opposed the reality of an actual operating plant. The analysis phase calculations, such as LOPA, can then be reviewed to identify gaps. In some cases, if the actual plant conditions resulted in higher demand rates and barrier failure rates than the design assumed, which results in a higher intolerable level of risk, the difference between design and actual could result in the need to change the original design, possibly by adding redundancy, increasing testing and maintenance, selecting different equipment, or even changing the operating conditions of the plant. This task can be greatly facilitated using connected enterprise software tools that allow for communication from operational phase tools that gather actual operating data back to the analysis phase tools. Once again, communication of these results is facilitated by presented in a bowtie diagram.

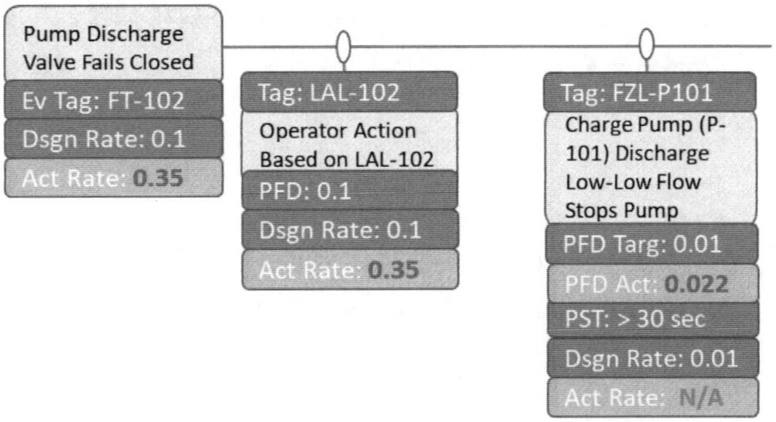

Figure 12 – Comparing Design and Actual Performance in Bowtie

Conclusions

Integrity management of process plant safety barriers is a critical activity that is suffering from lack of communication and lack of data. The process plant safety lifecycle contains three basic phases – analysis, design, and operation. During each phase, critical data is generated that must be passed back and forth from the different phases to ensure that designs are appropriate and tolerable levels of risk can be achieved. Enterprise software tools that utilizes standardized and open data structures facilitate this type of communication between applications and between phases using bowtie diagrams as the mechanism to communicate all this information in an easily understood format. Industry needs to work together to standardize the structure of the data that is passed between the phases and demand that software vendors adhere to these open standards. In this way, critical data can be passed from phase to phase facilitating the integrity management of barriers throughout the whole lifecycle, including the analysis phase where discrepancies between desired performance and actual performance can be assessed and, if necessary, design modifications can be introduced.

Conclusions

Marszal, E.M., *Unified Hazard Assessment – Bringing Together HAZOP, LOPA, Hazard Registers, and Bowtie in a Unified Structure*, Proceedings – Mary Kay O'Conner Process Safety Center 22nd Annual International Symposium, October 2019, College Station, Texas.

Marszal, E.M., *Applying Data Science to Hazard Analysis*, InTech Magazine, Jan-Feb 2020, International Society for Automation, Research Triangle Park, North Carolina.

586239 A Case Study on KOC Approach for PHA Revalidation of Existing Gathering Centre

Abstract:

Introduction:

A hazard is an inherent physical or chemical characteristic that has the potential to do harm (to people, property, or the environment). A hazard evaluation study is an organized effort to identify and analyze the significance of hazardous situations associated with a process or activity.

There are number of reasons such as changes in Process or Equipment, in on-site or off-site occupancy patterns and Gaps and Deficiencies in Prior PHA. In addition, new knowledge is now available to better understand the hazards potential and control, safeguard previously credited in the PHA have been removed / compromised or discredited, and Actual incidents have revealed scenarios not previously identified in the PHA. Positively, process changes have actually removed hazards from the process, prior PHA did not take proper credit for some risk reduction measures and new risks reduction measures have been added, which clearly reduced the original risk.

Objective:

The primary objective of a PHA revalidation is to produce an updated PHA that adequately identifies, evaluates, and proposes controls for the hazards of the process, as they are currently understood.

Methodology:

KOC has established KOC.PS.017 - Process Hazard Analysis (PHA) Revalidation Procedure to determine an appropriate approach for the revalidation (Redoing vs. Update and Revalidate), conducting revalidation in timely manner, establish responsibilities and confirming that the PHA is accurately reflects the hazards of the current process.

The revalidation process build upon the still pertinent portions of the prior PHA. Corrections are made and new content is added to the PHA as required (i.e., the PHA is Updated), and the results are documented to serve as the new PHA (i.e., the PHA is Revalidated).

KOC appointed well known third party consultant for onsite PHA revalidation and below are stages for PHA revalidation.

Team Selection

Selecting the Appropriate PHA Revalidation Approach

- Update and Revalidate.
- Retrofit, Update, and Revalidate
- Redo Review and Analyze Information Review of Operating Experience
- Evaluating of PHA Quality and Completeness
- PHA Revalidation Information Requirements

- Changes Review

- Review Previous Incidents
- Review the Status / Resolution of Previous PHA Recommendations
- Update Human Factors
- Update the Facility Siting Analysis
- Address Hazards Associated with Non-Routine Operating Modes
- Address Compliance of PHA with Regulatory Requirements Documentation of PHA Revalidation Recommendation Follow - up
- PHA Report Contents
- Wrap-Up Discussions

Results

Type of studies carried out and number of new recommendations will be discussed for Gathering Center 24.

Conclusion:

Continuous improvement in the control of newly introduced process safety risks.

Ensured the effective close out of pending actions raised during previous process hazard analysis.

Addresses all four pillars of Risk Based Process Safety.

Considerable time saving in comparison to Redo PHA.

Improvement in employee engagement.

Visual Dust Hazard Analysis – Understanding Threats and Assuring Controls

Michelle R. Murphy
Mica, LLC
Atkinson, NH
murphy@micadhm.com

David Hatch
Process Safety Integrity
Derby, UK
dhatch@psintegrity.com

Prepared for Presentation at
American Institute of Chemical Engineers
2020 Spring Meeting and 16th Global Congress on Process Safety
Houston, TX
March 30 – April 1, 2020

Visual Dust Hazard Analysis – Understanding Threats and Assuring Controls

Michelle R. Murphy
Mica, LLC
Atkinson, NH
murphy@micadhm.com

David Hatch
Process Safety Integrity
Derby, UK
dhatch@psintegrity.com

Keywords: Dust Explosions, Bowties, Flammable Atmospheres, Risk Management. Ignition Controls, Dust Hazard Analysis

Abstract

The potential harm and asset damage from a dust explosion have been known for many centuries. A wide variety of materials that are explosible in dust form exist in many industries, however knowing about an issue is not the same as properly understanding and adequately addressing the challenges.

The authors draw on their complementary experience in Dust Hazard Analysis (DHA) application (1) and BowTie implementation (2) to present a novel approach to scenario visualisation and asset management using bowties.

Just as a conventional HAZOP or Process Hazard Analysis (PHA) subdivides the process into nodes, the DHA breaks the plant into sections which are systematically challenged to determine if explosion or fire hazards exist. For dusts, this includes identification of competent ignition threats. Subsequent evaluation is conducted to determine which are prevented and/or mitigated by safeguards specific to each threat and consequence.

Risks are assigned and actions issued as per a normal hazard analysis, however the graphical presentation of the ignition scenarios and safeguards and the illustration of potential ignited particles which can travel and ignite within downstream equipment offers significant communication benefits. This can be particularly helpful in facilities or countries where English and/or Risk are not the primary language or familiar terminology. This paper will show how BowTies are an effective tool to engage less technical stakeholders and offer a robust framework to assure the health (presence and performance) of the assumed or planned human and hardware controls which must be sustained to provide the necessary risk reduction.

In summary, this novel approach evolves analysis into an operational framework to ensure that the ignition risks are better understood, and the controls are properly implemented, operated and maintained.

1 Combustible Dust Hazard Analysis

The potential harm and asset damage from a dust explosion have been known for centuries. A wide variety of materials that are explosible in dust form exist in many industries, however knowing about an issue is not the same as properly understanding and adequately addressing the challenges.

Figure 1 shows the progression from fire to flash fire to explosion for combustible dusts. These hazards exist in dust processing equipment and in building spaces where dust is escaping equipment. To prevent the realization of these unwanted events, one of the elements of the original fire triangle needs to be removed. While explosions are the most destructive, flash fires often result in fatalities, and having one of these events can trigger secondary events.

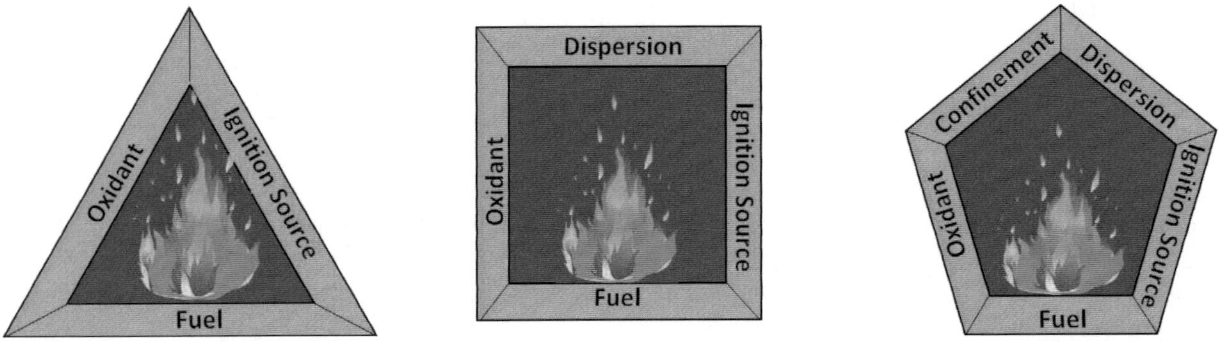

Figure 1 – Progression of combustible dust hazards from fire to flash fire to explosion

Dust hazard analyses (DHAs) are conducted to identify and assess dust flash-fire and explosion hazards in both process equipment and building spaces. Like PHAs, many methodologies are available for completing hazard identification and assessment of dust hazards. Most involve recording of the discussion in a spreadsheet format. Visual DHAs (bowties) are a novel approach to scenario visualization that can simplify communication of DHA results, elevate appreciation of safeguards (barriers), and improve the management of change (MOC) process.

At present, most applications involve development of bowties for higher hazard scenarios after a more traditional hazard analysis has been completed. This paper will present bowties as developed from traditional DHA records.

1.1 Regulations

In the US, the Occupational Health and Safety Administration (OSHA) does not have a comprehensive combustible dust standard. It relies on existing related regulations and recognized and generally accepted good engineering practices (RAGAGEPs). The most widely applied of these standards are from the National Fire Protection Association (NFPA®). Both the fundamentals standard NFPA 652 (3) and the industry-specific standards (4; 5; 6; 7), require dust hazards analysis (DHA) of dust handling operations. The NFPA does not specify a DHA technique to be adopted however one example is provided and several commonly applied hazard analysis techniques are listed in the appendix. In their recent paper, Murphy and Borene (8) detail the NFPA requirements for conducting DHAs and present a typical DHA process.

In Europe, ATEX (ATmosphères EXplosives) is the name commonly given to the two European Directives for controlling explosive atmospheres:

- Directive 1999/92/EC (9) (also known as 'ATEX 137' or the 'ATEX Workplace Directive') on minimum requirements for improving the health and safety protection of workers potentially at risk from explosive atmospheres.
- Directive 2014/34/EC (10) (also known as 'ATEX 95' or 'the ATEX Equipment Directive') on the approximation of the laws of Members States concerning equipment and protective systems intended for use in potentially explosive atmospheres.

Article 8 of the ATEX Workplace Directive requires the employer to draw up (and maintain) an Explosion Protection Document (EPD) which includes:

- Identification of hazards
- Evaluation of risks
- Definition of specific measures to safeguard the health & safety of workers at risk from explosive atmospheres

In the UK, the Dangerous Substances and Explosive Atmospheres Regulations 2002 (DSEAR) regulations (11) put into effect the requirements of ATEX 137 and require employers (duty-holders) to identify dangerous substances (including flammable gases, vapours, mists and combustible dusts), evaluate the associated risks, implement controls to remove/reduce flammable atmospheres and/or ignition sources, prepare emergency plans and ensure employees are properly informed about and trained to control the risks from dangerous substances. Although DSEAR does not require an EPD (since its contents are covered by Regulation 5 *Risk Assessment* and Regulation 7 *Places where explosive atmospheres may occur*) these are often prepared by duty-holders as a clear, concise, basis of safety that can be communicated with employees and other interested parties. This approach is intended to complement and/or supplement such documents.

1.2 Regulatory Deadlines

In the US the deadline for most facilities to complete a dust hazard analysis (DHA) is 7th September 2020 (later for agricultural and food manufacturers) in accordance with NFPA Standard 652 on the Fundamentals of Combustible Dust and the associated industry-specific standards.

In the UK and Europe, the ATEX Workplace Directive and DSEAR regulations are already effective and the obligations on duty-holders are long established. This approach aims to improve the awareness and understanding of those responsible for and/or affected by flammable atmospheres.

1.3 DHA Elements

NFPA states that a dust hazard analysis is intended to identify and evaluate areas of the process or building spaces where fire, flash fire, and explosion hazards exist. Once identified, they are to be evaluated for applicable fire and deflagration scenarios, including:

1. Safe operating ranges
2. Existing safeguards
3. Additional safeguards, as needed

Like all hazard analyses, an identification method is employed to identify hazards. Once identified, causes and consequences are postulated to define scenarios. In the case of combustible dust, these would focus on fire, flash fire and explosion scenarios. Existing safeguards are identified and evaluated for their ability to manage these events and remove or limit the consequences. Where existing safeguards are deemed inadequate, additional safeguards are recommended. Finally, a plan for implementation is put together. The process is depicted in Figure 2. An implementation plan is typically completed after the hazard analysis step and is not shown in the hazard analysis flow diagram.

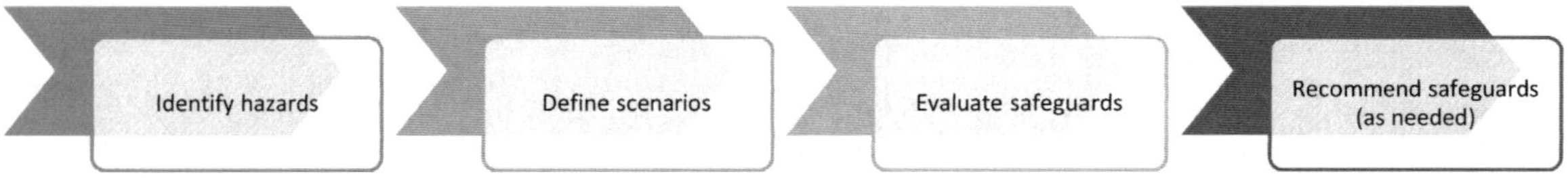

Figure 2 – Hazard analysis flow

1.4 Hazard Identification Methods

Many methods exist to identify dust hazards. Typical identification methods include:

- What-if/Checklist
- HAZOP
- Event Tree/Fault Tree
- Failure Modes and Effects Analysis

Due to the straightforward operation of combustible dust handling equipment, a checklist approach is often employed to identify hazards. Such a checklist can be based on specific requirements and recommendations found in several reference documents (e.g., NFPA codes and standards, CCPS Guidelines for Safe Handling of Powders and Bulk Solids (12) , CCPS Guidelines for Combustible Dust Hazard Analysis (13) and Lee's Loss Prevention in the Process Industries (14)).

CCPS Guidelines for Risk Based Process Safety (15) states "A company that uses its risk and understanding is better able to deal with the resultant risk, and subsequently, sustain long-term, accident-free, and profitable operations." DHAs are the first step in the risk assessment process.

2 Application of Bowties

2.1 Purpose of Bowtie

A 'bowtie' is a diagram that visualizes the risk you are dealing with in just one, easy to understand picture. The diagram is shaped like a bowtie, creating a clear differentiation between proactive and reactive risk management. The power of this type of diagram is that it gives you an overview of multiple plausible scenarios, in a single picture. In short, it provides a simple, visual explanation of a risk that would be much more difficult to explain otherwise.

Bowtie diagrams are now commonplace in understanding and managing process safety risks, particularly associated with major accidents. Historically they have been applied most frequently in the chemical, petrochemical and oil & gas industries but are equally applicable in all industries, e.g., transport, mining, energy, finance or healthcare and can be used to manage all risks (effect of uncertainty on objectives) including safety, environmental impact, asset damage or loss of reputation.

Bowties demonstrate how hazards are controlled and illustrate the links between controls (barriers) and the relevant components of the safety or risk management system.

2.2 Methodology

The bowtie diagram is based on the concept of the "Swiss Cheese Model" developed by James Reason (16). This model illustrates system failures which can be addressed by controls (barriers) represented by swiss cheese slices which are intended to provide 'defence in depth'.

Barriers have weaknesses which are either inherent i.e., always present or are active failures created during the scenario which, when aligned, allows the 'accident trajectory' to pass through the holes, as depicted in Figure 3.This model has evolved into the familiar and increasingly popular bowtie diagram which allows multiple trajectories (Cause-Consequence relationships) to be visualised on the same diagram.

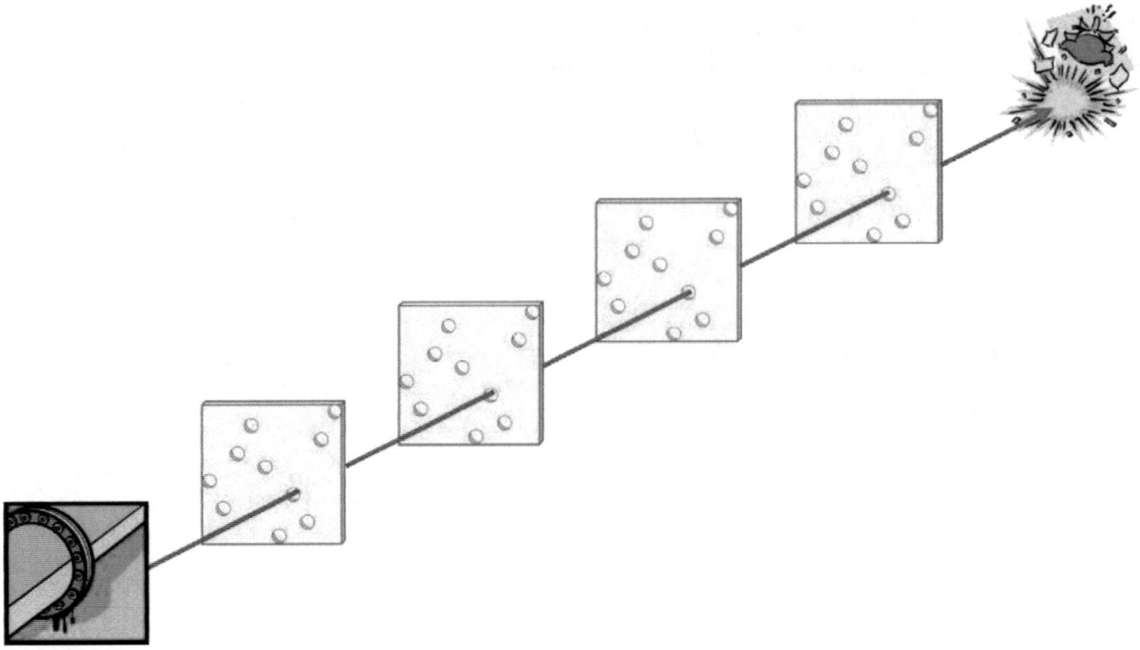

Figure 3 – Swiss Cheese Model

The components of a bowtie diagram are summarised as follows:

- Hazard = Operation, activity or material with potential to cause harm
- Top Event = Initial loss of control of Hazard
- Consequences = Negative effects (harm or damage) that could result from Top Event
- Threats = Potential reasons for loss of control of the Hazard leading to the Top Event
- Barrier = Engineering or Administrative risk reduction measure
    - Prevention
    - On its own can prevent a Threat from developing into a Top Event
        - Eliminate the Threat
        - Prevent the Top Event
    - Mitigation
    - On its own can mitigate the Consequences of the Top Event once it has occurred
        - Prevent consequence (likelihood)
        - Reduce impact (severity)

Barriers can either be hardware, human, or combinations of both which are implemented and operate as follows:

- Passive hardware
    - Always present e.g. structural or containment
- Active hardware

    - Automated response to detection e.g. control action or trip
- Active hardware + human
    - Human response prompted by hardware detection e.g. alarms
    - Hardware response prompted by human detection e.g. manual call point
- Active human
    - Human response prompted by observation e.g. walk around & intervention
- Continuous hardware
    - Always operating e.g. ventilation

These components are assembled into scenarios which are read from left to right as shown in the overview in Figure 4 below (with good practice guidance from the CCPS Bowtie book (17)):

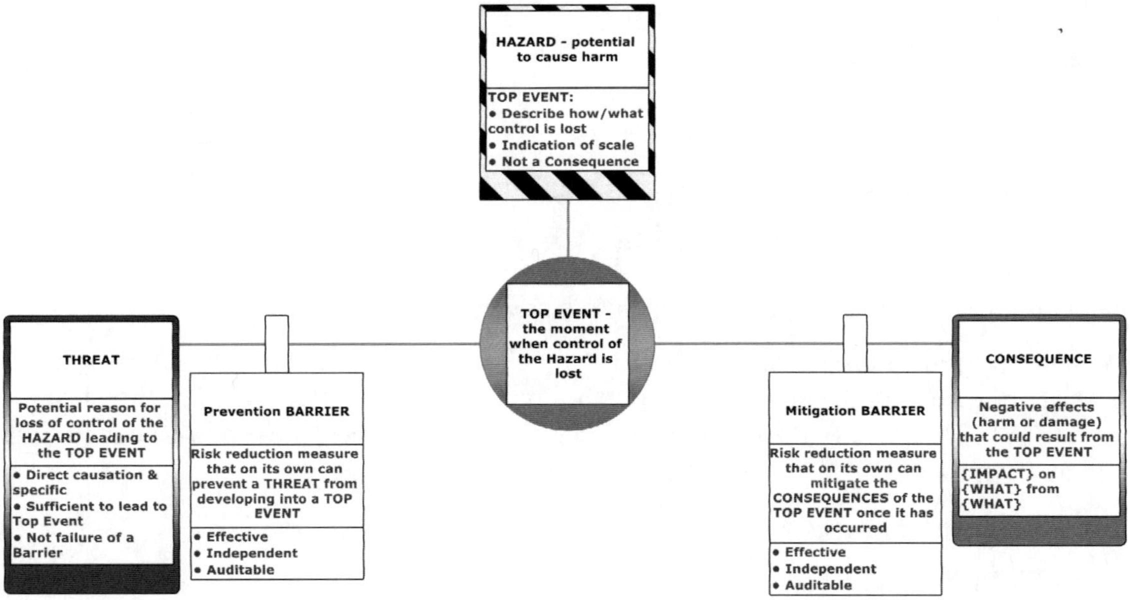

Figure 4 – Bowtie Model

A Top Event may have several Consequences (safety, environmental, asset, etc.) that result from it and several Threats (e.g., equipment failures, environmental influences or operational issues) that could cause it.

2.3 Application to Dust Hazard Analysis

Just as a conventional HAZOP or PHA subdivides the process into nodes, the DHA breaks the plant into systems which are systematically challenged to determine if explosion or fire hazards exist. For dusts, this includes identification of competent ignition threats. Subsequent evaluation is conducted to determine which are prevented and/or mitigated by safeguards specific to each threat and consequence.

Annex B of NFPA 652 (3) provides an example DHA of a wood milling process. Wood chips are brought in via rail car and trailer truck, unloaded and pneumatically conveyed into a storage silo, transported through a screw conveyer to a size reduction mill, and further processed for product delivery. The DHA is applied to each process component and is documented in paragraph format. Figure 5 shows how the results of the evaluation of the first node, offload duct to offload fan, would look in a simple bowtie representation.

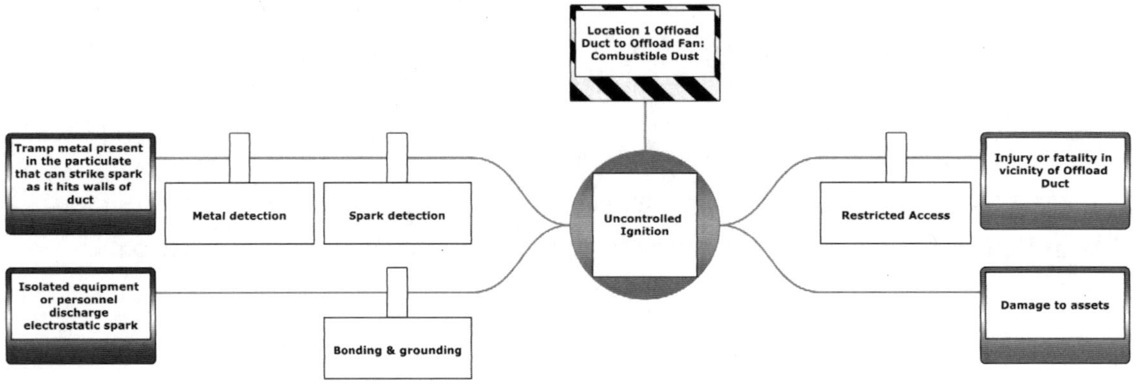

Figure 5 – Basic DHA of a process unit

A more detailed example is discussed later in this paper.

In the manner presented in Figure 5, each DHA node (major asset or sub-system) is represented by a separate bowtie with the credible/competent ignition sources represented as threats e.g. from EN 1127-1 (18) . To simplify the documentation of ignition sources, these can be coded to make the diagrams concise:

- AC Adiabatic compression and shock waves
- EA Electrical apparatus
- EC Stray electric currents, cathodic corrosion protection
- EW Electromagnetic waves
- EX Exothermic reactions, including self-ignition of dusts
- FL Flames and hot gases (including hot particles)
- HS Hot surfaces
- IR Ionizing radiation
- LT Lightning
- MS Mechanically generated sparks
- RF Radio Frequency (RF) Electromagnetic waves
- SE Static electricity

- US Ultrasonics

Hazard data can be added to the diagram including:

- Material and properties
- Safe Operating Range
- Deflagrable (Yes/No)
- Suspended in Air (Yes/No)
- Above Minimum Explosive Concentration (Yes/No)

The potential effects of an uncontrolled ignition (some processes involve controlled ignition e.g., biomass fuel) include fire, flash fire, or explosion that result in fatality, injury, and/or property damage (including business interruption). These can occur in the node/asset/system under review or in connected nodes/assets/systems e.g., ignited or smouldering particles passing downstream. Figure 6 shows the connection of consequences (upstream) that become threats (downstream). In the figure, uncontrolled ignition in the offload duct propagates to the offload fan. These have become threats to the downstream system.

NFPA specifically calls out the evaluation of potential deflagration propagation between parts of the process. The bowtie simplifies the application and visualization of this critical part of the analysis.

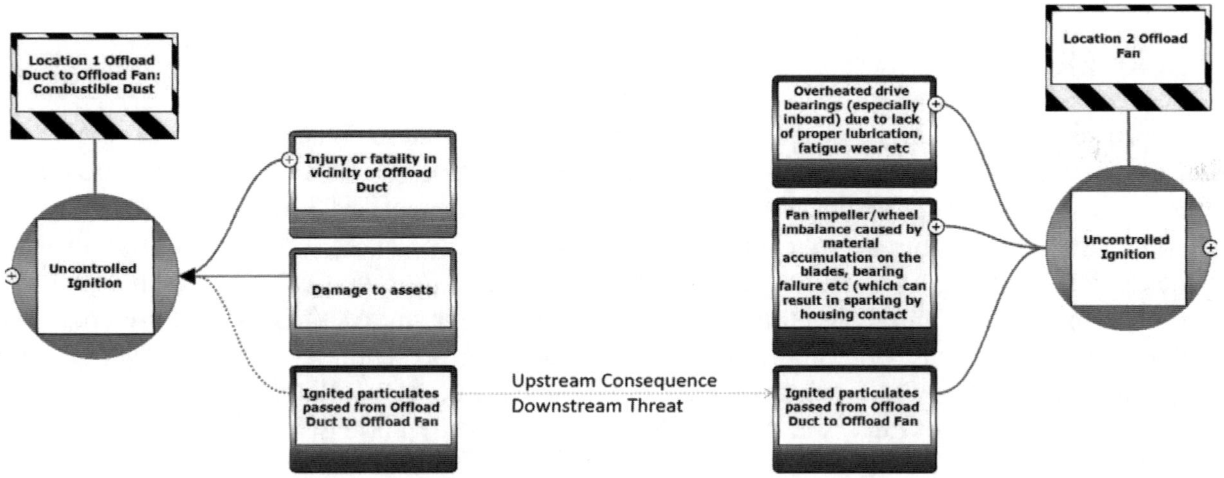

Figure 6 – Indirect Consequences (chained bowties)

Risks are evaluated based on the likelihood of a coincident flammable dust atmosphere (in ATEX, this is defined in hazardous area classification) and an ignition source (based on threat frequency or potential) and the severity of the fire or explosion impact. These are assigned and actions issued as per a normal hazard analysis. The graphical presentation of the ignition scenarios and safeguards and the illustration of potential ignited particles which can travel and ignite within downstream equipment offers significant communication benefits. This can be particularly helpful in facilities or countries where English and/or Risk are not the primary language or familiar terminology.

The visual nature of the bowtie helps confirm existing control measures (prevention & mitigation barriers) e.g. explosion vents, temperature sensors, etc. and identify where additional measures to meet risk target are needed.

2.4 Risk Management

Effective risk management requires an understanding of hazard scenarios and an appreciation (and respect) of the measures to prevent loss of control and mitigate the consequences. Bowtie diagrams provide clear, consistent information to a wide audience compared to detailed analyses which are limited to specialist interpretation.

Once barriers have been identified, a barrier-management program is necessary to ensure that controls are implemented and continue to operate and perform as required to provide ongoing risk reduction. This program involves monitoring the presence and measuring the effectiveness of hardware (technical) and human (organisational) barriers and repairing, replacing or refreshing them as and when appropriate in advance of any demands to act, i.e., before the scenarios occur. The visual nature of bowties can simplify the identification of barriers for program development.

3 Example

The approach of converting from a traditional spreadsheet format is presented in an example of a filter receiver (Figure 7). Filter receivers are one type of air material separator and are often found in dust handling operations as part of the pneumatic transport system. In this example, material is transferred from a silo to a day bin, through a filter receiver. The receiver has a rotary air lock before discharging by gravity into the bin. The material properties and safe operating ranges would also be documented for the DHA. The example employs the risk-ranking presented in the paper by Murphy & Borene (8), wherein C, L, and R represent a ranking of consequence, likelihood, and risk. The rank is determined from the consequence and likelihood assigned by the DHA team. The resultant rank is color-coded to depict a prioritization of risk, moving from green (acceptable) to red (highest priority).

Table 1 shows the results of an analysis of ignition sources, as part of an overall checklist review of the filter receiver. Note that the example is provided for demonstration purposes only and is not intended to represent a complete DHA.

The same record is presented in a bowtie format in Figure . The first impression is one of simplicity. The information is presented clearly and concisely, allowing a quick review of the barriers that are imperative to preventing the fatality. This allows for further discussion on the availability/reliability of the barriers. For example:

1. Equipment material is fixed so very effective barrier
2. Overload trip needs to be regularly tested

3. Ignition energy is a robust barrier as long as material doesn't change i.e., analysis only fit for specific dust
4. Explosion vent needs to be regularly inspected
5. Explosion vent size/capacity needs to match dust (explosion) properties i.e., if material changes vent may not be adequate.

The visual design can be used as a communication tool for design, operation, and maintenance to understand the importance of specific activities and the role they play in preventing and mitigating the event of an explosion in the receiver resulting in a fatality. Additionally, during the MOC process, it will be straightforward to see what changes will impact the barriers (here change of material, change of rotary valve speed, removal of explosion vent, for example).

In this example, the following bowtie features are exploited:

1. Threat types categorised from EN 1127-1
2. Barrier (safeguard) types
3. Consequence risk for 4 types of receptors:
    a. Safety = D
    b. Environment = not determined (grey box)
    c. Asset = not determined (grey box)
    d. Reputation = not determined (grey box)

It is also possible to show individual Threat likelihoods (with or without safeguards or Barriers) and the individual Barrier effectiveness to appreciate the risk reduction contribution that each makes.

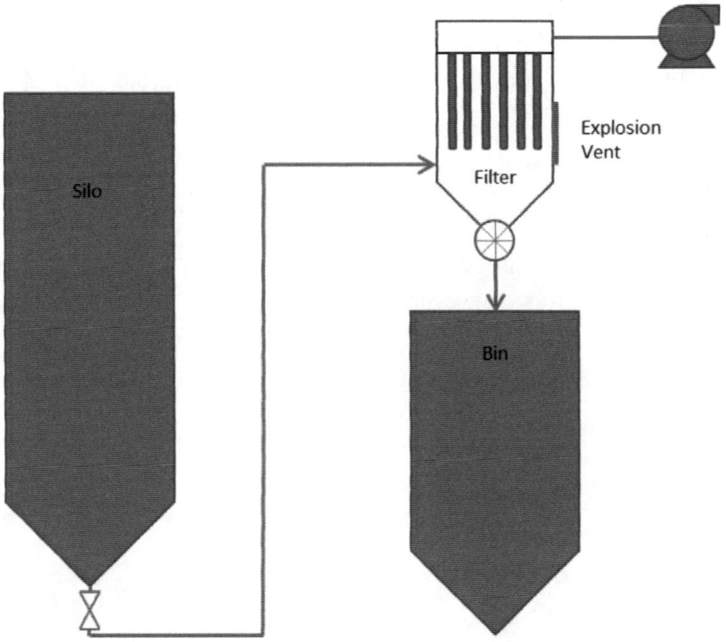

Figure 7 – Example receiving system

Table 1 – Excerpt of a traditional recording of a checklist DHA on a filter receiver

Item No.	Question	Answer	Cause	Consequence	Safeguards	C	L	R	Recommendation
1.1	Are competent sources of ignition present?	Bearings	Bearings cause ignition - N/A as bearings are outboard						
		Electrical spark	Electrical spark causes ignition - N/A as no electrical equipment in receiver						
		Impact spark	Impact spark caused by tramp material	Explosion in receiver, one fatality	Material of construction is stainless steel, explosion venting on receiver	1	1	D	
		Frictional spark	Frictional spark caused by tramp material getting caught in rotary valve	Explosion in receiver, one fatality	Rotary valve is low rpm, valve has an overload trip, explosion venting on receiver	1	1	D	
		Electrostatic spark	Electrostatic spark from unbonded flex boot causes ignition	Explosion in receiver, one fatality	MIE 100-300 mJ, explosion venting on receiver to safe location	1	1	D	1.1.1 Update grounding/bonding program to assure bonding is maintained across flex boots.
1.2	Do AMS with explosion hazards have isolation devices?	Yes, rotary valve on outlet of receiver	Electrostatic spark causes ignition in filter receiver and propagates forward to day bin	Explosion in day bin, one fatality	Rotary valve designed for material blocking deflagration isolation downstream	1	1	D	1.2.1 See Recommendation [1.1.1] to update grounding/bonding program.
		No isolation back to the silo	Electrostatic spark causes ignition in filter receiver and propagates back to the silo	Explosion in silo, one fatality		4	3	B	1.2.2 Add an isolation device (in accordance with NFPA 69) upstream of the filter receiver to prevent propagation back to the silo.
									1.2.3 See Recommendation [1.1.1] to update grounding/bonding program.

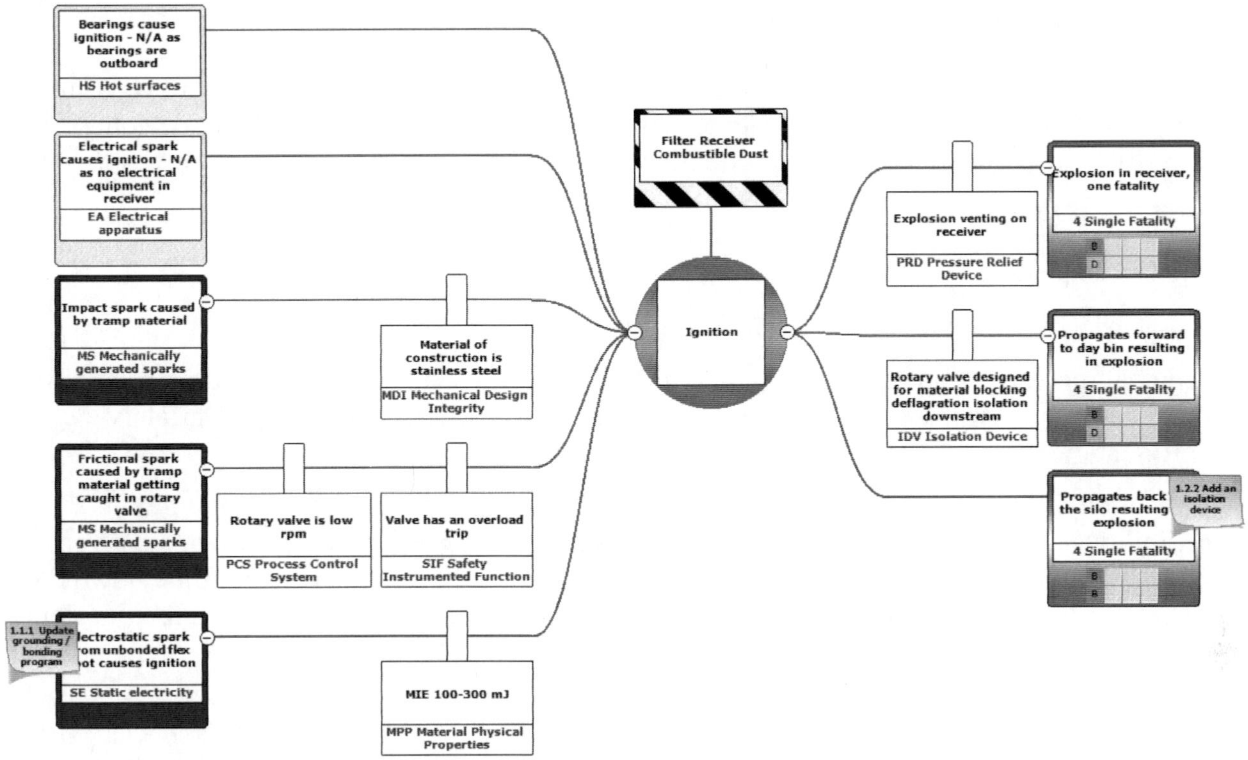

Figure 8 - Traditional DHA in bowtie format with ignition as the top event

4 Conclusion

4.1 Challenges of traditional hazard identification methods

The consequences and risks of a dust explosion have been known for many years and assessments have been conducted to address these. A traditional hazard analysis format (e.g., Checklist, HAZOP) is often applied which may have the following weaknesses:

1. Controls wrongly assigned to causes/threats/ignition sources
2. Controls missing from appropriate causes/threats/ignition sources
3. Failure to distinguish between prevention and mitigation/recovery measures
4. Failure to identify and evaluate connections between nodes/equipment where smouldering particles can ignite downstream dust

Furthermore, the presence (availability) and performance (reliability) of barriers is not part of the assessment and therefore a static snapshot of assumed or planned controls.

4.2 Benefits over traditional hazard identification methods

There are a number of identification methods ranging from simple checklists to more rigorous, systematic techniques which are well publicised (19) and employed. Bowties provide context to the explosion scenarios in a simple but effective visualisation of the relationships between:

- Causes of uncontrolled ignition;
- Consequences of those events, and;
- Barriers that prevent the event and/or mitigate the effects

A graphical representation offers a language that facilitates a common understanding of hazards, unwanted events, their risks, and controls to help direct or deploy resources appropriately and proportionately.

Classification of barriers, either by type or owner, allows dependencies or common mode vulnerabilities to be evaluated and addressed. A common mode failure occurs when a single event causes multiple barriers to fail (either within the same bowtie or across multiple bowties) and therefore barriers should be independent of the threat that they prevent and of other barriers on the same threat to top event or top event to consequence pathway.

4.3 Communication

The aim is not just to make existing assessments more engaging i.e. to involve all stakeholders but also to provide a life-long dynamic framework where threats (competent ignition sources) and barriers (control measures) are monitored, evaluated and actions taken to ensure that protection is sustained and risk targets are maintained. Visualisation enables duty holders not only to analyse their dust hazards but also to communicate the analyses to front line personnel to ensure that they understand the risks they are responsible for managing and sustain the protection measures for which they are accountable.

The current aim is to enhance not replace traditional methods.

4.4 Operational vs design focus

Bowties are not a static snapshot of assumed/planned controls but a live asset/risk management platform that can be updated (manually or automatically) to show the current state of health (presence and performance) of controls and the current risk exposure.

Barriers degrade over time and their performance must be monitored, measured and sustained at the required level to achieve the necessary risk reduction.

4.5 Better management of change implementation

Identification of barrier criticality and ownership makes management of change (MOC) more robust since the potential impact of defeating, degrading or deleting barriers within a single or across multiple scenarios (bowties) is more apparent. Suitable bowtie software with a barrier database can be filtered/sorted to focus on the deployment of barriers. This allows proper addressing in all scenarios where an affected barrier or barriers is/are implemented. Because barriers can be hardware, human or a combination of both, technical and organisational changes can be evaluated to ensure that the risk reduction is not unduly compromised.

The authors believe this approach highlights and addresses existing deficiencies whilst providing the same risk assessment functionality expected or required. For combustible dust this means consideration of probability of a flammable atmosphere (using ATEX (20), NFPA classification of dust flash-fire or explosion hazard area (3), or NFPA classification as a hazardous location for electrical installations (21)), probability of competent ignition source, and severity of fire, flash fire, or explosion consequences.

The goal is to operationalise scenarios and demonstrate that duty-holders are and remain in control through ongoing barrier maintenance and robust change management.

An initial step for duty-holders would be to migrate their existing tabular (worksheet) assessments into bowties to expose and address weaknesses. Visual DHA is designed to be different (evolution rather than revolution) with a format where existing information is not lost but knowledge is gained.

References

1. **Murphy, M.** Making Sense of Combustible-Dust Hazard Analysis. *Chemical Engineering Progress.* 2016, April 2016.

2. **Hatch, D., McCulloch, P., Travers, I.** Enhancing PHAs: The Power of Bowties. *Chemical Engineering Progress.* 2019, February 2019.

3. **NFPA.** *NFPA 652 Standard on the Fundamentals of Combustible Dust 2019 Edition.* May 2018.

4. **NFPA.** *NFPA 654, Prevention of fire and Dust Explosions from the Manufacturing, Processing, and Handling of Combustible Pariculate Solids.* Augst 2019.

5. **NFPA .** *NFPA 61 Standard for the Prevention of Fires and Dust Explosions in Agricultural and Food Processing Facilities.* November 2019.

6. **NFPA.** *NFPA 484 Standard for Combustible Metals.* May 2018.

7. **NFPA.** *NFPA 664 Standard on the Prevention of Fires and Explosions in Wood Processing and Woodworking Facilities.* June 2016.

8. **Murphy, M. and Borene, M.** How to Conduct a Dust Hazards Analysis. *Process Safety Progress.* 2018, October 2018.

9. **European Commission.** *ATEX 137 workplace directive 1999/92/EC.* 1999.

10. **European Commission .** *ATEX 95 equipment directive 2014/34/EC.* 2014.

11. **UK Government.** *Dangerous Substances and Explosive Atmospheres Regulations (DSEAR).* 2002.

12. **CCPS.** *Guidelines for Safe Handling of Powders and Bulk Solids.* November 2004.

13. **CCPS .** *Guidelines for Combustible Dust Hazard Analysis.* May 2017.

14. **Elsevier.** *Lees' Loss Prevention in the Process Industries 4th Edition.* 2012.

15. **CCPS.** *Guidelines for Risk Based Process Safety.* March 2007.

16. **Reason, J.** *Managing the risks of organizational accidents.* 1997.

17. **CCPS.** *Bow Ties in Risk Management: A Concept Book for Process Safety.* September 2018.

18. **British Standards Institute.** *BS EN 1127-1:2011, Explosive atmospheres. Explosion prevention and protection. Basic concepts and methodology.* 2011.

19. **CCPS.** *Guidelines for Hazard Evaluation Procedures 3rd Edition.* April 2008.

20. **British Standards Institute.** *BS EN 60079-10-2: 2015, Explosive atmospheres. Classification of areas. Explosive dust atmospheres.* 2015.

21. **NFPA.** *NFPA 70 National Electrical Code.* January 2020.

585578 Vinyl-Siloxane Cleavage: Applying Thermo-Analytical and Hybrid Calorimetry Techniques to Inform Process Safety and Design

Abstract:

The use of homogeneous and heterogeneous acid catalysts is widespread in the processing of siloxanes as they protonate Si-O bonds, enabling a vast array of rearrangement reactions. Therefore, developing a fundamental understanding how acid catalysts interact with various siloxane moieties is crucial to safe and sustainable operation. In this study, we present our findings on the interaction of acid catalysts with vinyl pendant siloxanes, specifically focusing on the exothermic acid-catalyzed cleavage of the C-Si bond with gas evolution. Thermo-analytical techniques including Differential Scanning Calorimetry (DSC) and Thermogravimetric Analysis (TGA) were used to screen various moieties to evaluate gas and heat generation. Adiabatic Accelerating Rate Calorimetry (ARC) measurements were performed to quantify the kinetics of C-Si cleavage and evaluate effect of operational conditions on the kinetics of the cleavage reaction. Then, a hybrid technique, ARCxGC/MS was utilized to identify the various reaction products and the correlate them with the detected onset temperatures. Using molecular modelling and additivity techniques like secondary Benson Group estimation, the theoretical Heat of Reaction for the proposed reaction was determined and compared to predictions from the afore-mentioned calorimetry techniques. From the above effort, it was concluded that cleavage of the vinyl-silicon bond due to activation by the acid catalyst is an exothermic and gas-generating reaction. The findings were used to generate, for the first time, a detailed quantitative thermo-kinetic model of the vinyl-Si cleavage reaction pathway useful for design safety. Our findings were supported by a cursory review of available literature that qualitatively suggest that scission on the C-Si bond is activated in the presence of electrophiles and protic acids.

586618 Safety Critical Elements (SCE): Rationalizing Our Barriers

Abstract:

Safety Critical Elements (SCE) can be described as part of an installation and such of its structure, equipment and systems or any part thereof, whereby failure of which could cause or contribute substantially to the release of a hazard with major risks. In other words, SCEs are the barriers which are in place to prevent or limit the effect of a major accident.

In early 2013, the organization has embarked on the adaptation of SCE as part of Hazard and Effects Management Process (HEMP) and subsequently, business units within the organization has adopted the concept of SCE within their respective processes. Offshore business unit, particularly, had incorporated SCE identification process as part of Safety Case and subsequently, SCE management process within their maintenance system. The SCE identification and management protocols for offshore business units were established based on available standards and practises, from both within the organization as well as from external sources.

As part of the Safety Cases revalidation, which needs to be done every 5 years, the review of the SCEs that were previously identified for operating facilities under offshore business units was also triggered. These reviews highlighted the real challenges or pain points encountered by operating facilities in managing SCEs as part of their day-to-day operation and maintenance. The challenges or pain points can be summarized as follows:

1. The number of items / equipment / system identified and included as SCE for a particular operating facility is high. On an average, it approximately constitutes about 30-40% of the overall asset register.
2. The high number of items / equipment / system tagged as SCEs has resulted in high number of SCE related work orders i.e. inspection, testing and preventive maintenance (ITPM) task being generated through the maintenance management system, which overwhelms the maintenance personnel at site.
3. This causes overdue and subsequently leads to backlog of the SCE specific ITPM task to be carried out, which is reflected in the Facility Status Monitoring (FSM) system.
4. FSM is being directly monitored by higher management and it raises a serious concern on the healthiness status of the SCEs at operating units in relation to its specified function.

Based on the feedback gathered from operating facilities, it is agreed that there is a need for rationalization of the SCEs currently listed in the respective facilities asset register. This paper details the approach for SCE rationalization which was established and implemented for operating facilities, in view of streamlining the list of SCEs to reflect only those items / equipment / system which are critical in-line with the purpose and definition of SCE.

586170 Advances in Layers of Protection for Combustible Dust Hazards

Abstract:

Often considered as two separate scenarios, combustible dust fires and explosions can often be related. Layers of protection are a benefit in the mitigation of both of the potential hazards presented by combustible dust. New solutions in the last few years for hazard assessment, prevention systems, and mitigation systems have helped create new layered approaches to combustible dust. Dust hazard analysis are the starting point with more elaborate alternatives to prescriptive approaches like risk-based analysis. New test protocols for quick water mist suppression systems have provided an improved layer to the traditional fire protection solution for combustible hazards in process equipment. Intelligent and connected mitigation equipment now provides improved reaction times, the ability to remotely monitor, and more reliable communication networks between components.

This presentation will examine the layers of mitigation protection for both fire and explosion hazards. Specifically, it will discuss how these layers relate and when designed together can provide a higher level of safety. Factors such as reliability (SIL), hazard assessment methodology, intelligent “smart” mitigation solutions, the impact of human error, and active vs passive safeguards will be discussed. A case study of a process area will be walked through during the presentation to show how layers of mitigation protection for both combustible dust fire and explosion hazards can improve safety.

587094 Leveraging the Success of Accelerative Learning and PHAs

Abstract:

The concept of Accelerative Learning has been known and practiced for several decades. This technique has proved to be a very welcome tool for adult learning, time and time again. This concept has been widely applied in the industries in lieu of the conventional classroom training methods, and its results have benefitted in achieving success for various corporations. The success has meant that the audience has improved their engagement in the class, embraced the challenging topics, increased the awareness and helping in bringing about a mind shift in some cases, through this learning exercise

On the other hand, Process hazard analysis (PHA), have also been in existence for the past few decades. Some of its methodologies are preferred over the other for level of details, and are preferred over the other for its brevity and time saving feature. These are always conducted in a workshop environment, and its success is dependent on the level of engagement and preparation. Meanwhile, the sessions can also become a victim on team that has been reluctant to participate for various reasons.

The last decade 2010-2019 has come with its set of challenges. Industries cut back on various workshops due to the financial crunch, due to their ever shrinking budgets and also the economic downturn in the oil & gas industry, which led to a cascading effect in other industries. This led to changes in industry approaches being altered to focus on "More with Less". One emergent outcome of this need, was the philosophy of Smart PHAs or Smart HAZOPs

The author has applied the concepts of AL to PHAs, and has changed the way the PHAs are facilitated in a workshop environment. This has resulted in higher level of personnel engagement, that has fostered the brain storming needs of a PHAs, and has left the audience highly energized after the PHA sessions, such that they want to return for me. Through this paper, the author will share the success stories while highlighting the benefits of this hybrid technique.

586044 Initiative for Advanced Risk Assessment to Prevent Accidents in Our Complex

Abstract:

Against recent accidents at chemical plants in japan, we have implemented improvements on tangible and intangible elements in our company. It is based on Guidelines for Risk Based Process Safety（RBPS） for the purpose of properly managing risk sources than before.

On the intangible elements, we made a new risk assessment (RA) system for quantitative evaluation in order to manage properly risk source. In addition, to operate the RA system, we created qualification system to raise the knowledge and skills of employees who conduct RA.

With regard to the tangible elements, we are considering the practical use of advanced technologies in order to grasp the quantitatively risk which causes the leakage of hazardous material.I will introduce about details next.

592114 Process and adsorbent for separating ethanol and associated oxygenates from a biofermentation system

Abstract:

Abstract for poster presentation or talk:

Abstract Title: Process and adsorbent for separating ethanol and associated oxygenates from a biofermentation system

Presenter Information: Deepak Sharma; Sr. Process Engineer, Engineering and Technology, Bayer AG

Patent number: US8658845B2

Patent Authors:

Deepak Sharma - Sr. Process Engineer, Bayer Ag, St. Louis; (Presenting Author)

Anil Oroskar – CTO, Orochem Technologies, Napervielle, IL;

David W. House – Scientist, Orochem Technologies, Napervielle, IL;

Alice M. Havill – Scientist, Lanzatech, New Zealand

The principles of process intensification if used on existing processes can lead to development and redesign of existing processes to become more sustainable. This presentation will discuss a process and an adsorbent for the separation of ethanol associated oxygenates from a dilute mixture of ethanol and associated oxygenates in water in the presence of organic compounds derived from a biofermentation process. After pretreatment, the separation is carried out in a simulated moving bed adsorption system employing a stationary phase adsorbent comprising fluorinated carbon or modified C18 silica gel selective for the adsorption of ethanol and associated oxygenates, such as 2,3-butanediol, with a mobile phase desorbent selected from the group consisting of methanol, ethanol, propanol, and methyl tertiary butyl ether. The process is useful for removing water from dilute aqueous mixtures of organic compounds comprising ethanol in dilute concentration in water and produced by fermentation, biomass extraction, biocatalytic, and enzymatic processes which are not economically recoverable by conventional distillation methods.

Keywords: Separation; Adsorption; SMB; Purification

Patent Assignee: Orochem Technologies LLC

592111 Pump safety, the undiscussed topic in chemical industry leading to several fatal incidents

Abstract:

Abstract Title: Pump safety, the undiscussed topic in chemical industry leading to several fatal incidents

Presenter Information: Deepak Sharma; Sr. Process Engineer, Engineering and Technology, Bayer AG

Pumps are one of the equipment's in the process industry that require a thorough review during HAZOP studies. The constant source of energy given to the pump, if not converted to kinetic energy to flow the fluid, can lead to heating of the pumped fluid. If the fluid keeps accumulating the energy provided by the motor, in a blocked-in scenario, it can lead to very high temperatures eventually leading to loss of containment of the pump and the pump can get damaged. There have been instances in the industry where fragments of damaged pumps have flown several feet away from the point of loss of containment leading to injuries. The poster will outline a scenario where a pump is run blocked in on suction and discharge for water and a chemical which has a low thermal decomposition temperature. Arc data is used to evaluate the onset temperature of thermal decomposition. The poster will outline safeguards that can be employed to avoid catastrophic loss of containment of the pump. Outlined below are safeguards used to avoid loss of containment of a pump in case of a blocked-in scenario:

- Install PSV on suction of pump
- Employ a low amp/power interlock on the pump to turn off the pump in case of blocked in situation
- Have a high-pressure interlock on the discharge line of the pump to direct flow back to the tank in case of pump deadhead
- Install an Interlock to turn off the pump in case the feed forward valves (automatic valves) close inadvertently

Pump deadhead scenarios are not only catastrophic when in flammable service there are several salts that have low thermal decomposition temperatures. Pumps in water service left blocked in could also blow catastrophically leading to reportable incidents. I feel this is an important topic to bring to the attention of students and professionals in the chemical industry. Poster will include pictures of actual installed pumps along with required safeguards in the field along with supporting P&ID depictions.

589440 New Kind of Linear Nanocomposite Applied in Improving Rheological Behavior of Crude Oil and Its Mechanism Studied through Molecular Simulation Method

Abstract:

In this work, a new kind of linear nanocomposite produced by carbon nanotube(CNT) as pour point depressant (C2-100) was developed by blending with MCNT nanoparticle and copolymer named Ethylene-vinyl acetate (EVA). Its effect of reducing yield stress and modifying the morphology of wax agglomerate were also evaluated. Besides, the synergetic effect of C2-100 nanocomposite with the magnetic field was also conducted. The mechanism of modifying rheological behavior of waxy crude oil was studied through molecular simulation method, which aims at understanding the interaction between this nanocomposite and wax molecule. The results indicated that C2-100 nanocomposite could reduce the yield stress of waxy crude oil more effectively than the copolymer (EVA) even at low concentration, as well as changing the morphology of wax crystals and agglomerate. Moreover, its effect of reducing yield stress becomes more obvious when the alternate magnetic field was applied. In addition, the results of molecular simulation also presented that the effect of improving rheological behavior of C2-100 nanocomposite is related to the adsorption behavior of wax molecules on the surface of MCNT nanoparticle, which is adsorbed with EVA molecules. In addition, the adsorption behavior of different fragments of EVA molecules can also influence the adsorption behavior of wax molecules on the surface of EVA-attached MCNT nanoparticle, which may influence the effect of C2-100 to improve the rheological behavior of waxy crude oil.

589683 Solar Energy Conversion and Noise Characterization Sensing in Building Integrated Solar Photovoltaic Devices with Ventilation

Abstract:

A study on solar energy conversion and noise characterization sensing in solar photovoltaic devices with ventilation is conducted. A building integrated parallel plate solar photovoltaic (PV) device was built with a pair of PV modules, a ventilated air cavity and an insulating back panel of plywood board filled with polystyrene installed in an outdoor test-room. Photovoltaic amplification was attained with power output from a potentiometer through rotation of its circular knob. A parallel plate PV device was studied for its
electrical parameters as Resistance-Capacitance (RC) electrical analog circuit. The effect of inductive and capacitive heating losses was considered in evaluating electrical characteristics of a PV device exposed to solar radiation. The characterization of noise interference, due to power difference of two intensities in a wave on PV Device is presented. Noise characterization of a wave is obtained depending on type of wave. The difference of two power intensities is due to transmission of light, sound, heat, electricity, fluid, and fire into a parallel plate PV device. Based on analysis of a solar PV device, a definition with measurement expression and unit for an electrical noise were coined. Standard definitions of noise sources, their measurement equations, their units and their origins under limiting reference conditions were also devised. A review for sensing noise of physical agents is conducted for sensing noise of physical agents (sun, light, sound, heat, fluid, electricity and fire). With integrated environmental control, these environmental parameters are to be characterized through various environmental monitoring sensors.

589945 Structural Characterization and Model Representation of Vitrinite-Rich India Coal By Using Advanced Analytical Techniques

Abstract:

The detailed investigation of the chemical structure of Karharbari coal (vitritine rich subbituminous coal from India) by utilizing advanced instrumental analytical techniques such as solid state ^{13}C nuclear magnetic resonance (^{13}C-NMR) spectroscopy, Fourier transform infrared (FT-IR) spectroscopy, X-ray diffraction (XRD), Flash-pyrolysis gas chromatography mass spectroscopy (Py-GC/MS), Laser desorption ionization time-of-flight mass spectroscopy (LD-TOF-MS), X-ray photoelectron spectroscopy (XPS), and high-resolution transmission electron microscopy (HRTEM). The Karharbari coal was an amorphous loosely organized lower rank coal with aromaticity 0.79. This was indicated by XRD results and also confirmed with ^{13}C-NMR results (aromaticity =0.84) and FTIR results (aromaticity = 0.82). The average molecular weight of the aromatic cluster is 507 amu by NMR result which is closer to the result obtained by HRTEM (MW = 530.46 amu). The presence of different heteroatoms like carbon, oxygen, nitrogen and sulphur with their functionalities, were determined by using a XPS technique. Different carbon/oxygen forms present in the Karharbari coal are (C-C) and (C-H) 68.53%, (C-O) 23.46%, (COO^-) 1.96%, (C=O) 6.05% and different nitrogen forms such as pyridine, pyrrolic, quaternary, oxidized and their distribution (mol %) are 19.36%, 45.68%, 31.25%, 3.71 respectively, and different sulphur forms were present i.e. corresponding to thiophenes, sulphones, sulphuric acid and sulphates with the molar content of 16.47%, 41.61%, 21.37%, 20.55% respectively. The construction of the individual molecules will be discussed in detail in the rest of this paper. The individual molecules were geometry-optimized and Materials Studio Amorphous Builder module was used to construct a three dimensional associated structure.

589479 Life Management of High Energy Piping with Highlights from ASME B31.1

Abstract:

High energy piping (HEP) systems – or as defined by the ASME B31.1 Power Piping Code, Covered Piping Systems (CPS) – are critical assets with respect to production facilities. A failure can have catastrophic consequences from a safety perspective and in terms of equipment loss and downtime. In fact, loss of high-pressure steam due to a failure in a main run of steam supply piping could shut an entire production unit down. Because of this and the requirements of the Code, HEP maintenance and inspection programs are established to ensure that the integrity of the systems is maintained in a safe and reliable condition throughout their lifecycle. There is no one-size-fits-all for the life management of these critical assets because there is a great variety in age, design, materials, and owner objectives. Life management of these assets is beneficial in that it provides a structured, systematic approach to creating and implementing a program including evaluating the current program (whether formal or informal), what the objectives of the program are (short and long term), what equipment is in place, and its history.

A HEP program put in place for overall life management is not synonymous with an inspection program, although inspections are an important part of an overall program. Inspections should be used in conjunction with analytical tools, real time monitoring, and laboratory examinations. The overall approach is life management – not inspection management – of the equipment.

This paper describes a HEP life management approach which includes identifying locations of interest, risk-ranking those locations, implementing a plan for determining the health of the highest risk locations, and taking actions based on the system health that has been determined. This paper also presents and discusses some of the requirements of the B31.1 code, which at the highest level states that each operating company shall develop operation and maintenance procedures for piping systems deemed necessary to ensure safe facility operations based on the provisions of the code, relevant industry experience, the operating company's experience (and knowledge of its facility), and the conditions under which the piping systems are operated.

590131 Optimization of C3MR LNG Liquefaction Process Using a Parallelized Simulator

Abstract:

Controlling the cost of energy-intensive processes such as C3MR (propane/mixed refrigerant) LNG liquefaction is a prime opportunity for optimization. Our work combines different C3MR models to develop an optimization that uses multiple independent variables, relaxed constraints, and flowsheet alterations to achieve an optimal operating cost. This work uses a rigorous model of the process in a process simulator rather than a reduced model of the process response. Use of a first-principles process simulator increases the accuracy of the model and therefore allows a more realistic optimization, at the cost of increased computing time. Parallelization (using multiple CPU cores) in the process simulator greatly increases the speed of the optimization calculations. CHEMCAD's optimization tools and multi-core technology make it possible to easily define the optimizations, run the optimizations efficiently, and generate global optima with a high degree of confidence.

590196 Modifying Fischer-Tropsch Reaction Sequence with Addition of Zeolite to Iron-Based Activated-Carbon Supported Catalyst

Abstract:

In addition to previously presented scientific work, modified reaction sequence of Fischer-Tropsch (FT) process was studied by addition of Zeolite ZSM-5 to an iron-based activated-carbon-supported catalyst. Main reactions involved due to FT catalyst are Fischer Tropsch Synthesis (FTS) and Water Gas Shift (WGS) reactions. Main reactions due to ZSM-5 are cracking, oligomerization, isomerization, cyclization, aromatization, conversion of alcohols to hydrocarbons, and coke formation. The waxy and linear structure FT products are converted to gasoline range, lighter hydrocarbons, iso-paraffins, and aromatics using ZSM-5. The catalyst ratio of 1:2, FT catalyst: ZSM-5, is used at 300 PSIG and 300°F. Separate-bed configuration, i.e. FT and ZSM-5 are separated in a reactor, arrangement is used for the reaction sequence analysis. Alcohols with carbon-chain from 3-5 produced by FTS are converted into hydrocarbons over ZSM-5. Waxy hydrocarbons with carbon-chain greater than 23 produced by FTS are cracked and isomerized to small n-paraffins and iso-paraffins using ZSM-5. The hydrocarbon with carbon chain from 7-10 produced by ZSM-5 cracking reactions are converted to aromatic products over ZSM-5. Produced gasoline to kerosene range n-paraffins are further cracked into smaller chain (1-5) iso- and n-paraffins. Some of the iso-paraffins are oligomerized and aromatized over ZSM-5 to form more aromatics. Further, multiple aromatic rings combine to undergo cyclization and dehydrogenation reactions over ZSM-5 to form coke. Water produced as a byproduct of FTS is combined with some of the carbon monoxide, which is the WGS, to form carbon dioxide and hydrogen.

590427 Lateral Retention Force Required for Sliding Saline Droplets over Silianized Silicon Surface

Abstract:

The lateral retention force required to slide a liquid drop on a surface is expected to increase with the time that the drop rests on the surface prior to sliding till it reaches a plateau. The molecular reorientation of the molecules at the periphery of the solid surface subjected to the unresolved vertical component of surface tension force causes this phenomenon. In this system, the surface tension of deionized water is increased by adding NaCl and is expected that the lateral retention force is increased proportionally with concentration. The retention force increased with increase in concentration reaching a maximum only to drop to a minimum with concentrations closer to saturation. That is, the lowest sliding force is required for both lower and higher concentrations of NaCl. This sudden decrease in observed lateral force is not a function of surface tension and the surface on which the drop slides. This phenomenon is associated with adhesion of salt to the surface and is supported by the drop sliding motion and depinning of the liquid from the solid.

590433 Study of Droplet Motion across Polymeric Substrates By Centrifugal Adhesion Balance (CAB)

Abstract:

On a given surface, the contour of the drop changes accordingly based on factors of the force applied, chemical structure of the fluid, drop size, and surface topography. With increasing force that is exerted on the drop, we determine the comprehensive trend for drop motion with respect to the controlled factors mentioned above. Drop motion is categorized into four phases and the point at where the initial displacement of receding line is observed is defined as the onset of motion. In this study, the onset of motion is affected by the factors mentioned above and aids in the determination for another value correlation known as Gs, known as Interfacial Modulus. The resistance to interaction towards the droplet as observed is quantified with respect to change in surface topography. This study therefore characterized drop motion and specifically onset motion based on controlled variables (factors). Work of adhesion is also quantified across different substrates to characterize the detachment of liquid from its substrate. All these parameters such as Gs, onset of motion, contact angles were analyzed with different polymeric substrates, drop sizes, and various fluids used.

590665 Population Balance Modeling and Spatially-Oriented Control for Crystal Size and Shape in Crystallization Processes

Abstract:

Crystallization is an important separation and purification process with relevant applications in the food, pharmaceutical, fine chemical and fertilizing industries. In these processes, modeling and process control for crystallization are focused on crystal size distribution (CSD) and shape of the crystals in order to make it possible to achieve the desired quality and purity of the product of interest. The control of crystal size and shape is of considerable importance for industrial crystallization processes, as crystal morphology has great influence on downstream steps and also from the point of view of product specification criteria. The consideration of crystal shape in modeling of crystallization processes is also important due to its impact on product effectiveness, such as the bioavailability and tablet stability in pharmaceuticals. Control of the crystallization process is can be facilitated with online measurements, which provide information to be used in the modeling of the process. However, if online measurements are not available, the strongly nonlinear dynamics of these processes makes difficult to establish relations between some inputs (e.g. temperature) and the final desirable characteristics of the crystal product. In this context, image analysis has been shown to be a suitable technique to monitoring the process [1].

In our present work, a population balance model-based scheme for controlling the size and shape of crystals is presented. The modeling was performed using a two-dimensional population balance (2D-PBM) in terms of the two characteristic dimensions of the crystals, describing the evolution of crystal size and shape distributions along the batch time. To identify the nucleation, growth and dissolution kinetics parameters using the 2D-PBM approach along the length and width directions, QICPIC-LIXELL's high speed dynamic image analysis equipment was used for the online measurement of the CSD, shape parameters and aspect ratio distribution of crystals during the experiments. In this way, the developed model allowed for the prediction of the crystal size and shape distributions during growth-dissolution cycles.

The developed framework was employed to obtain optimal temperature policies for controlling the crystal size and shape distributions that meet target criteria for size, morphology and productivity using the dynamic image analysis along the runs. Based on the concept of using measurements along batch crystallization processes as a trajectory in a phase space [2], a trajectory-endpoint control problem is presented with an oriented control scheme that is applied to the shape of the crystals. The scheme was used to control the batch cooling crystallization of potassium dihydrogen phosphate. The reported experimental data demonstrate the application of the control policies for temperature to produce crystals of the desired average shape and size in different batch times. Thus, using the first principle modeling coupled to the developed trajectory tracking control, the proposed framework has the potential to be applied to industrial crystallization processes for shape control and to be extended to polymorph control investigations.

References:

[1] Eisenschmidt, H.; Bajcinca, N.; Sundmacher, K. (2016) Optimal Control of Crystal Shapes in Batch Crystallization Experiments by Growth-Dissolution Cycles. *Crystal Growth & Design*, v. 16, p. 3297-2803.

[2] Griffin, D. J., Grover, M. A., Kawajiri, Y., Rousseau, R. W. (2015) Mass-count plots for crystal size control. *Chemical Engineering Science*, v. 137, p. 338-351.

590450 Auditing the Process Hazard Analysis - Key Characteristics to Scrutinize in a Good-Quality HAZOP Study and Supporting Effective HAZOP Study Updates

Abstract:

Although Process Safety Management (PSM), requiring good-quality Process Hazard Analyses, was promulgated in the United States in 1992, and HAZOP Study techniques date back to the mid-1970's, the HAZOP Study technique is still not uniformly applied in a high quality fashion. The objective of this paper is to provide experience-based tips on Process Hazard Analysis auditing, and some key characteristics that the auditor can scrutinize to assure quality implementation. Analogously, a similar set of characteristics that can facilitate future HAZOP Study updates can also contribute to a quality HAZOP Study that supports the spirit of Safety Management System objectives for the identification of hazards potentially impacting plant personnel, the community, and the environment. This paper will also review tips for structuring the HAZOP Study to facilitate later updates and revalidation.

590342 Development of High Performance Organic-Based Cathodes

Abstract:

For lithium-based batteries, structure/property relations in terms of cell voltage are desired for new organic-based cathode active materials. In this initial pursuit of structure/property relations, ab-initio calculations were performed on a few organic-based molecules to show the effect on the theoretical cell voltage by inclusion of nitrogen and/or oxygen. Preliminary test results are presented on a newly developed molecule and methods to potentially improve on the cell performance are given.

590705 Electrochemical Kinetics of Degradation of Graphite Anodes in Lithium Ion Batteries

Abstract:

Growth of the solid electrolyte interphase (SEI) is a major driver of capacity fade in LIBs. Despite its importance, the fundamental mechanisms remain unclear, primarily because of the complicated reaction pathways involved[1–3]. SEI growth can be both electrochemical and chemical in nature[4], and thus, it is a strong function of the potential and degree of lithiation of the electrode. We model the early-stage and long-term growth of SEI by accurately capturing the potential dependence of its formation kinetics as well as long term rate limiting steps, and validating it against the world's largest open source battery cycling data, generated in-house[5]. This is done using the Multiphase Porous Electrode Theory (MPET) framework[6] on graphite (phase separating) and carbon black (non phase separating) particles.

Lithium plating is another key degradation phenomenon that has been elusive, and it becomes important while trying to fast-charge batteries, i.e., 0% - 80% state-of-charge in 30 mins. We show that lithium plating is a key function of electrode morphology, phase-separation dynamics and potential. Phase-separation in graphite is modeled in the electrode using the Cahn-Hilliard Reaction framework described by Bazant[7]. We understand the electrochemistry of the onset of lithium plating with in-situ measurements connected to real time cell potential in a phase-separating electrode for the first time[8].

Results indicate that the peak SEI-forming currents are higher for higher driving currents. Also, we find that SEI only grows during electrode lithiation, i.e. the battery only degrades while being charged. We also find that onset of lithium plating is correctly captured only when phase separation in active material is accounted for. Further, the onset of plating is delayed on electrodes with a thick SEI layer – understanding SEI/plating coupling is integral to predicting fast charging manufacturing protocols for LIBs. This work holds promise for the predictive design of procedures[9] for manufacture and formation of LIBs.

[1] Cohen, Y. S.; Cohen, Y.; Aurbach, D. Micromorphological Studies of Lithium Electrodes in Alkyl Carbonate Solutions Using in Situ Atomic Force Microscopy. J. Phys. Chem. B 2000, 104, 12282–12291. https://doi.org/10.1021/jp002526b.

[2] Horstmann, B.; Single, F.; Latz, A. Review on Multi-Scale Models of Solid-Electrolyte Interphase Formation, Current Opinion in Electrochemistry 13, 62-69 2019.

[3] Nie, M.; Abraham, D. P.; Seo, D. M.; Chen, Y.; Bose, A.; Lucht, B. L. Role of Solution Structure in Solid Electrolyte Interphase Formation on Graphite with LiPF6 in Propylene Carbonate. J. Phys. Chem. C 2013, 117 (48), 25381–25389, https://doi.org/10.1021/jp409765w.

[4] Das, S.; Attia, P. M.; Chueh, W. C.; Bazant, M. Z. Electrochemical Kinetics of SEI Growth on Carbon Black: Part II. Modeling. J. Electrochem. Soc. 2019, 166 (4), E107–E118. https://doi.org/10.1149/2.0241904jes.

[5] Attia, P. M.; Das, S.; Harris, S. J.; Bazant, M. Z.; Chueh, W. C. Electrochemical Kinetics of SEI Growth on Carbon Black: Part I. Experiments. J. Electrochem. Soc. 2019, 166 (4), E97–E106. https://doi.org/10.1149/2.0231904jes.

[6] Smith, R. B.; Bazant, M. Z. Multiphase Porous Electrode Theory, J. Electrochem. Soc. 2017, 164 (11). https://doi.org/10.1149/2.0171711jes.

[7] Bazant, M. Z., Theory of chemical kinetics and charge transfer based on nonequilibrium thermodynamics, *Accounts of Chemical Research*, *46*(5), 1144–1160. https://doi.org/10.1021/ar300145c

[8] T. Gao, Y. Han, S. Das, T. Zhou, D. Fraggedakis, N. Nadkarni, C. N. Yeh, W. Chueh, J. Li, M.Z. Bazant, Interplay of lithium intercalation and plating on graphite using in-situ optical measurements, submitted.

[9] Huang, W.; Attia, P. M.; Wang, H.; Renfrew, S. E.; Jin, N.; Das, S.; Zhang, Z.; Boyle, D. T.; Li, Y.; Bazant, M. Z.; McCloskey, B. D.; Chueh, W. C. and Cui, Y.; Nano Letters 2019 19 (8), 5140-5148. DOI: 10.1021/acs.nanolett.9b01515

589787 Model Development of Solid Oxide Fuel Cell Thermal Performance Using Artificial Neural Network

Abstract:

NASA's Johnson Space Center has recently begun efforts to eventually integrate air-independent Solid Oxide Fuel Cell systems, with landers that can be propelled by LOX-CH4, for long duration missions. Using landers that utilize such propellants, provides the opportunity to use such systems as a power option, especially since they will be able to process methane into a reactant through fuel reformation. To ensure fuel reformation in the systems, Steam Methane Reformation (SMR) will be employed. Various lead-up activities, such as hardware testing and computational modelling, have been initiated to assist with this developmental effort including the development of dynamic empirical models using artificial neural network. Such a modeling approach has shown to be accurate ($R^2>0.9$) to predict the thermal and electrical performance for SOFC systems. Hence, such models could be used for system optimization and control design.

Impact of Spacer Nature on Rheological Properties of Cationic Gemini Surfactant

Abstract ID: 589650

Shams Kalam[1], Muhammad Shahzad Kamal[2], Shirish Patil[1] and S. M. Shakil Hussain[2]

(1) KFUPM, Dhahran, Saudi Arabia

(2) Center for Integrative Petroleum Research, King Fahd University of Petroleum & Minerals, Dhahran, Saudi Arabia

Surfactants play a vital role in Chemical Enhanced Oil Recovery (cEOR) techniques. Gemini surfactants are a unique style of surfactants having more water-soluble hydrophilic head and oil-soluble hydrophobic tail units. Both groups are connected by a spacer at or near to the head groups. Polymers are added into the surfactants to increase the sweep efficiency of the flooding processes. This paper presents the rheological properties of Surfactant Polymer (SP) solutions in which two recently synthesized cationic gemini surfactants with different spacer nature are used.

The rheological properties of novel SP solutions were determined using Discovery hybrid rheometer (DHR-3) from TA Instrument. The geometric configuration set for this work was concentric cylinder. All experiments were carried out at 80 °C, while the shear rate for steady shear viscosity measurements ranges from 0.001 to 1000 s^{-1}. 0.25 weight % of a polymer FO 4290 SSH was used with different concentrations of newly synthesized gemini surfactants for Rheological measurements. Two hours magnetic stirring and twenty hours retention time were provided to make uniform surfactant polymer solutions.

Both surfactants exhibited good thermal stability in seawater with no precipitation. Shear viscosity and storage modulus were measured as a function of shear rate and angular frequency respectively at different surfactant concentrations ranging from 0 to 0.5 mM, where 0 mM shows pure polymer. We discovered that counter ions also influence the rheological properties significantly. Research outcomes revealed that the novel SP solution with a benzene ring and chlorine counterion produces better performance in comparison with benzene ring and bromine counterion. Moreover, the effect enhances if we increase one more benzene ring in the spacer.

This study reports two recently synthesized SP solutions for cEOR processes in carbonate reservoirs. It presents the effect of counterion on rheological properties and improves the understanding of tuning the surfactant structure in making SP solutions to increase the flooding sweep efficiency.

590257 Real Time Energy Management Systems Implementation for Continually Optimized Performance

Abstract:

It is well known that energy management is key for the optimal operation of refining and petrochemical sites. Industrial energy systems are constituted by complex utilities networks that are interconnected and have complex interactions that in order to achieve it best performance need to comply with increasing environmental rules, push against operating constraints and adapt to altering power prices and rapidly changing equipment availability.

After the initial installation, management and continuous improvement of these systems present challenges not only from a technological but also from an organizational point of view.

The implementation of an advanced digital Energy Management System (EMS) which includes capabilities for multi-period optimized scheduling, real time optimization and monitoring, deployed either as cloud-based or on-premise solutions that could be acting on open or closed loop is known to provides large benefits.

Critical is the sustainability of the EMS in the ever-evolving energy supply and demand of refining and petrochemical sites. Commercial, asset and organizational factors all change and the EMS needs to adapt to the developing environment of each enterprise meaning the type of implementation, including appropriate management structure, culture, best practices and Key Performance Indicators (KPIs) needs to be continually improved.

The EMS implementation at the large and complex Eastman site in Longview, TX, will be presented as an example. It will provide insight on the initial successes achieved in real time energy management for both, cost optimization and energy related KPIs performance tracking. The case study evolves to discuss how technical asset improvements, once replicated in the EMS, provided alternative operating modes, large additional benefits and sustained optimal

Andrew Lammers, NN2

Eastman Longview

Pablo Montagna*, Cristian Rojas

KBC Advanced Technologies, Inc

589818 Analysis of Fuel Cell System through Project Based Learning for Application of Classical Thermal Fluid Concepts in Mechanical Engineering Laboratory Course

Abstract:

A hands-on project involving analysis of air noncircular channels in Fuel Cell to measure the pressure was introduced in a thermal fluid laboratory course. Although the students studied fluid mechanics as a theoretical course, this project gave them the opportunity to build an experimental setup from scratch to help showcase their hands-on skills and to apply their understanding of relevant concepts of fluid sciences. The results of the project included a successful test of a conceptual procedure, designed to measure and calculate pressure drop and flow rates across the anode and cathode sides of a fuel cell. The students enjoyed completing and learning from the project. They also provided very useful feedback on the project by incorporating a functional experimental system for the future laboratory project. This project based learning opportunity allowed students to build their confidence in what they are learning as well as provide a glimpse of what the real world may hold.

589438 Measured Versus Calculated Sour Gas Properties

Abstract:

Measured versus calculated sour gas properties
Abstract:
Recent Environmental agencies have restricted standards for sulfur emission from gas chemical processing. Sour gas can extremely impact the delivery and processing by corroding the equipment. Gas properties such as compressibility, density and viscosity are a major factor controlling prediction of future production pipelines, and production tubing and pipelines in gas industry.
Sour gas effect on all gas industry aspects usually gas properties are measured in laboratories but occasionally experimental values are not available due to complexity and time consuming hence a reliable method would be proposed.
We are assessing the accuracy of the industry standard methods in calculating sour gas properties, compressibility factor using the following correlations:
Drunchuck - Abukkasem, Hall – Yarbarough, Drunchuck- Purvis- Robinson, Londono et-al, Alanazi et-al and Azizi et-al.
Viscosity using empirical correlations such as Carr-Kobayashi Burrows, Lea-Gonzalez –Eakin and Dean- Stiel.
Density using equations of state Redlich Kwang and Soave -Redlick - Kwang.
This study shows that the best correlation for calculating compressibility factor and density of high concentration sour gas is Londono-et al, all other considered methods are not accurate. All other methods investigated include high error especially for high H2S concentration.
However the error in estimation of Z factor for most of the considered methods has great impact on estimated sour gas density. Therefore we recommend that sour gas properties be measured experimentally at high pressure and high temperature conditions.

587967 The Study of a S-CO2 Oxy-Fuel Combustion Power Plant for Low Grade Coals

Abstract:

Recently, it is very important to have reliable solutions that can maximize power generation efficiency of current power generation processes and decrease emissions of CO2 and other pollutants because there are increasing concerns about the global warming and the depletion of fossil fuels. Among the available carbon capture technologies and power generation technologies, a S-CO2 (supercritical carbon dioxide) cycle based oxy-fuel combustion process appears to be a promising mid-term solution for the production of electricity from fossil fuels. Since supercritical carbon dioxide (S-CO2) involves a rapid density change at small temperature and pressure ranges, the S-CO2 cycle can extract large amounts of energy from relatively small equipment and it is possible to minimize or simplify a power generation block and to obtain higher thermal efficiency than existing steam cycles used in power generation plants. Therefore, an indirect S-CO2 oxy-fuel combustion power plants for low grade coals was designed and analyzed by using ASPEN-PLUS. Also, the operating characteristics of the indirect S-CO2 Oxy-fuel combustion power plant were compared with those of conventional air combustion power plants.

590099 Characterization and Performance Assessment of Oil Entrapment By Commercially Available Surfactants

Abstract:

Oil spills occur frequently and are a threat to the health and safety of human society and the well-being of the environment and ecosystem. Remediating oil spills is an ongoing challenge. Surfactant based dispersing agents, such as the Corexit® product line, are widely used for remediation of oil spills mainly through direct application on top of the oil slick. This work encompasses the investigation of variables that influence the emulsification process and therefore the encapsulation of oil within the surfactant micelles. Of central importance are the conditions under which oil entrapment in emulsions even occurs. A central finding is that there appears to be a threshold shear rate for emulsifying oil with surfactants, below which no emulsification occurs. Different schemes for applying shear stress in this work included simply stirring, using a high energy colloid mill, and flowing the mixture through a high shear capillary. Also, the sequence in which the oil/surfactant/water (O/S/W) components are mixed impinges strongly on whether and how much oil entrapment occurs. To try to correlate shear stress and velocity profiles throughout the sample volume with entrapment efficiency and kinetic stability of these O/S/W systems, Computational fluid dynamics (CFD) simulations were conducted. Another important factor is the ionic strength of the aqueous medium, which greatly impacts kinetics of these emulsions, but with a smaller impact on the threshold at which any emulsion is obtained at all. Surfactants tested include Corexit® EC9500A, sodium dodecyl sulfate (SDS), Triton X-100, and cetyltrimethyl ammonium bromide (CTAB). The quantification of oil entrapment was performed through either thermal gravimetric analysis (TGA) or differential scanning calorimetry (DSC).

Designing a Native Model for Assessment of the Effectiveness of Advertising

Sorush Niknamian
PhD, military Medicine Department, Liberty University, USA
Email: so.niknamian@gmail.com

Abstract

Reviewing and evaluating the effectiveness of advertising can be used as a criterion to convert weaknesses into strengths and budgeting. Also, Identification of the factors influencing the effectiveness of advertising can lead to more professional design and implementation. In the banking industry, with the increasing number of banks and increased right of customers to choose, the role of advertising has become increasingly important to succeed in this field of competition; so it should be said that advertising can make the bank close to its goals when it is sufficiently effective. The current research aimed to identify and rank the effective factors on the design of a native model for evaluation of the effectiveness of advertising of the Eghtesad Novin bank in Iran. The statistical population of this research is all managers and experts of the Eghtesad Novin bank. According to the research literature, the criteria and sub-criteria affecting the effectiveness of advertising have been extracted. By using the fuzzy Delphi technique and based on the experts' opinion, the final indices were extracted. Then, using the fuzzy AHP method, the rank of criteria has been addressed. The derived criteria are demographic characteristics, individual emotions, and attitudes of the customer, apparent features of the message, message content, market characteristics, cultural, economic features, design and nature of the service, advertising methods. The results of the ranking showed that the methods of advertisement, the design, and nature of the service, the apparent characteristics of the message, the emotions and attitudes of the customer, the demographic characteristics, the content of the message, economic, market characteristics, were placed in the first to ninth ranks, respectively.

Keywords: banking services, banking services marketing, advertising, advertising effectiveness,

Introduction:

Advertising is one of the promotional mix elements of marketing mixes, which play an important role in attracting customers. Advertising is a bridge between the organization and customers and can introduce the organization and its products to customers, inform consumers about the goods and services of the organization, convince them about the benefits of the product, and ensure them to have the best choice even after the purchase when the customer is uncertain (Buck, 2016: 172).

Given the increasing competition among the brands in the marketing sector, advertising is a necessary tool to introduce the characteristics of goods and services to customers. Leong et al. (1999) pointed out the fact that consumers overcome advertising information, and many of them do not remember the advertisements that they are exposed to, and it is necessary to consider the advertising and its attractive methods to attract customers' attention. The ability to attract customers' attention is one of the main indicators of advertising effectiveness that should be taken into consideration. The attention of customers is a key component of the overall effectiveness of advertising (Pileliene et al., 2017: 2). Any professional marketing without advertising will not be effective. Despite relatively high advertising costs, advertising is mentioned as an investment rather than the cost. Since many advertisements cause that the advertising companies do not benefit their advertising because of the wrong way of advertising, the competitive companies benefit from the advertising of that company (Hartnet, 2016: 23)

Advertising should be in line with the goals and plans, as well as the strategies of the organization. Therefore, we should see to what extent the advertising has met such goals. Allocating funds without evaluating the effectiveness of past advertising is the waste of resources of the organization. By evaluating the effectiveness of advertising, changes can be made to the form, the content of messages, type of media, budget, communication channels, and even the time and conditions for conducting advertising to create more useful and effective advertising than the past (Yazgirudo-Yasteh, 2015: 359). In the current competitive world, the importance of advertising is increasing. As a result, businesses have devoted a great deal of money to customers' attraction, but it is important to see whether the company advertising has been designed and implemented in such a way to meet the goals of the company. This issue has been often neglected in our country. (Kim, 2014: 2107).

The amazing effect of advertising on the lives of individuals requires that this scientific, social and communicative branch be carefully studied and advertising is made with a scientific, artistic, specialized and expert view. Professional advertisement can be a sharp sword for sales promotion, both inside and outside the country, and otherwise, it is a waste of time, money and energy. A professional advertising lever can, with the help of new technology, penetrate humans, and cause more desire for consumption (Arribeh, 2016: 463). Today, banks have turned their attention to the widespread use of mass media advertising, given the highly competitive structure governing on them. Also, given the relationship among a wide range of businesses of organizations and households and banks through various deposit accounts, loans, and other financial services, banks can benefit from advertising in a proper way to meet their goals, namely the introduction of their services, the communication with customers, their effectiveness on them in choosing a bank to do their banking business and ultimately attracting them. Unfortunately, in most banks, despite the massive advertising and the allocation of large amounts of energy and budget, the effectiveness of such advertisements is less evaluated. Since advertising is one-way, it is necessary to receive feedback from the customers (Rabiei, 2011: 19).

Regarding marketing, the effectiveness of advertising depends on the wide range of coverage of various consumers and the cost of advertising campaigns. Economists have defined the effectiveness of advertising as a change in the profit derived from promotional activities. Therefore, the effectiveness of bank advertising is the potential quality of information reported to consumers, and in promotional campaigns, it aims at creating a bank image and informing banking products and services, as well as obtaining a specific economic result before and after the transmission of the message (Denis et al., 2012: 85).

Eghtesad Novin Bank (EN bank), which is one of the oldest banks in Iran, due to the addition of new banks, should use new tools to protect and even increase its market share for the assessment of the effectiveness of its advertising. For this purpose and given that the bank uses various advertisements, it is necessary to design a model to evaluate the effectiveness of the bank advertising; therefore, the present study seeks to answer the following questions:

- What indicators are used to assess the bank native advertising model?

- How are the indicators for bank advertising evaluation weighed and ranked?

In this research, we have tried to identify the criteria using the Fuzzy Delphi process; then we will use the Fuzzy AHP method to rank the identified criteria. In the next section of the research, the literature has been studied. Then the research design has been expressed, and in the last section the conclusion has been presented.

1-Research literature:

1-1-Marketing of banking services

The growing trend in service activities has led to the emergence of new markets and large service companies. The complexity of service industries, such as information systems, management consultancy, and logistics, has increased, as service providers have entered the market with a variety of business backgrounds and capabilities to provide a broader range of services. Facing this complexity requires knowledge and skills to understand the services. The definition of the service includes all activities that have been done (for example, writing document specifications) (Selviaridis et al., 2013: 1398).

In the industrial literature, there are various ways to define the service and product requirements. Van Weele (2010) has distinguished technical and applied definitions from each other. The first requirement expresses the exact technical specifications of the goods with the required services and inputs while it focuses on performance demands and demand-oriented performance orientations. In this regard, Axelsson and Wynstra (2002) have proposed four methods for definition of services: input definitions are based on the sources and capabilities of the service provider while the process-oriented approach focuses on how services are presented regarding activities and processes. Output-oriented definitions focus on service capabilities and performance levels, as well as the interaction between the suppliers and recipients. Ultimately, they focus on stress-oriented results in the economic value of services (Selviaridis et al., 2013: 1399).

The financial sector forms an essential part of the service sector. In this regard, banks have used marketing to improve the strategic performance of banks. In particular, since the 1980s, this trend has been confirmed and criticized. Marketing researchers and experts are interested in researching the various aspects of marketing in banking. The quality of services is essential for customers who use banking services, and the provision of accurate and high-quality services leads to the survival of banks in the competitive market. Improving the quality of services will increase the access and response to the demands of the bank's customers, which will make the bank more distinct than other competitors. Therefore, measuring the quality of services in the banking industry is essential. Providing quality services is a hard and difficult task, in this regard, the features of services in general and banking services, in particular, should be explained. Banking services require a high degree of appropriate communication between service providers and customers. If services are not provided properly, they will be at high risk.

About the five features that distinguish them from material services (intangibility, non-allocation of materials for production, diversification, non-corruption, and lack of ownership), financial services have two other specific features, that have been firstly developed by McKechnie (1997). These characteristics include trusteeship and two-way information. These special features mean that the evaluation of services is more difficult than material products, and they not only create a challenge in the development of marketing activities, but also they are the source of the problem in providing an optimal level of quality services. Banking services require a high level of communication between service providers and customers, as well as a high degree of customer participation in the process of providing services. Angurr et al. (1999) emphasized that due to the high standardization and the development of technologies in the banking sector, human factor has been considered as the most important one in making distinctions among competitors. The expertise and motivation of the employees, as well as the level of customer involvement in the provision of services, significantly affect the quality of services. Due to the wide range of banking activities, attention should be paid to cultural and social changes that affect the attitudes of customers (Durdana et al., 2015: 100).

Marketing of banking services has three components as follows: if any of them were properly executed, banking services would change the country.

1- Market study: Identification of the banking services market system, including the status of current banking services and their weaknesses, strengths, expectations, demands, ideas and opinions of customers, the conditions of environmental instruments affecting the market system, in particular government policies and governing laws, and even global changes in the banking system and successful models in this system. Indeed, market study is the foundation of the evolution of the banking system, because market study (which is the result of constant communication with the market and conducting marketing studies and research in various forms) leads to comprehensive, up-to-date, relevant and necessary information for decision-makers and bank managers (Munsung Rheea & Satish Mehrab, 2006: 508).

2. Market creation: It involves the use of a variety of simple and complex marketing techniques and competitive strategies, as well as the growth, development and market creation in the banking system; it needs to understand the internal customers (employees) and external

customers, to communicate with them and meet their needs and demands. Market creation in thc competitive environment requires appropriate strategies (Moch, 2013: 2908).

3. Market protection: One of the most important tasks of marketing in the banking system of the country is to maintain market share, a stable relationship with customers, and to convert ordinary customers to loyal and supportive ones. The more intense the competition, the more valuable the loyal customers will be, as marketing research in the world has shown that the cost of finding a new customer in firms is 3 to 33 times higher than the cost of maintaining current customers regarding the type of products and services. Therefore, market protection prevents the costly advertising, and attraction of new customers and provides a kind of confidence and serenity for market survival, especially in competitive conditions (Fredriksson & Moro, 2014: 69).

1-2-Effectiveness of bank advertising

One of the indicators of success in any organization is the extent to which the organization achieves its goals. Here, the success of any organization is determined by its own goals. This approach has been called effectiveness in the management literature. The effectiveness is the degree to which the organization achieves its goals. Goals must first be defined to measure effectiveness in any organization. Then, plans are set up to fulfill these goals, and thus the plans will be implemented. In the end, the results are compared with predetermined goals (Manjegani et al., 2015: 124).

About 100 years ago, William Heketh statement, "Half of the advertisements were wasted, we just do not know which half it is" attracted everyone to the accountability of advertising. What matters to advertisers is that how much the amount of money spent on advertising has returned? Has the cost spent on making and broadcasting the advertising been effective? Has the effect of advertising been strong or weak? Knowing the answer to these questions is important to the advertiser because today companies pay much money to make and broadcast advertising. According to Jack Trout, the money we use to advertise is very much. Every second of the World War II was $ 9,000, and Vietnam cost $ 22,000 per minute. But in the American Football Final, every minute of advertising cost $ 2 million, so the advertiser has the right to know about the effectiveness of his advertising (Aksoy, Atilla, 2005, Yeni Reklamcilik, Istanbul Bilgi Universitesi, 95). It is very difficult to measure the impact of long-term advertising on many occasions. Also, academically in the world of advertising, it is impossible to create a comprehensive theory that can be used to determine the effectiveness of advertisements on people at different levels under different circumstances. The effectiveness of media is its capacity to create desired results, or special conditions in a particular position and content, that is, attaining the desired communication goals (Wery, 2012)

Evaluating the effectiveness of advertising indicates how much advertising has succeeded in achieving the goals. The evaluation of the effectiveness of advertising acts as a feedback, informs us about the weaknesses and strengths of our advertising by which we find our shortcomings in the advertising. The marketer may choose an appropriate stage regarding product life or brand position, and use it to define his advertising goals and ultimately evaluate its effectiveness (Eckler & Bolis, 2011: 2).

The effectiveness of advertising is a function of its content (message), the execution (how it is transmitted), and repetition (the number of times each consumer sees the message) (Cutler, 1997, quoted by Moorthy & Hawkins, 2005). Some of the effectiveness criteria are the call and recognition of the brand name (Park, 2006), awareness, willingness and intention to purchase (Decrop, 2007), attitude towards the ad, attitude toward the brand name and intent to purchase (Muda et al., 2014)

The control of the communication element has a direct effect on customer perception of different dimensions of the brand name, such as service quality and loyalty. It means that the elements of communication, advertising and promotion must be carefully implemented and controlled (Suvittawat, 2015). Informative advertising is one of the factors that shape the behavior of the bank customers. Studies conducted both on attracting, retaining and loyalty of customers, show the need to examine the impact of advertising on attracting and loyalty of customers (Ioannou, Boukas and Skoufari, 2014: 25).

An ideal model of world-class advertising can: 1. transfer a consistent meaning to the consumer all over the world. 2. Transfer positive global capacity for the brand. That is, the results of advertising among different peoples should produce a positive effect and transfer high-quality signals (Strebinger et al., 2018: 410).

Advertisers want to prepare consumers to like goods or services, but sometimes it harms them through unwanted and intrusive advertising. It leads to the creation of a series of negative emotions in advertising. The purpose of ads is to create positive feelings in consumers about advertised goods and services (Bell & Buchner, 2018: 1).

On the other hand, research shows that commercial advertising can be positively evaluated and considered, and this positive evaluation leads to a positive evaluation of the brand with the product. However, advertising can have negative effects that lead to the formation of negative attitudes towards that product and service. Therefore, by increasing the volume of messages, it can have positive, negative or neutral effects. Therefore, marketers should evaluate advertising messages and their behavioral outcomes. With this regard, the effectiveness of advertising, based on a hierarchical approach, should be fully investigated (Tafani et al., 2018: 125).

There are various factors which influence effectiveness of advertisement like media, time, market life stage, advertisement life stage (new or old advertisement) and also advertisement creative signs (like content and administrative elements) (Talis et.al, 2005). Also using popular cartoon personages (Ranjbarian, 2009), size (Rousat, Venous & Ebrahimi, 2004, Baltas, 2003), color (Park, 2008, Moriarty, Mitchell, Wells, 2009, Cheng & Kao, 2011), complexity or simplicity of the advertisement (Morrison & Dainoff, 1972, Di Res, 2008), using irony factor (Mukherjee, Hober, 2004), using famous artists (Rousta, Venous, Ebrahimi , 2004) and consumer features (Matrin, Phaymi, Agee, 2002, Park, 2008) are among these variables. Advertisement repetition is also a factor which helps in reminding its content to the audiences (Rousta, Venous, Ebrahimi, 2004). One of considered factors in the debate of evaluating effectiveness of advertisement is the attitude towards advertisement. Attitude towards the advertisement includes public interest or lack of interest in an advertising stimulus in encountering a special topic (Moaen & Minor, 2009). Mental imagery (B one & Elen, 1992), content transparence and emotions called through the advertisement are among variables influencing the attitude towards the advertisement. Some of these factors may lead to strong impacts while some of them may create weak impacts or don't include any impact. Awareness of the impact of each of these factors makes a considerable help in making management decisions (Talis et.al, 2005).

Fotini et.al (2009) provided a model of effective advertisements for advertising websites of big companies. The results of the study showed that measurements, activities, priorities and intentions of people have a considerable impact on evaluation and qualitative factors of relation-oriented marketing have a great impact on advertisements effectiveness indicators. The extracted criteria include:

- Advertisement policies: product information, company information, product general display, consistency with traditional advertisement

- Medium features: message features, information structure, responding time, facilitating research, beauty, interactions, communications in real-time, amusement, information control

- Relation- oriented marketing policy: communications, feedback, supporting customers

Today, organizations for competition in the business world and selling their goods and services require purposeful and efficient marketing activities. One of topics which has a very important role in the process of marketing of an organization and assign a high volume of budget to it is advertisement. Advertisement causes some changes in the knowledge, attitude and behavior level of audiences, so attention to evaluating advertisement effectiveness is a necessity for the organizations that consider advertisement as requirement of continuous life, growth and development. By assessing advertisement effectiveness and investigating consistency with the organization objectives and its expected impacts, some changes may be created in its dimensions, content and methods so that a more effective and useful advertising is performed. Regarding the research literature and internal and external background which has been stated, here a general categorization of the model proposed criteria has been provided in the following table:

Table 1- summation and extracted variables

scholar	sub standard	criterion
) Dargi & Mohammadian (2010), Ashouri (2016), Mansouri Moayed (2005), Aghapour (2016), Mirzadh & Mehrmanesh (2016), Keramforn (2011), Mac Ki- Nesbiat et.al (2011), Rosemary et.al (2012), Wang (2015), Maryion et.al(2017)	awareness of sexuality differences	individual factors
Dargi & Mohammadian (2010), Ashouri (2016), Samsam Shariat Keramforn (2011)(2007), Mirzadeh & Mehrmanesh (2016), Mac Ki- Nesbiat et.al (2011), Rosemary et.al (2012), Wang (2015), Maryion et.al(2017)	attention to age requirements	
Dargi & Mohammadian (2010), Dehdashti Shahrokh (2009)	various consumers (permanent, ordinary and case consumers)	

Khashizadeh (2010), Danaei & Dasar (2016), Pillay et.al (2012), Marion et.al (2017)	advertisement richness	
Danaei & Dasar (2016), Dargi & Mohammadian (2010), Pillay et.al (2012), Olson et.al (2012)	advertisement accuracy	
Ashouri (2016), Rabiei (2011), Taeini Harari (2014)	positive mentality of the organization	
Dargi & Mohammadian (2010), Shahmohammadi (2011), Olson et.al (2012)	honesty feeling in advertisement	
Fakhari et.al (2012)	threatening feeling	
Ashourizadeh (2016), Hamidizadeh (2013)	attitude towards brand	
madhoushi (2013), Olson et.al (2012)	being convinced	
Ahmadi (2013), Mac Ki – Nesbiat et.al (2011), Olson et.al (2012)	convicting audiences	
Ashouri (2016), Hosivar (2014), Olson et.al (2012)	mental imagery	
Siadat & Sakhi Jourashk (2014)	positive feeling	
Navaei et.al (2015), Rozmari et.al (2012)	audience being active	
Mosavi Esmaeilzadeh (2015), Olson et.al (2012)	knowledge level	
Dargi & Mohammadian (2010), Ashouri (2016), Shafiei Nikabadi & Zarei (2017)	attitude towards advertisement	
Dargi & Mohammadian (2010), Hosseinzadeh et.al (2016), Sanei et.al (2016)	familiarity with methods and forms of advertisement	message
Ashouri (2016), Olson et.al (2012), Alan (2012), Patric Ro (2014)	attention to reduction of expectation time of the user for observing advertisement	
Danaei & Dasar (2016), Forentebury & Mac Gldric (2011)	beautiful and appropriate	

	combination of colors	
Aghapuor (2016), Sanei et.al (2016)	site and location of advertisement placement	
Hassanzadeh & Danaei (2016), Sanei et.al (2016)	access to sufficient and required information	
Dargi & Mohammadian (2010), Darzban (2016)	starting the advertisement with a challenge	
Navaei et.al (2015), Salimi Fard & Pourdehghan (2016), Bargar et.al (2016)	starting the advertisement with a statement and explanation	
Dargi & Mohammadian (2010), Mirzadeh & Mehrmanesh	using helping words and terms for decision making	
Dargi & Mohammadian (2010), Daneshvar et.al (2016), Alan (2012)	using stimuli with diversions more than usual and expected extent	
Sanei et.al (2016)	using images of women and children	
Sohrabi Ranani (2016), Fortenbury & Mac Goldric (2011)	creativity in the message	
Shafei Nikabadi & Zarei (2017), Denis et.al (2012)	the text of advertisement message being a predicate sentence	

Khashizadeh (2010), Dargi & Mohammadian (2010), Sohrabi Ranani (2016), Patric Ro (2014), Abdolsalam et.al (2016)	continuation and repetition of message	
Daneshfar & Nabizadeh (2015), Fortenbury & Mac Goldric (2011)	using trustable and famous people	
Salimifard & Pourdehghan (2016)	common dialects and terms	
Daneshfar & Nabizadeh(2015), Forentnburi & Mac Goldric (2011)	loveliness of the popular person for the observer	
Daneshfar & Nabizadeh (2015), Forentnburi & Mac Goldric)(2011	familiarity of famous person with the observer	
Mansouri Moayed (2005), Aghapour (2016)	emotional music	
Khonsazadeh (2010), Hosseinzadeh et.al (2016), Aghapour (2016)	using various methods	
Siadat & Sakhi Jourashk (2014), Sohrabi Ranani (2016)	high coverage and being universal	
Navaei et.al (2016), Mosavi Esmaeilzadeh (2015), Hosseinzadeh et.al (2016), Mirzadeh & Mehrmanesh (2016), Rosemary et.al (2012), Mikit et.al (2012), Spenkar (2013)	media advertisement	
Navaei et.al (2015), Mosavi Esmaeilzadeh (2015), Hosseinzadeh et.al (2016), Ashouri (2016), Hassanzadeh & Danaei (2016), Mikit et.al (2012)	multimedia advertisement	
Navaei et.al (2015), Mirzadeh & Mehrmanesh (2016)	radio advertisement	
Navaei et.al (2015), Mosavi Esmaeilzdeh (2015), Ashouri (2016), Hassanzadeh & Danaei (2016)	cell phone advertisement	
Navaei et.al (2015), Hosseinzadeh et.al (2016), Mirzadeh & Mehrmanesh (2016)	creating TV advertisement	
Ahmadi (2013), Salimi Fard & Pourdehghan (2016), Hoang et.al (2013)	attraction of rewards	
siadat & Sakhi Jourashk (2014), Koumra et.al (2016)	being benevolent	

Darzaian (2016), Zabkouich & Louksten (2011), Abdolsalam et.al (2016)	coordination in speaking and behavior	
Ahmadi (2013), Salimi Fard & Pourdehghan(2016), Honari et.al (2015)	reality of messages	
Madhoushi (2013), Salimi Fard & Pourdehghan (2016), Houang et.al (2013)	graphic-designing	
Shahmohammadi (2011), Madhoushi (2013), Salimi Fard & Pourdehghan(2016)	psychological – geographical division	
Mirzadeh & Mehrmanesh (2016), Zabkouich & Lokesten (2011), Menroz et.al (2012), Choubhai et.al (2013), Mokan & Oltz (2014), Ayono et.al (2014)	advertisement on web	
Dehdashti Shahrokh (2009), Shahmohammadi (2011), Mirzadeh & Mehrmanesh (2016), Menroz et.al (2012), Choubhai et.al (2013), Mokan & Oltz (2014), Ayono et.al (2014), Henri et.al (2015), Yazghiro- Basteh (2015), Lin (2016)	internet & cell phone advertisement	
Shahmohammadi (2011), Sanei et.al (2016), Menroz et.al (2012), Chobai et.al (2013), Ayono et.al (2014), Henri et.al (2015)	selecting advertising media	
Shahmohammadi (2011), Sanei et.al (2016), Howang et.al (2013), Henri et.al (2015), Yazghirdo-Yasteh (2015), Martina Santa et.al (2017)	artistic aspect of advertisement	
Kashizadeh (2010), Hoshivar(2014), Pelpin et.al (2017)	gender of speaker in the advertisement	
Hosseinzadeh et.al (2016), oribeh (2016), Madra et.al (2017)	text repetition	
Shahmohammadi (2011), Hassanzadeh & Danaei (2016), Abri & Joumar (2014), Ayouno et.al (2014), Oribeh (2016), Pachman (2016), Bak (2016), Madra et.al (2017)	advertisement	
Samsam Shariat (2007), Sohrabi Ranani (2016), Daneshvar et.al (2016), Bak (2016), Kiji-Kali (2017)	factors relating to usage	environment

Fakhari et.al (2012), Mirzadeh & Mehrmanesh (2016), Howang et.al (2013), Bak (2016), kiji-Kali (2017)	factors relating to facilitation	
Dargi & Mohammadian (2010), Hamidizadeh (2013), Darzban (2016), Belman et.al (2012)	awareness of different status of the environment status	
Dargi & Mohammadian (2010), Mirzadeh & Mehrmanesh (2016)	awareness of access location of objective group to the product	
Dargi & Mohammadian (2010), Salimifard & Pourdehghan (2016)	awareness of norms	
Dargi & Mohammadian (2010), Hamidizadeh (2013), Darzban (2016), Belman et.al (2012), Koumra et.al (2016)	awareness of religions and religious values	
Dargi & Mohammadian (2010), Hassanzadeh & Danaei (2016)	awareness of method of evaluating services in relation to other services	
Hassazadeh & Danaei (2016), Radmiker (2015)	attention to security	
Samsam Shariat (2007), Mirzadeh & Mehrmanesh (2016), Belman et.al (2012)	economic beliefs	
Salimifard & Pourdehghan (2016), Pilay et.al (2012), Howang et.al (2013), Komra et.al (2016)	ethical beliefs	
Salimifard & Pourdehghan (2016)	Islamic teaching	
Samsam Shariat (2007), Daneshvar et.al (2016), Radmiker (2015)	strategic look to advertisement plan	
Dargi & Mohamadian (2010), Ansari & Rasei (2016), Abri & Joumar (2014)	market share	
Dargi & Mohamadian (2010), Darzain (2016)	obviating primary needs of audiences	

Salimi Fard & Pourdehghan (2016), Kim et.al (2012), Howang et.al (2013)	reference groups	
Sohrabi Ranin (2016), Fontini et.al (2009), Belman et.al (2012)	awareness of services	factors relating to services
Salimi Fard & Pourdehghan (2016), Abri & Joumar (2014), Radmiker (2015)	stating potential benefits of purchase ad services consumption	
Dargi & Mohamadian (2010), Hosseinzadeh et.al (2016), Fontini et.al (2009)	rate of services complexity	
Samsam Shariat (2007), Darzaian (2016), Fontini et.al (2009), Radmiker (2015)	display of expected experience	
Khashizadeh (2010), Darzain (2016)	rate of familiarity of consumer with services	
Ansari & Rasaei (2016), Belman et.al (2012)	market researches	
Hosseinzadeh et.al (2016), Mirzadeh & Mehrmanesh (2016)	record of exploiting services	
Dargi & Mohammadian (2010), Ansari & Rasaei (2016), Lousi (2012), Park (2015), Marion et.al (2017)	being unique	
Dehdashti Shahrokh (2009), Daneshvar et.al (2016), Radmiker (2015), Jankoski et.al (2016)	price, benefits	
Daneshvar et.al (2016), Fontini et.al (2009), Jankoski et.al (2016)	relation with customer	
Samsam Shariat (2007), Daneshvar et.al (201), Fontini et.al (2009), Jankoski (2016)	supporting customer	
Dehdashti Shahrokh (2009), Ansari & Rasaei (2016), Fontini et.al (2009), Radmiker (2015), Jankoski (2016)	feedback	

2- Research plan

The present study is of descriptive applied type which has been performed with survey method. The statistical population of this study includes all managers and experts of Eghtesad

Novin Bank. The statistical population of the present study contains all people who are exposed to advertisements of Eghtesad Novin Bank across the country. For selecting samples, Cochran formula is used for unlimited societies that based on this formula, the samples number were determined 400 persons for each province. The sample people were selected based on the following sampling method. At first, using stratified sampling method , the samples were randomly selected from five geographical regions of the country (north, south, east, west and center) of nine provinces (Mazandaran, Khorasan Razavi, Zanjan, Chahar Mahal Bakhtiari , Kerman, East Azarbaijan , Kermanshah, Ilam and Tehran).

3- Variables and research model

In this study, fuzzy Delphi technique has been used for identifying main indicators. For ranking the final criteria and sub standards, multi-criteria decision making models and AHP model have been used with fuzzy approach. For analysis of obtained data, Excel software and coding in the environment of this software have been used. The results obtained from data analysis have been provided in the present chapter.

3-1 Identifying criteria and sub standards of factors effective on advertisement effectiveness

At the first step, the primary indicators were identified and selected. In the first stage, by reviewing the literature and performed studies, a large number of indicators were extracted from the literature. For screening the indicators and identifying the final indicators, fuzzy Delphi approach has been used. The expert views have been collected in the field of rate of significance of main indicators. Fuzzy average and defuzzification output of rates relating to indicators are presented in table 2. The rate of defuzzification greater than 7 is acceptable and every indicator which has the score higher than 7 is confirmed (Wu& Fang, 2011).

Table 2- fuzzy average and fuzzy screening of indicators of factors effective on advertisement effectiveness

Result	Crisp	mean	U	M	L	
acceptance	7.15	(2, 7.97, 9)	9	7.97	2	Q01
acceptance	7.84	(6, 8.01, 9)	9	8.01	6	Q02

Result	Crisp	mean	U	M	L	
acceptance	7.92	(6, 8.13, 9)	9	8.13	6	Q03
acceptance	7.39	(4, 7.83, 9)	9	7.83	4	Q04
acceptance	7.70	(6, 7.8, 9)	9	7.80	6	Q05
rejection	6.68	(2, 7.28, 9)	9	7.28	2	Q06
acceptance	7.71	(6, 7.82, 9)	9	7.82	6	Q07
acceptance	7.66	(6, 7.75, 9)	9	7.75	6	Q08
acceptance	7.02	(2, 7.78, 9)	9	7.78	2	Q09
acceptance	7.27	(2, 8.15, 9)	9	8.15	2	Q10
acceptance	7.09	(2, 7.88, 9)	9	7.88	2	Q11
rejection	6.85	(2, 7.52, 9)	9	7.52	2	Q12
acceptance	7.43	(4, 7.9, 9)	9	7.90	4	Q13
rejection	6.71	(3, 7.07, 9)	9	7.07	3	Q14
acceptance	7.17	(2, 8.01, 9)	9	8.01	2	Q15
acceptance	7.13	(2, 7.95, 9)	9	7.95	2	Q16
rejection	6.85	(4, 7.02, 9)	9	7.02	4	Q17
acceptance	7.43	(4, 7.89, 9)	9	7.89	4	Q18
acceptance	7.70	(6, 7.8, 9)	9	7.80	6	Q19
rejection	6.80	(3, 7.21, 9)	9	7.21	3	Q20
acceptance	7.75	(6, 7.88, 9)	9	7.88	6	Q21
rejection	6.62	(3, 6.93, 9)	9	6.93	3	Q22
acceptance	7.61	(6, 7.66, 9)	9	7.66	6	Q23
acceptance	7.22	(2, 8.08, 9)	9	8.08	2	Q24
acceptance	7.04	(2, 7.81, 9)	9	7.81	2	Q25
acceptance	7.71	(6, 7.82, 9)	9	7.82	6	Q26
rejection	6.65	(3, 6.97, 9)	9	6.97	3	Q27
acceptance	7.98	(6, 8.22, 9)	9	8.22	6	Q28
acceptance	7.70	(6, 7.79, 9)	9	7.79	6	Q29
rejection	6.54	(3, 6.81, 9)	9	6.81	3	Q30
acceptance	7.19	(4, 7.53, 9)	9	7.53	4	Q31
acceptance	7.79	(5, 8.18, 9)	9	8.18	5	Q32

Result	Crisp	mean	U	M	L	
acceptance	7.53	(4, 8.05, 9)	9	8.05	4	Q33
acceptance	7.05	(2, 7.82, 9)	9	7.82	2	Q34
acceptance	7.43	(4, 7.89, 9)	9	7.89	4	Q35
rejection	6.92	(3, 7.38, 9)	9	7.38	3	Q36
acceptance	7.02	(2, 7.78, 9)	9	7.78	2	Q37
rejection	6.91	(3, 7.37, 9)	9	7.37	3	Q38
rejection	6.96	(2, 7.68, 9)	9	7.68	2	Q39
rejection	6.99	(2, 7.74, 9)	9	7.74	2	Q40
acceptance	7.42	(4, 7.88, 9)	9	7.88	4	Q41
acceptance	7.10	(2, 7.9, 9)	9	7.90	2	Q42
rejection	6.59	(2, 7.13, 9)	9	7.13	2	Q43
rejection	6.55	(2, 7.08, 9)	9	7.08	2	Q44
acceptance	7.77	(6, 7.9, 9)	9	7.90	6	Q45
rejection	6.91	(2, 7.62, 9)	9	7.62	2	Q46
acceptance	7.09	(2, 7.88, 9)	9	7.88	2	Q47
acceptance	7.68	(6, 7.78, 9)	9	7.78	6	Q48
acceptance	8.19	(6, 8.53, 9)	9	8.53	6	Q49
rejection	6.88	(2, 7.57, 9)	9	7.57	2	Q50
acceptance	7.75	(5, 8.12, 9)	9	8.12	5	Q51
acceptance	7.03	(2, 7.8, 9)	9	7.80	2	Q52
rejection	6.96	(4, 7.19, 9)	9	7.19	4	Q53
acceptance	7.04	(2, 7.8, 9)	9	7.80	2	Q54
rejection	6.96	(3, 7.45, 9)	9	7.45	3	Q55
acceptance	7.02	(2, 7.78, 9)	9	7.78	2	Q56
acceptance	7.92	(5, 8.38, 9)	9	8.38	5	Q57
acceptance	7.00	(2, 7.76, 9)	9	7.76	2	Q58
rejection	6.90	(1, 7.86, 9)	9	7.86	1	Q59
acceptance	7.10	(4, 7.4, 9)	9	7.40	4	Q60
acceptance	7.86	(6, 8.04, 9)	9	8.04	6	Q61
rejection	6.97	(2, 7.71, 9)	9	7.71	2	Q62

Result	Crisp	mean	U	M	L	
acceptance	7.02	(2, 7.79, 9)	9	7.79	2	Q63
acceptance	7.73	(6, 7.84, 9)	9	7.84	6	Q64
acceptance	7.01	(2, 7.76, 9)	9	7.76	2	Q65
rejection	6.65	(3, 6.98, 9)	9	6.98	3	Q66
acceptance	7.01	(2, 7.77, 9)	9	7.77	2	Q67
acceptance	7.90	(6, 8.11, 9)	9	8.11	6	Q68
acceptance	7.55	(5, 7.83, 9)	9	7.83	5	Q69
acceptance	8.02	(6, 8.28, 9)	9	8.28	6	Q70
acceptance	7.95	(6, 8.17, 9)	9	8.17	6	Q71
acceptance	7.03	(2, 7.79, 9)	9	7.79	2	Q72
acceptance	7.65	(5, 7.98, 9)	9	7.98	5	Q73
acceptance	7.00	(1, 8, 9)	9	8.00	1	Q74
acceptance	7.04	(2, 7.8, 9)	9	7.80	2	Q75
acceptance	7.03	(2, 7.8, 9)	9	7.80	2	Q76
acceptance	7.30	(4, 7.7, 9)	9	7.70	4	Q77
rejection	6.88	(2, 7.57, 9)	9	7.57	2	Q78
acceptance	7.63	(4, 8.19, 9)	9	8.19	4	Q79
acceptance	7.65	(5, 7.98, 9)	9	7.98	5	Q80
rejection	6.59	(3, 6.88, 9)	9	6.88	3	Q81
acceptance	7.09	(2, 7.88, 9)	9	7.88	2	Q82
rejection	6.83	(2, 7.5, 9)	9	7.50	2	Q83
acceptance	7.12	(2, 7.93, 9)	9	7.93	2	Q84

From the 84 indicators which were assessed, 24 indicators which have acquired the average score less than 7 are omitted. So, 60 indicators are finally confirmed (Adelazaret.al, 1995).

The remained indicators will be finally examined in the format of a 60-question questionnaire with analytical hierarchy process.

3-2 determining the priority of factors effective on effectiveness of advertisement using fuzzy AHP technique

For determining the priority of major criteria, fuzzy analytical hierarchy (FAHP) has been used. The analysis process is as follows.

1- Dual comparison of major criteria based on objective and determining the weight of major criteria

2- Dual comparison of sub standards of each criterion and determining sub standards of each cluster

3- Multiplication of the weight of sub standards in the weight of respective criterion and determining the final weight of sub standards

For dual comparison of elements, 9-degree Saaty scale has been used. 9-degree Saaty scale has been provided by Thomas Saaty , codifier of analytical hierarchy theory. Also, in this study fuzzy approach has been used for making the values quantitative. So, Saaty fuzzy range has been used.

Table 3- fuzzy range equal to 9-degree Saaty scale in AHP technique. Habibi et.al , 2014, 77

reversed fuzzy equivalent	fuzzy equivale nt	the word term of comparison status of i to j
(1,1,1)	(1, 1, 1)	Preferred Equally
$\left(\frac{1}{3},\frac{1}{2},1\right)$	(1, 2, 3)	intermediate
$\left(\frac{1}{4},\frac{1}{3},\frac{1}{2}\right)$	(2, 3, 4)	Preferred moderately
$\left(\frac{1}{5},\frac{1}{4},\frac{1}{3}\right)$	(3, 4, 5)	intermediate
$\left(\frac{1}{6},\frac{1}{5},\frac{1}{4}\right)$	(4, 5, 6)	Preferred Strongly
$\left(\frac{1}{7},\frac{1}{6},\frac{1}{5}\right)$	(5, 6, 7)	intermediate
$\left(\frac{1}{8},\frac{1}{7},\frac{1}{6}\right)$	(6, 7, 8)	very strongly Preferred
$\left(\frac{1}{9},\frac{1}{8},\frac{1}{7}\right)$	(7, 8, 9)	intermediate
$\left(\frac{1}{9},\frac{1}{9},\frac{1}{9}\right)$	(9, 9, 9)	Extremely Preferred

(source in papersL Khorshid & Zabihi, Lee my H.I; Wen-Chin Chen, Ching-Jan, Chang, 2008, 101)

The pattern of criteria and sub standards has been drawn in the figure using AHP technique. Also, the criteria and sub standards of the study have been named with numeric index so that the research flow is easily detectable.

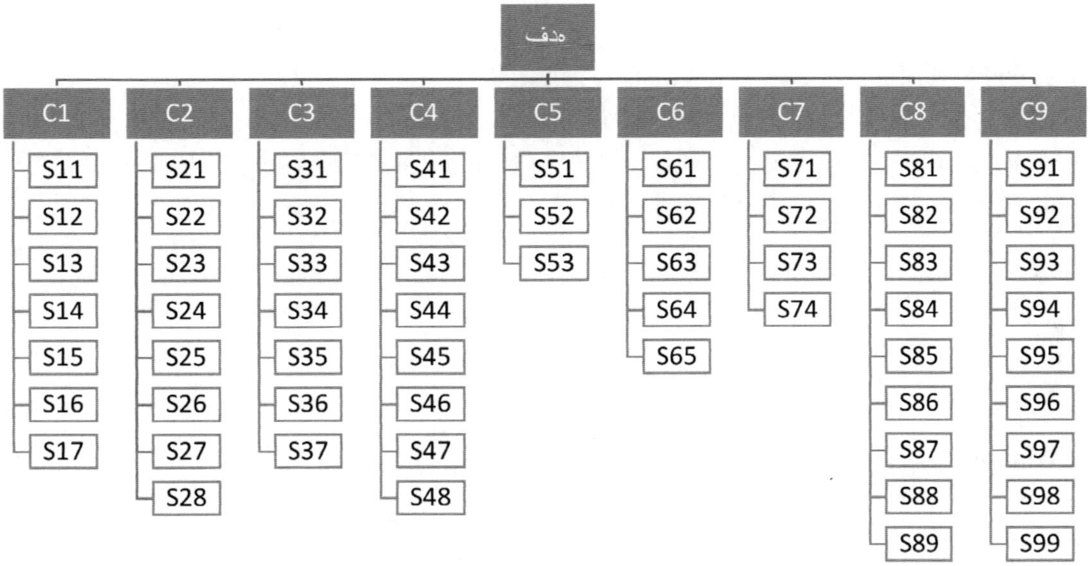

Figure 1- hierarchical model of criteria and sub standards of factor effective on advertisement effectiveness

1- Determining the priority of major criteria based on objective

In the first step, the major criteria based on the objective have been compared dually. Dual comparison is very simple and all elements of each cluster should be compared two by two. Therefore, if there is n element in a cluster, $\frac{n(n-1)}{2}$ comparisons will be performed. Since there are four criteria, so the number of performed comparisons is equal to:

$$\frac{n(n-1)}{2} = \frac{9(9-1)}{2} = 36$$

So, 36 dual comparisons have been performed from group perspective of experts. The experts view has been made quantitative using fuzzy scale. Collecting experts view has been collected with 9-degree Saaty range. So, 6 dual scales of four major four criteria have been performed based on the experts view. Using geometric average the experts views have been aggregated. Fuzzy aggregation will be calculated as follows:

Equation 1

$$F_{AGR} = (\prod(l), \prod(m), \prod(u))$$

After matrix formation of the obtained dual comparisons, the special vector has been calculated. At first, the fuzzy extension of each row is calculated. Every entry of dual comparison matrix $\widetilde{X}$

Is displayed as $\tilde{x}_{ij}$.

The fuzzy extension of every row is also displayed with symbol $\tilde{S}_i$. Therefore, the fuzzy extension of each row is calculated as follows:

Equation 2

$$\tilde{S}_i = \sum_{j=1}^{n} x_{ij}$$

Then, the fuzzy extension of every row elements will be as follows:

Fuzzy extension of row 1 (1,1,1) ⊕ (1,27,1,47,1,75) ⊕ (0.47, 0.55, 0.66) ⊕ (1.52, 1.81, 2.14) ⊕ (0.93, 1.14, 1.39) ⊕ (1.26, 1.61, 2.04) ⊕(1.25, 1.51, 1.84) ⊕(0.78, 0.97, 1.24) ⊕ (0.59, 0.75, 0.96) ⊕ (9.08, 10.8, 13.02)

Fuzzy extension of row 2 (0.57, 0.68, 0.79)⊕(1, 1, 1)⊕(0.58, 0.7, 0.89)⊕(1.84, 2.28, 2.75)⊕(0.68, 0.82, 1.02)⊕(1.42, 1.81, 2.34)⊕(1.64, 2.04, 2.53)⊕(0.71, 0.88, 1.12)⊕(0.73,
fuzzy extension of row 3 (1.52, 1.82, 2.11)⊕(1.13,)0.9, 1.1)⊕ = (9.17, 11.11, 13.55
1.42, 1.72)⊕(1, 1, 1)⊕(1.89, 2.34, 2.88)⊕(0.7, 0.86, 1.09)⊕(1.45, 1.78, 2.22)⊕(1.55, 1.88, 2.29)⊕(0.57, 0.71, 0.93)⊕(0.67, 0.8, 0.96)⊕ = (10.47, 12.61, 15.2)

Fuzzy extension of row 4 (0.47, 0.55, 0.66)⊕(0.36, 0.44, 0.54)⊕(0.35, 0.43, 0.53)⊕(1, 1, 1)⊕(0.97, 1.21, 1.51)⊕(0.96, 1.33, 1.8)⊕(2.14, 2.7, 3.37)⊕(0.94, 1.21, 1.56)⊕(1.3, 1.69, 2.16)⊕ = (8.49, 10.57, 13.14)

Fuzzy extension of row 5 (0.72, 0.88, 1.07) ⊕ (0.98, 1.22, 1.48)⊕(0.92, 1.17, 1.43)⊕(0.66, 0.82, 1.03)⊕(1, 1, 1)⊕(0.67, 0.83, 1.04)⊕(0.23, 0.29, 0.36)⊕(0.28, 0.32, 0.39)⊕(0.3, 0.38, 0.48)⊕ = (5.76, 6.91, 8.29)

Fuzzy extension of row 6 (0.49, 0.62, 0.79)⊕(0.43, 0.55, 0.7)⊕(0.45, 0.56, 0.69)⊕(0.55, 0.75, 1.04)⊕(0.96, 1.2, 1.48)⊕(1, 1, 1)⊕(0.34, 0.41, 0.51)⊕(0.27, 0.34, 0.44)⊕(0.24, 0.31, 0.41)⊕ = (4.74, 5.75, 7.08)

Fuzzy extension of row 7 (0.54, 0.66, 0.8)⊕(0.4, 0.49, 0.61)⊕(0.44, 0.53, 0.64)⊕(0.3, 0.37, 0.47)⊕(2.81, 3.5, 4.3)⊕(1.95, 2.43, 2.96)⊕(1, 1, 1)⊕(0.34, 0.42, 0.54)⊕(0.29, 0.36, 0.47)⊕ = (8.06, 9.76, 11.78)

Fuzzy extension of row 8 (0.8, 1.03, 1.28)⊕(0.89, 1.14, 1.41)⊕(1.08, 1.41, 1.76)⊕(0.64, 0.82, 1.06)⊕(2.55, 3.1, 3.63)⊕(2.25, 2.97, 3.72)⊕(1.86, 2.39, 2.94)⊕(1, 1, 1)⊕(0.58, 0.69, 0.83)⊕ = (11.66, 14.56, 17.64)

Fuzzy extension of row 9 (1.04, 1.33, 1.69)⊕(0.91, 1.11, 1.37)⊕(1.04, 1.25, 1.5)⊕(0.46, 0.59, 0.77)⊕(2.08, 2.63, 3.3)⊕(2.42, 3.23, 4.1)⊕(2.15, 2.74, 3.43)⊕(1.2, 1.46, 1.72)⊕(1, 1,)1)⊕ = (12.29, 15.35, 18.88

Then, the fuzzy sum of total priorities column elements is calculated

Equation 3

$$\sum \tilde{S}_i = \sum_{i=1}^{n} \sum_{j=1}^{n} x_{ij}$$

The sum of elements of the priorities column of major criteria will be as follows:

$$\sum_{i=1}^{4} \sum_{j=1}^{4} M_g^j = (79.72, 97.41, 118.59)$$

For normalization of each criterion priorities, that criterion total rates sum should be divided into sum of all priorities (column elements), since the rates are fuzzy, so fuzzy sum of every row is multiplied in summation reverse. The summation reverse should be calculated.

Equation 4

$$if\ \tilde{F} = (l, m, u)\ then\ \tilde{F}^{-1} = \left(\frac{1}{u}, \frac{1}{m}, \frac{1}{l}\right)$$

Then, based on the equation 4, we will have

$$\left(\sum_{i=1}^{n} \sum_{j=1}^{n} M_g^j\right)^{-1} = (0.01, 0.01, 0.01)$$

$$P_k = \sum_{i=1}^{n} M * \left(\sum_{i=1}^{n} \sum_{j=1}^{n} M_g^j\right)^{-1}$$

So, the results obtained from normalization of obtained rates will be as follows:

$\widetilde{W}_{C1} =$ (0.08, 0.11, 0.16)
$\widetilde{W}_{C2} =$ (0.08, 0.11, 0.17)
$\widetilde{W}_{C3} =$ (0.09, 0.13, 0.19)
$\widetilde{W}_{C4} =$ (0.07, 0.11, 0.16)
$\widetilde{W}_{C5} =$ (0.05, 0.07, 0.1)

$\widetilde{W}_{C6} =$ (0.04, 0.06, 0.09)

$\widetilde{W}_{C7} =$ (0.07, 0.1, 0.15)

$\widetilde{W}_{C8} =$ (0.1, 0.15, 0.22)

$\widetilde{W}_{C9} =$ (0.1, 0.16, 0.24)

Each of the obtained rates is fuzzy and normalized weight relating to major criteria. There are various methods for defuzzifcation of the obtained rates. In this study, Bojadziev proposed method has been used for defuzzification.

Table 4- defuzzification and computation of final and normal weight of criteria of factors effective on advertisement effectiveness

Normal	Deffuzy	X3max	X2max	X1max	Crisp
0/111	0/117	0/113	0/115	0/117	C1
0/114	0/120	0/115	0/118	0/120	C2
0/103	0/137	0/133	0/135	0/137	C3
0/108	0/113	0/112	0/113	0/113	C4
0/070	0/073	0/072	0/073	0/073	C5
0/060	0/063	0/062	0/063	0/063	C6
0.101	0/107	0/103	0/105	0/107	C7
0.149	0/157	0/153	0/155	0/157	C8
0.158	0/167	0/163	0/165	0/167	C9

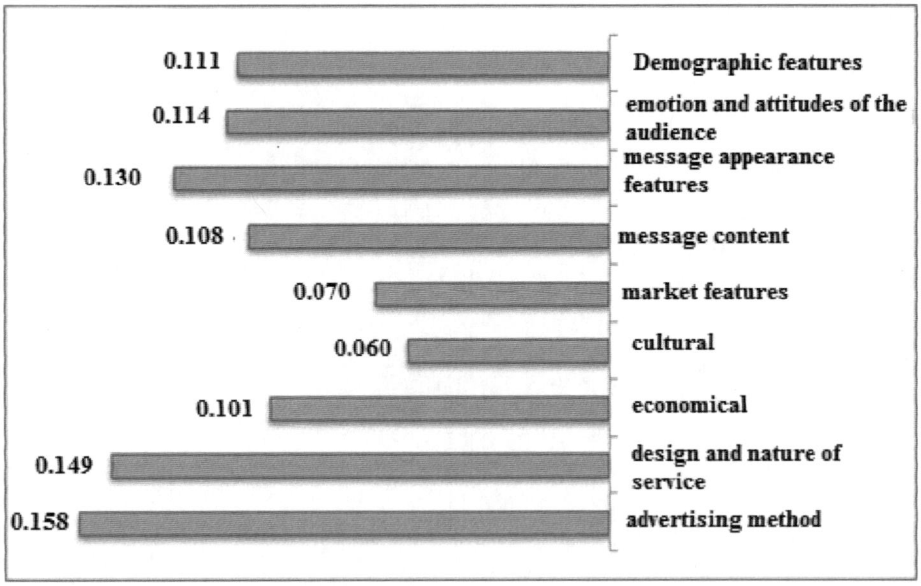

Figure 2- graphical display of significance of criteria of factors effective on advertisement effectiveness

Based on the obtained special vector:

- Demographic features with weight 0.111 have been located in rank five

- Individual feelings and attitude of the addressee with weight 0.114 has been located in rank four.

- The message surface features with weight 0.13 has been located in rank three.

- The message content with weight 0.108 has been located in rank six

- The market feature with weight 0.07 has been located in rank eight.

- Cultural with weight 0.06 has been located in rank nine.

- Economic with weight 0.10 has been located in rank seven.

- Plan and nature of services with weight 0.149 has been located in rank second.

- Advertisement methods with weight 0.158 has been located in rank first.

The inconsistency rate of performed comparisons has been obtained 0.030 which is smaller than 0.1 and so the performed comparisons is sufficed.

2- Comparison and determining priority of sub standards

At first, the experts' views have been collected in the dual comparison of sub standards. Then, the performed calculations for fuzzification of the experts views average have been provided for determining the priority of sub standards.

Priority of sub standards of demographic features

The audience job with weight 0.151 has been located in the third rank.

The audience income with weight 0.093 has been located in the seventh rank.

The audience religion with weight 0.16 has been located in the second rank.

Social class with weight 0.141 has been located in the fifth rank.

Awareness of gender differences with weight 0.173 has been located in the first rank.

The individual knowledge level with weight 0.135 has been located in the sixth rank.

Priority of sub standards of the audience individual feeling and attitude

Positive mentality with weight 0.124 has been located in the fifth rank.

Feeling of honesty in advertisements with weight 0.089 has been located in the eighth rank.

Threatening feeling with weight 0.14 has been located in the third rank.

Mental image with weight 0.102 has been located in the seventh rank.

Attitude towards brand with weight 0.121 has been located in the sixth rank.

Convincing audiences with weight 0.149 has been located in the first rank.

Attitude towards advertisement with weight 0.143 has been located in the second rank.

The individual perception of accuracy of advertisement with weight 0.133 has been located in the fourth rank.

Priority of sub standards of the message surface features

Then, based on the results of performed dual comparisons:

Beautiful and appropriate combination of colors with weight 0.143 has been located in the fourth rank.

Attention to reduction of the user waiting time with weight 0.159 has been located in the second rank.

Graphics- design with weight 0.131 has been located in the sixth rank.

Creativity in the message with weight 0.137 has been located in the fifth rank.

Emotional music with weight 0.099 has been located in the seventh rank.

Text repetition with weight 0.153 has been located in the third rank.

Access to required and sufficient information with weight 0.178 has been located in the first rank.

Determining the priority of sub standards of the message content

Benevolent with weight 0.093 has been located in the eighth rank.

Coordination in speech and behavior with weight 0.125 has been located in the fourth rank.

Reality of the messages with weight 0.119 has been located in the sixth rank.

Using women and children images with weight 0.131 has been located in the third rank.

Gender of speaker in the advertisement with weight 0.106 has been located in the seventh rank.

Using trustful and famous people with weight 0.16 has been located in the first rank.

Loveliness of famous people for the observer with weight 0.122 has been located in the fifth rank.

Familiarity of the famous person with the observer with weight 0.144 has been located in the second rank.

Priority of sub standards of the market features

Then, based on the results of the performed dual comparisons:

Factors relating to its application in the market with weight 0.326 have been located in the second rank.

Awareness of different aspects of status and environment with weight 0.283 has been located in the third rank.

Awareness of evaluation method of service receiver with weight 0.391 has been located in the first rank.

Priority of cultural sub standards

Then, based on the results of performed dual comparisons:

Awareness of norms with weight 0.199 has been located in the third rank.

Awareness of religion and religious values with weight 0.238 has been located in the first rank.

Ethical beliefs with weight 0.18 have been located in the fourth rank.

Obviating primary needs of the audiences with weight 0.164 has been located in the fifth rank.

Reference groups with weight 0.219 have been located in the second rank.

Determining the priority of economic sub standards

Then, based on the results of the performed dual comparison:

Economic beliefs with weight 0.261 have been located in the second rank.

Strategic look at an advertising program with weight 0.215 has been located in the fourth rank.

Market share with weight 0.274 has been located in the first rank.

Attention to economic security with weight 0.251 has been located in the third rank.

Priority of the plan and services nature sub standard

Awareness of services with weight 0.104 has been located in seventh rank.

Expressing potential benefits of purchase and services consumption with weight 0.117 has been located in the third rank.

Services complexity rate with weight 0.089 has been located in the ninth rank.

Uniqueness with weight 0.108 has been located in the sixth rank.

Rate of familiarity of the services consumer with weight 0.12 has been located in the second rank.

Price and allowances with weight 0.114 has been located in the fourth rank.

Relation with the customer with weight 0.111 has been located in the fifth rank.

Displaying the expected experience with weight 0.098 has been located in the eighth rank.

Supporting customers with weight 0.139 has been located in the first rank.

Priority of sub standards of the advertising methods

Beginning the advertisement with a challenge with weight 0.079 has been located in the ninth rank.

Beginning an advertisement with a statement and explanation with weight 0.107 has been located in the fifth rank.

Using stimuli with diversions more than usual and expected with weight 0.097 has been located in the seventh rank.

The advertisement message text being predicating with weight 0.126 has been located in the third rank.

Using diverse methods with weight 0.088 has been located in the eighth rank.

Media advertisement with weight 0.113 has been located in the fourth rank.

Dimensions of TV advertisement with weight 0.151 have been located in the first rank.

Psychological, geographical division with weight 0.107 has been located in the fifth rank.

Advertising media selection with weight 0.132 has been located in the second rank.

Determining the final priority of major criteria with FAHP technique

For determining final priority of major criteria using FAHP technique, the weights relating to major criteria (W1) and weight of indicators based on each criteria (W2) are needed. The results of comparison of the research sub criteria relating to them forms matrix W2. For determining the final priority of indicators with AHP technique, it is sufficient that the weight of indicators based on every criterion (W2) is multiplied in the major criteria (W1). Any of these matrices has been calculated in the previous steps. Based on the obtained output, it is obvious that the market share item, one of economic factors with weight 0. 277 is the most important indicator in the existing indicators. Item of awareness of evaluation method of service receiver in relation to other services with weight of 0.0273 has the second priority. Item of economic believes, one of economic factors, has the third priority.

So, the research model is as follows:

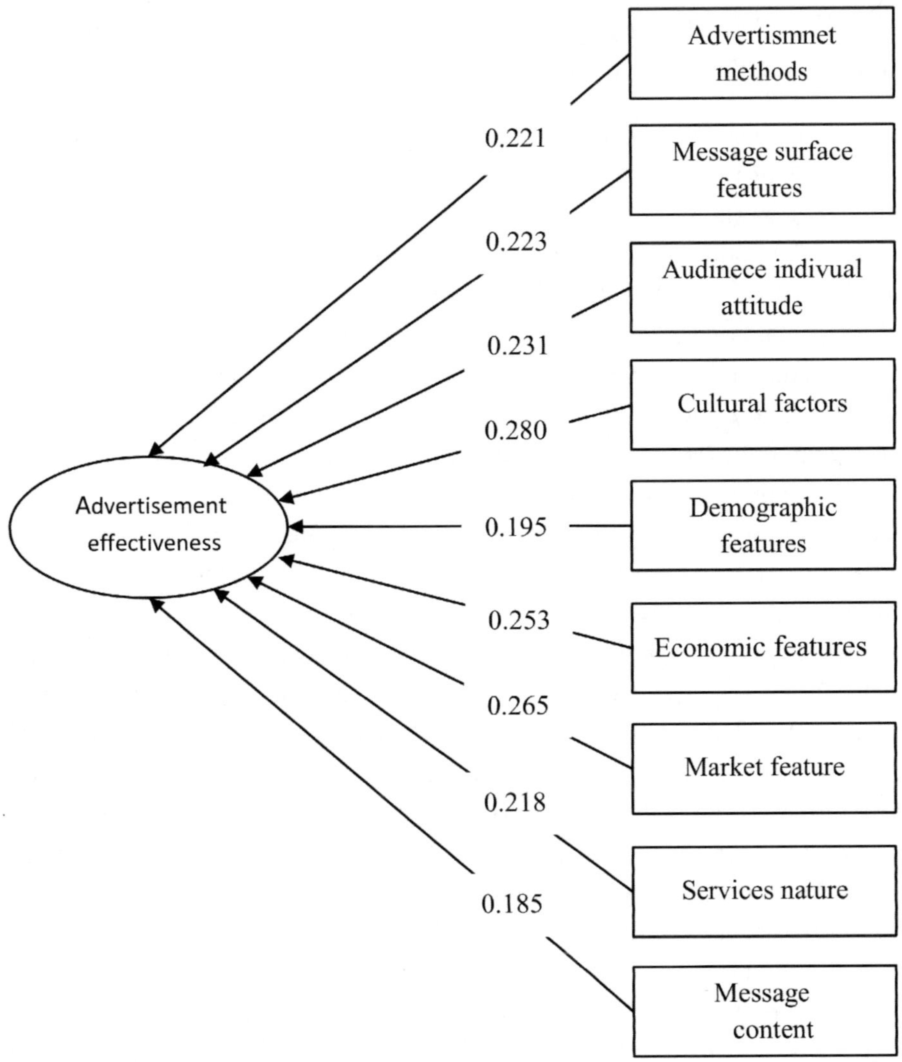

4- Conclusion:

In this study, identification and ranking of factors effective on designing the native model of advertisement in Eghtesad Novin bank has been addressed. In this first step, the primary indicators were identified and selected. In the first stage, by reviewing the literature and performed studies, a great number of indicators were extracted from the literature. For screening the indicators and identifying final indicators, fuzzy Delphi approach has been used. The experts' views have been collected in the field of significant rate of major indicators. From 84 indicators which were assessed, 24 indicators which have acquired score less than 7 are omitted. So, 60 indicators were finally confirmed. Then, regarding the performed calculations, the final weight of any of the model indicators have been calculated with fuzzy AHP. Based on the obtained output, it is obvious that the market share of economic factors with weight of 0.027 is the most important indicator in the existing indicators. The item of awareness of evaluating services in relation to other services is one of market features factors with weight 0.0273 which has second priority. The item of economic beliefs with weight 0.0264 is in the third priority.

Today, considering extension of competition in the international level and formation of world competition, all countries of the world may be successful in this arena to coordinate them with this trend and create more value for their customers by services and products they provide. In the today modern world, almost all people are exposed to advertisements and other communicational activities. Effectiveness of advertisement is one of important issues in the advertisement domain. In this regard, Mansouri Moyed (2005) showed that attention attraction, commercial advertisements of Maskan Bank are effective but in respect of other variables, that is, creating interest, stimulation of tendency and measurement to purchase, commercial advertisements of Maskan Bank are not effective. Samsam Shariat (2007) showed that TV channels and cartoon animations , attractive package, colors relating to the product and innovation, noticing safety feeling and associated symbols, informing of quality, background and price, dialects and common phrases, diversity in the message content, in TV, broadcasting from 18-23 and before beginning programs (in media) have been effective factors in effectiveness of commercial effectiveness from people viewpoint. Mosavi Esmaeilzadeh (2015) showed that oral advertisements (in 4 dimensions) and after that TV advertisement (just in 2 dimensions) are effective. Also, radio, internet and billboard advertisement are not effective. And also Friedman test was used for ranking advertisement tools. Salimi Fard & Pourdehghan (2016) showed that among factors effective on attitude to cell phone advertisement, components of general attitude towards advertisement, knowledge level and value have respectively the greatest significance and components of time of presentation and reference groups with the least significance in effecting the individuals attitude towards cell phone advertisement. Ansari & Rasaei (2016) showed that the creativity of advertisement message , market research, competition, market share, uniqueness and relation with customer are effective on the effectiveness of brand advertisement. Louis (2012) showed that advertising in the urban public spaces will have a great impact in the consumers' mentality. Since some advertisements are transitory and are often displayed in a short period of time and just in several spaces, but advertisement in public spaces exist everywhere and is similar to advertising through print and media, but since they have lower cost for production and distribution, have lower cost in this respect. Zabckowich & Loukston (2011) showed that in necktie advertisements use more of sensory requests and information signs. Findings showed that in western and European countries, people establish more contact due to globalization increase. They declared that content of printed advertisements, includes visual and text advertisements which are effective in establishing relation with many audiences. Also, ethnic identity as an important cultural variable increases attention to consumer purchase behavior. Besides that, they found out that the image of people, models, ceremonies and managers constitute an important part of images and background image plays an important role in effectiveness of visual factors in advertisement. Mac Ki- Nesbiat et.al (2011) showed that younger adults remind emotional messages specially negative ones better than others and older adults more prefer positive and logical messages to negative and emotional ones, but attitude towards advertisement among younger adults is not different. Also, they found that older adults have the most positive attitude towards advertisements.

Perzouti (2015) showed that advertising models are different in various ages and each age is consistent with a special model and using various and consistent models with people conditions may make the advertisement more effective. Abri & Joumar (2014) showed that advertising messages are effective on the consumer behavior and the advertiser may use various media for communicating consumer and media schedule is effective on the advertisement success. Attention to environmental factors where the customer lives are important factors in the advertisement success. Madra et.al (2017) showed that the advertisement text, text repetition impact the advertisement effectiveness. Maryion et.al (2017) showed that simultaneous use of several media like TV and internet may increase effectiveness of the advertisement and considering two variables of gender and multimedia create a better individual attitude towards the message effectiveness.

In ranking the identified factors, advertisement methods have assigned the first rank among factors. It is suggested that in designing advertisement and selecting advertising instruments, the dimensions of TV advertising, selecting median and predicative message text should be noticed. Using advertising instrument like internet advertisement, adverting on public vehicles like buses and metro due to great number of people who use them can increase stability of the advertisements in the customers mind.

In codifying and designing bank advertisements, the design and services nature should be noticed. It is proposed that supporting customer, familiarity of consumer with services are considered and during the advertisement process, it is required that the potential benefits of purchase and services consumption are expressed.

In designing the advertisement and content of bank advertising message, it is required to use famous and trustful people for introducing the services. Also, children as future customers and audiences should be noticed since they are future potentials. Total advertisements provided by the bank through various media should move in a unit line so that the audience can establish a logical relation between the observed advertisements in various media that is total advertising efforts lead to synergy of advertisement effectiveness. This issue won't be acquired unless through a media planning and efficient advertisement.

In advertisements, the message surface features should be noticed. It is required that in advertisement, access to required and enough information for the customers, attention to reduction of expectation time of user for observing the advertisement and suitable repetition of the text are considered. Providing large information in an advertisement which has a short time and providing information that require special mathematical analysis should be avoided.

References:

Hosseinabadi, Maryam, Mahmoud Nematian, Mohammad Besharati, investigating the relation between mixed elements of marketing and customers satisfaction with Tehran

Sarmayeh bank branches, management and entrepreneurship studies, fall 2016, 2nd period, No. 3 (page 162-169)

Dargi Parviz, 2008, subjects and topics of marketing management with Iran market attitude, Tehran, Rasa publication

Sarmad, Zohreh, Bazargan Abbas, Hejazi, Elaheh, 2014, research methods in behavioral sciences, Tehran, Agah. 27-13

Salimi Fard, Khodakaram, Pourdehghan, Adel, 2016, identifying and ranking factors effective on attitude towards cell phone advertisement using AHP, Brand management, period 3, no. 6, summer 2016, page 133-160

Seyed Akbar Manjgani, Vahideh Tabasi Lotfabadi, 2015, factors effective on effectiveness of advertisement on financial stoop of legal clients of Mashhad City with AHP approach, finance journal, fall 2015, No. 75 (page 121-146)

Sanayei, Ali, Majid Mohammad Shafiei, Maedeh Amini Valashani, 2016, effectiveness of bank advertisement in customer attraction, commercial management searches, spring and summer 2016, No. 15 (page 185-210)

Mahmoudian, Mahmoud, Rabiei, Ali, Jamili, Bita. 2011, evaluation of effectiveness of Parsian bank advertisement and identifying the most important factors in increasing its effectiveness in Tehran city, modern marketing studies, summer 2011, No. 2 (page 17-40).

Manjagani, Seyed Akbar, Tabasi Lotfabadi, Vahideh, 2015, factors effective on advertisement effectiveness on financial stoop of legal clinets of Mashhad city with AHP approach, finance journal, fall 2015, No. 75, 1-121

Venous, Davar, Safaeian, Mitra, (2005), marketing bank services in Iran and bank services features, Tehran, Negah Danesh, 4th edition

Angulo, N. (2016). 10 digital marketing trends to watch in 2016 and beyond. Retrieved from: http://www.marketingdive.com/news/10-digital-marketing-trends-to-watch-in-2016-./and-beyond/411457

Bell Raoul, Buchner, Axel.(2018). Positive Effects of Disruptive Advertising on Consumer Preferences, Journal of Interactive Marketing 41 (2018) 1–13

Corvi. Elisabetta.(2010).The effectiveness of advertising: a literature review, 10th Global Conference on Business & Economics, ISBN : 978-0-9830452-1-2

Cramphorn Michael F.,(2011), Forum Gender effects in advertising, International Journal of .Market Research Society Vol. 53 Issue 2

Cronin, J. J. & Fox , G. L. (2010). "The implications of third party customer complaining for .advertising efforts". Journal of advertising. 39(2): 21-23

Decrop, A. (2007). The influence of message format on the effectiveness of print advertisements for tourism destinations. International Journal of Advertising, 26(4), 505–.525

Denis .A. Shakhov and ,Anna A. Panasenko.(2012).Evaluating Effectiveness of Bank Advertising in the Internet: Theory and Practice, World Applied Sciences Journal 18 (Special Issue of Economics): 83-90, 2012 ISSN 1818-4952

Denis 1,2 A. Shakhov and 3Anna A. Panasenko.(2012).Evaluating Effectiveness of Bank Advertising in the Internet: Theory and Practice, World Applied Sciences Journal 18 (Special Issue of Economics): 83-90, 2012

Durdana Ozretic-Dosen Ines Zizak , (2015),"Measuring the quality of banking services targeting student population", EuroMed Journal of Business, Vol. 10 Iss 1 pp. 98 - 117

Eckler, P., and Bolls, P. (2011). Spreading the Virus: Emotional Tone of Viral Advertising and Its Effect on Forwarding Intention and Attitudes. Journal of Interactive Advertising, .11(2): 1-11

Eric W.T. Ngai Ka-leung Karen Moon S.S. Lam Eric S. K. Chin Spencer S.C. Tao , (2015),"Social media models, technologies, and applications", Industrial Management & Data Systems, Vol. 115 Iss 5 pp. 769 – 802

Fotini Patsioura Maro Vlachopoulou Vicky Manthou, (2009),"A new advertising effectiveness model for corporate advertising web sites", Benchmarking: An International Journal, Vol. 16 Iss 3 pp. 372 - 386

Frederic Crane, , G. (2010). «Marketing FO Entrepreneurs: ConceptandApplication for new .Venture. New York: SAGE Pablication, Inc

Frolova. Svetlana.(2014).THE ROLE OF ADVERTISING IN PROMOTING A PRODUCT, Thesis CENTRIA UNIVERSITY OF APPLIED SCIENCES ,Degree Programme in Industrial Management May 2014, p"1-61

Henry Boateng Abednego Feehi Okoe , (2015),"Consumers' attitude towards social media advertising and their behavioural response", Journal of Research in Interactive Marketing, Vol. 9 Iss 4 pp. 299 -312

Henry Boateng Abednego Feehi Okoe , (2015),"Consumers' attitude towards social media advertising and their behavioural response", Journal of Research in Interactive Marketing, Vol. 9 Iss 4 pp. 299 -312

Huang. Jinsong, Song Sub, , Liuning Zhouc, , , Xi Liu .(2013).Attitude Toward the Viral Ad: Expanding Traditional Advertising Models to Interactive Advertising, Journal of Interactive Marketing, Volume 27, Issue 1, February 2013, Pages 36–46

Idris, I., M.S.A. Yajid and A. Khatibi, 2009. Personal and Psychological Factors-Does it .Impact the Choice of Advertising Medium? Journal of Social Sciences, 5(2): 104-111

Ioannou Myria, Nikolaos Boukas, Evi Skoufari .(2014).Examining the role of advertising on the behaviour of co-operative bank consumers, Journal of Co-operative Organization and Management, Volume 2, Issue 1, June 2014, Pages 24-33

Jankowski. Jarosław, Przemysław Kazienko, Jarosław Wątróbski, Anna Lewandowska, Magdalena Zioło.(2016).Fuzzy multi-objective modeling of effectiveness and user experience in online advertising, Expert Systems with Applications, Volume 65, 15 December 2016, Pages 315-331

Kim, C., Park, S., Kwon, K., & Chang, W. (2012). An empirical test to measure the effectiveness of online advertising in online marketplaces using a hierarchical Bayes model. .Expert Systems with Applications (39), 117–128

Kim, ji-Hern., Hyun, yong j., (2010). " A model to investigate the influence of marketing – mix efforts and corporate image on brand equity in the IT software sector " . industrial . Marketing Management. No 40,2011, pp 424-438, Emerald Group Publishing Limited

Kim, S., & Jun, J. (2016). The impact of event advertising on attitudes and visit intentions. .Journal of Hospitality and Tourism Management, 29, 1-8

Kirby, J. and Marsden .(2006). Connected marketing : the viral, buzz and word of mouth .revolution, Oxford : Butterworth-Heinemann, 2006

Madera. Quetzali, Oscar Castillo ⁎, Mario García-Valdez, Alejandra Mancilla..(2017). A method based on Interactive Evolutionary Computation and fuzzy logic for increasing the effectiveness of advertising campaigns, Information Sciences 414 (2017) 175–186

Marion Garaus and Udo Wagner , Anna-Maria B¨ack.(2017). The Effect of Media Multitasking on Advertising Message Effectiveness, Psychology & Marketing, Vol. 34(2):)138–156 (February 2017

Martín-Santana J.D., E. Reinares-Lara , P. Reinares-Lara .(2017).Influence of radio spokesperson gender and vocal pitch on advertising effectiveness: The role of listener gender, Spanish Journal of Marketing – ESIC, Volume 21, Issue 1, February 2017, Pages 63-71

McKay-Nesbitt, J., V. Manchanda, R., C. Smith, M., & A. Huhmann, B. (2011). Effects of age, need for cognition, and affective intensity on advertising effectiveness. Journal of Business Research (64), 12–17

Moch, Nils (2013). Competition in fragmented markets: New evidence from the German banking industry in the light of the subprime crisis. Journal of Banking &; Finance, Volume 37, Issue 8, Pp 2908–2919

Moorthy, S. and Hawkins, S. (2005). Advertising repetition and quality perception. Journal of .Business Research, 58, 354– 360

Mucan. Burcu, Özeltürkay. Eda Yaşa.(2014).Social Media Creates Competitive Advantages: How Turkish Banks Use This Power? A Content Analysis of Turkish Banks through Their Webpages, Procedia - Social and Behavioral Sciences, Volume 148, 25 August 2014, Pages 137-145

Muda, M.; Musa, R.; Niana-Mohamed, R. and Borhan, H. (2014). Celebrity Entrepreneur .Endorsement and Advertising Effectiveness. Social and Behavioral Sciences, 130, 11-20

Oladele. Kehinde Joseph, Ogunnaike, Akinbode Mosunmola, Aka Deborah .(2016). Effective Advertising: Tool For Achieving Client-Customer Relationships, Researchjournali's Journal of Media Studies Vol. 2 | No. 1 February | 2016

Olsen, G., Pracejus, J. W., & O'Guinn, T. (2012). Print advertising: White space. Journal of .Business Research (65), 855–860

Park, M. (2006). The compensatory effects of pictorial and verbal information for haptic information on consumer responses in non-store shopping environments. Unpublished Ph.D, .Ohio, Ohio state University

Park, T. (2008). Effective design of advertisement in handheld devices (Doctoral dissertation .)PURDUE UNIVERSITY

Park, T. (2008). Effectiveness Design of Advertisement in Handheld Devices. Dissertation for .doctoral degree, Purdue University, West Lafyette India

Pergelova, A., D. Prior and J. Rialp, 2010 Assessing Advertising Efficiency Does the Internet Play a Role? Journal of Advertising, 39(3): 39-54

Pilelienė Lina Viktorija Grigaliūnaitė , (2017)," The effect of female celebrity spokesperson in FMCG advertising: neuromarketing approach ", Journal of Consumer Marketing, Vol. 34 - .Iss 3 pp

Rosemary J. Averya, Matthew D. Eisenbergb, Kosali I. Simon .(2012).The impact of direct-to-consumer television and magazine advertising on antidepressant use, Journal of Health Economics, Volume 31, Issue 5, September 2012, Pages 705–718

Selviaridis. Kostas, Martin Spring b, Luis Araujo b .(2013). Provider involvement in business service definition: A typology, Industrial Marketing Management 42 (2013) 1398–1410

Strebinger .Andreas, Xiaoling Guo, Ferdinand Klauser, Peter Grant-Hayd .(2018).Is Multi-Ethnic Advertising a globally viable strategy for a Western luxury car brand? A mixed-method cross-cultural study, Journal of Business Research 82 (2018) 409–416

Suvittawat, A., (2015). "Commercial Bank Service Quality: Chonburi Province, Thailand". .International Journal of Management Sciences and Business Research. 4(2): 2226-8235

Tafani. Eric, Elyette Rouxa, Rainer Greifeneder .(2018).In the mood for action: When negative program-induced mood improves the behavioral effectiveness of TV commercials, Journal of Business Research 84 (2018) 125–140

Te'eni-Harari, T. (2014). "Clarifying the relationship between involvement variables and .advertising effectiveness among young people". Journal of Consum Policy. 37: 183–203

Thomas, S. and Kohli, S.C. (2011). Can Brand Image Move Upwards after Sideways? A .Strategic Approach to Brand Placements. Business Horizons, 54 (1): 41-49

Yahyaoui . Hamdi, Own .Hala S..(2018). Unsupervised clustering of service performance behaviors, Information Sciences 422 (2018) 558–571

Yuksel, A., Yuksel, F. & Bilim, Y. (2010) "Destination Attachment: Effects on Customer Satisfaction and Cognitive, Affective and Cognitive Loyalty", Journal Of Tourism Management, Vol.31, pp: 274-284.

STUDY OF NATURAL ANTIOXIDANT INHIBITORS FOR CORROSION PROTECTION OF MILD STEEL IN ACIDIC MEDIUM

Ms. Ibtihal Hassan AL-Bahrani, National University of Science and Technology, Sultanate of Oman.

ABSTRACT

Corrosion is a natural phenomena and it is a very big loss for structure as well as business; so corrosion minimization is one of the biggest areas of research. In this project; corrosion inhibition efficiency of berries juice extract was tested in mild steel sample against the sulfuric acid medium which was prepared from four types of berries; blackberries, blueberries, cranberries, and raspberries.

The experiment was designed by using Design Expert V 11 where immersed time, acid and berries juice extract concentrations were set as factors and a limited range was selected. The corrosion rate was set as a response. 20 runs were under taken and the experiments were done in the lab with maximum care.

The corrosion rate of 1020.374 mpy minimized to 597.659 mpy in presence of 5ml of extract and the efficiency of berries juice extract reached to 41.427%. The RSM gave 232.022 mpy corrosion rate as a minimum value and the optimum equation given by RSM was used for Genetic Algorithm analysis which gave 188.866 mpy corrosion rate at the optimizing conditions.

The surface and chemical characterization of the mild steel sample in presence and absence of berries juice extract were analyzed along with the relations between temperature, immersed time, acid and berries juice extract concentrations with response. The mixing tank was designed, calculations and drawing were presented.

KEYWORDS

Corrosion, Mild steel, Berries, Inhibitor, Weight loss, Sulfuric acid, RSM, SEM.

INTRODUCTION

Mild steel is used in most of our daily life applications as well as in industrial applications. In other hands, mild steel has low corrosion resistance which corrodes easily once it reacts with the surrounding environment; corrosion is a phenomenon where the mild steel reacts with the deteriorates and surrounding environment. Corrosion of mild steel is a continuous and constant problem where it severe financial and life loses as it leads to loss in the product, efficiency of equipment and failures in some cases (Roberge,2012).

The corrosion phenomenon controlled by adding a small amount of component to the corrosive environment which is a corrosion inhibitor. There are organic and inorganic inhibitors. The inorganic inhibitors are used in limit amount as they are toxic such as nitrites, molybate, nitrate, silicate, chromate, and phosphate (Cafferty,2010). Now a day, the corrosion engineers looking for cheap, renewable, friendly, readily available and ecologically acceptable inhibitors which are organic inhibitors which comes from the plants such as antioxidant components consisting of tannins, saponins, flavonoids, alkaloids and amino acids. Organic inhibitors consist of some heteroatoms such as oxygen, nitrogen, multi-bound, and sulphur atoms with the conjugated system which allows the absorption in the metal surface. Those atoms attract the absorption of some components on the metal surface and that done by the interaction of metals and by the electron of the conjugated system or by forming a bond with the free electron of heteroatom which forms a layer in the metal surface, as a result, it decreases the corrosion rate (Roberge,2012).

Berries are small pulpy fruits which have up to 9.2 mmol of antioxidant per 100g of berries. Berries can be used as a corrosion inhibitor as they are rich in antioxidant. The studies show that the berries juice extract of blueberries, cranberries, raspberries, and blackberries rich with antioxidant where the antioxidant capacity of them reach to 10 mg/ 100 g of blueberries (Michalska&Lysiak,2015), 58 mg/100g of cranberries (Blumberg,J.,et al,2013), 26.2mg/ 100 g of raspberries (Bobinaite,R.,Viskelis&Venskutonis,2015), and 21 mg/100 g of blackberries (Kaume,L.,et al,2012).

OBJECTIVES

To prepare anticorrosive extract (berries), to study corrosion in sample in presences and absences of the inhibitors, to determine the optimum conditions to inhibit corrosion and to design mixing tank.

METHADOLOGY, DESIGN AND EXPERIMENTATION

PREPARATION OF MILD STEEL SAMPLES

The mild steel sheets were cut into 2.5*3 cm^2 by using steel cutter and cleaned by surface cleaning machine to remove any corroded products in the metal surface then 4mm hole done by using drilling machine. Bunches for all metal samples (1-20) were made to differentiate between them. The drilled samples were washed by distilled water then rinsed with benzene and acetone respectively and dried with a hot air dryer. They were weighted using sensitive balance and kept in the desiccator until the time of usage.

PREPARATION OF CORROSIVE MEDIUM

The corrosive medium of different concentrations of H_2SO_4 prepared by using 36.8 N of the sulfuric acid.

PREPARATION OF BERRIES JUICE EXTRACT

137g, 209g, 282g, and 200g of raspberries, cranberries, blueberries, and blackberries respectively have been used to produce 309.5 ml of berries juice extract. The different types of berries washed by cold water, crashed in the electrical crasher and compressed mechanically by hand then filtered by cloth filter. The filtrate filtered again by using filter paper. The berries juice extract kept in the freezer at 0°C till the time of usage.

The efficiency of berries juice extract has been tested by running three experiments. The cleaned and weighted mild steel samples immersed in 0ml(blank), 2ml and 5ml of berries juice extract respectively with 50 ml of 1M H_2SO_4. After 2 hours; the samples have been taken out and cleaned by distilled water, benzene and acetone then dried by air dryer and weighted again. The corrosion rate and inhibitor concentration have been calculated by using equations (1) and (2) respectively.

$$Cr(mpy) = 22300\times\frac{\Delta W}{\rho\times A\times t} \quad \textbf{equation(1)}$$

$$IE(\%) = \frac{\text{initalcorrosionrate-finalcorrosionrate}}{\text{initalcorrosionrate}}\times 100 \quad \textbf{equation(2)}$$

DESIGN THE EXPERIMENT CONDITIONS

The experiments conditions have been designed by using Design Expert software-version 11 and following Response Surface Module(RSM); the range of the parameters have been given as inputs to the software which are inhibitor concentration (5-15 ml), immersed time (180-240 min) and acid concentration (1-2 M) and the response set as corrosion rate in mpy. Table(1) shows the experiment conditions given by the software.

Table(1): Designed runs given by RSM.

Run.No	Factor(A): inhibitor, ml	Factor(B): Time, min	Factor(C): Acid concentration, M
1	10	180	1
2	15	240	2
3	5	120	1
4	5	240	1
5	10	180	1.5
6	15	180	2
7	10	120	1.5
8	5	120	1.5
9	10	180	2
10	10	180	1.5
11	10	180	1.5
12	15	240	1
13	15	120	1
14	5	240	2
15	10	180	1.5
16	15	180	1.5
17	5	120	2
18	10	180	1.5
19	10	180	1.5
20	10	240	1.5

APPLYING THE DESIGNED RUNS IN THE LABORATORY

The weighted mild steel sample has been hanged by using a plastic thread in a 50 ml of sulfuric acid in presence of berries juice extract of known concentration and contact time as per the software; the batch has been kept in the shaker. After the contact time the mild steel sample cleaned by using 120 grade emery paper to remove any corroded products, washed by distilled water, rinsed by benzene and acetone respectively and dried by hot air dryer, weighted by using sensitive balance. The corrosion rate in mpy was calculated by using equation(1) where the weight losses results have been used.

STUDYING THE IMPACT OF TEMPERATURE IN CORROSION RATE

To study the impact of temperature in the corrosion rate; mild steel samples weighted and hanged for 2 hours in 50 ml of 1M sulfuric acid with 10ml of berries extract at different temperatures(30, 40, 50 and 60 °C) as shown in Table(2); each batch has been kept in the water bath shaker with heater to provide perfect contact between the solution and mild steel sample and to maintain the required temperature. Corrosion rate has been calculated by equation(1).

Table(2): Experiment conditions of temperature into corrosion rate.

SI.NO	Temperature (°C)	Inhibitor concentration (ml)	Acid concentration (M)	immersed time, (hours)
1	30	10	1	2
2	40	10	1	2
3	50	10	1	2
4	60	10	1	2

SURFACE AND CHEMICAL CHARACTERIZATION ANALYSIS

The mild steel surface of blank, absence and presence of berries juice extract samples have been studied by using Scanning Electron Microscopy (SEM) and Energy Dispersive Spectroscopy (EDS).

OPTIMIZATION STUDY BY GENETIC ALGORITHM

Matlab has been used for optimization where the genetic algorithm method has been followed. The actual equation form Design expert(RSM) has been used where "Optimtool" followed. The same limit of the conditions from RSM has been selected. The same previous procedure has been followed in the lab with maximum care by using those optimum conditions which given by the software, 1.001M of H_2SO_4 with 15 ml of berries juice extract immersed for 237.208 minutes.

DESIGN OF MIXING TANK

Design of mixing tank is a part of the project where the industrial scale mixing tank has been designed, the tank will be formed from high resistance material such as S.S to be able to resist the corrosive medium of sulfuric acid which will be mixed with berries juice extract. It is a container which used to mix various material (heterogeneous materials) together to form homogenous materials even the different viscosity materials can be mixed easily which provides continuous process, good control, easy cleaning, and simplest construction. It can be made from S.S material which provides a smooth surface and it can be easily cleaned (Petropedia,2018).

Parameters to be considered to design mixing tank are length, torque, weight, stiffness and hydraulic loading.(Dynamixianc,2015). The important factors in design mixing tank are tank type, tank volume, specific gravity, process and environment.(Ibryamoval,Y.et al,2009)

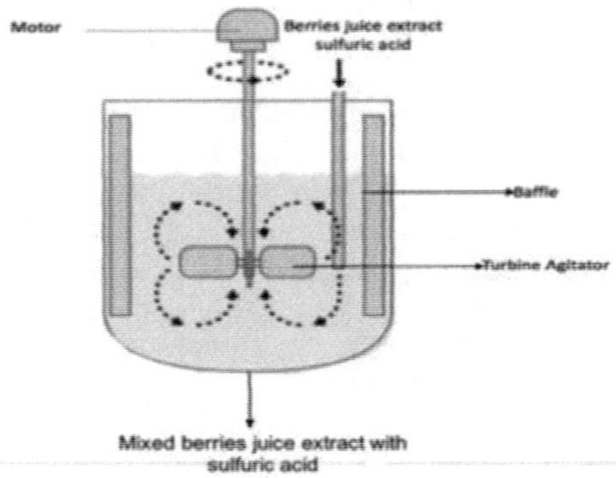

Figure(1): Mixing Tank(*Mixing Tank*,2018)

From the graph in Figure(2) Turbine agitator has been selected as it is suitable with the viscosity of the fluid and the tank volume. (Sinnot&Towler,2009)

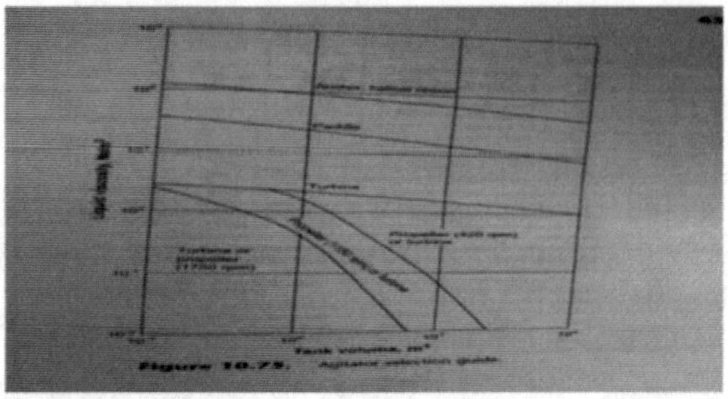

Figure(2): Graph of liquid viscosity Vs tank volume. (Sinnot&Towler,2009)

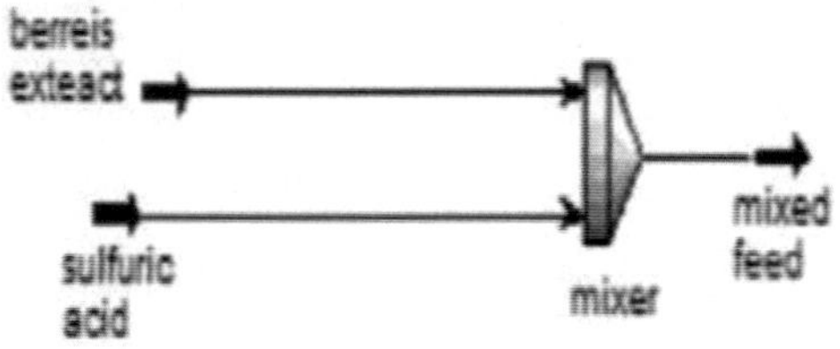

Figure(3): Result of Mixing Tank by Aspen Hysys.

Table(3): Calculated parameters for designing mixing tank

Tank diameter	m	3.5
Tank Height	m	1.81
Across sectional area	m^2	9.62
Volume of the tank	m^3	17.5
Impeller diameter	m	1.16
Length of paddle	m	0.29
Width of paddle	m	0.23
power	watt	204085
Reynolds Number	-	136486.6391

RESULTS AND ANALYSIS

EFFECT OF BERRIES JUICE EXTRACT CONCENTRATION INTO CORROSION RATE AND INHIBITOR EFFICIENCY

As shown in Table(4) the corrosion rate in the absence of berries juice extract inhibitor reach to the maximum (1020.374mpy); the corrosion rate has been minimized to 898.117mpy and 597.659mpy in presence of 2 and 5 ml of berries juice extract inhibitor respectively which means that berries juice extract inhibits the corrosion as the concentration of the berries juice extract increases. As well as the inhibitor efficiency reaches to 41.427% which obtained at 5ml of berries juice extract inhibitor in 1M of sulfuric acid immersed for 2 hours. The adsorption of the organic compounds such as ascorbic acid that present in the berries juice extract is blocking the active sites on the mild steel samples as it is forming a protective film on the metal surface as what kumar.A,2012 mention in his studies so the corrosion rate keeps decreasing as the amount of the berries juice extract is increasing. (Kumar.A,2012) Those compounds have heteroatoms such as oxygen which is a center of adsorption. In addition; It forms a coordinate bond as it has the ability to transfer the electrons to the vacant d-orbital of mild steel from the heteroatoms. (Ostovari, L. et al,2009)

Table(4): Experimental results of testing the efficiency of berries juice extract.

SI.NO	Inhibitor concentration (ml)	Corrosion rate (mpy)	Inhibitor efficiency (%)
1	0	1020.374	-
2	2	898.117	11.976
3	5	597.659	41.427

EFFECT OF IMMERSED TIME INTO CORROSION RATE AND INHIBITOR EFFICIENCY

The maximum corrosion rate reaches to 765.005mpy which obtained at 5ml of berries juice extract in the presence of 50 ml of 2 M of H_2SO_4 at 2 hours. Since the minimum corrosion rate of 232.022mpy obtained in presence of 5ml of berries juice extract with 50 ml of 1 M of H_2SO_4 immersed for 4 hours. The effect of immersed time into corrosion rate can be observed at low acid concentration and in the presence of berries juice extract, the corrosion rate is minimized at high immersed time. The decrease in the corrosion rate with the increase in immersed time is related to the adsorption of the corrosion product ($FeSO_4$) as shown in equation(3); According to Xianghong,L. et al ; the corrosion product of $FeSO_4$ becomes thicker with the immersed time, so it form a protective film (barrier) in the surface of mild steel (Xianghong, L. et al,2009).

$$Fe_{(s)}+H_2SO_{4(aq)} \rightarrow FeSO_{4(aq)}+H_{2(g)}$$, **equation(3)**

EFFECT OF ACID CONCENTRATION INTO CORROSION RATE

The effect of acid concentration into corrosion rate has been studied by using 1, 1.5 and 2 M of H_2SO_4 at different inhibitor concentrations and immersed time where the corrosion rate increase with the acid concentration at low inhibitor concentration and immersed time which reach to 765.005mpy at 2M H_2SO_4, 5ml of berries juice extract which immersed for 120 minutes as a maximum corrosion rate as shown in Table (3). In presence of the corrosive medium of H_2SO_4 with less amount of berries juice extract; the mild steel surface was exposed to H_2SO_4 medium where the corrosion reaction started as shown in equation(3) and $FeSO_4$ was formed with H_2 bubbles.

EFFECT OF TEMPERATURE INTO CORROSION RATE

From Table(5) can be noted that the rate of corrosion increases with temperature where it reaches to maximum 2407.994mpy at maximum temperature of 60°C. According to Ambrish, S. et al, at high temperature the berries juice extract molecules were partially absorbed in the mild steel surface so it was exposed to the corrosive medium of H_2SO_4, as a result, increasing the corrosion rate (Ambrish,S. et al,2012).

Table(5): Results of studying the effect of temperature into corrosion rate

SI.NO	Temperature,° C	Corrosion rate, mpy
1	30	701.507
2	40	1666.404
3	50	2008.861
4	60	2407.994

ANALYSIS BY RSM

Once the experimental results and corrosion rate calculated the RSM analysis has been done where the corrosion rate has been given to the software as a response. The response must be modified where the results fit with linear. The fit summary shows that the model F-value and P-value of 12.66 and 0.0001 respectively imply that the model is significant where A, B, C, and B^2 are the significant terms in this model only. 1.48 is a lack of fit P-value which indicates that the fit is not significant as there are pure errors.

0.5562 of the predicted R^2 is a reasonable agreement with 0.7105 of the adjusted R^2. The signal to noise ratio of 13.7252 of Adeq precision that the ratio grater than 4 is desirable. (Mycourses,2016)

The process variables are studied using ANOVA. The estimate coefficients in term of coded factors are A-Inhibitor, B-time, C-Acid and B^2. It is creating an accompanying condition for the sample with 1.0000 correlation coefficient. The coded and actual equations given by software are shown in equation(4) and (5) respectively.

- Corrosion rate(mpy)= 370.52-40.52(A)-126.97(B)+ 80.85(C) +66.58$(B)^2$ **equation(4)**
- Corrosion rate(mpy)= 1189.15487-8.10479(Inhibitor) -8.77421(Time) +161.69302 (Acid Conc)+0.018495 $(Time)^2$ **equation(5)**

The various relations can be studied from one factor graph; it can be noted that the corrosion rate increases with acid concentration and decreases with the increase in the amount of berries juice extract and immersed time as can be seen from the plots in Figure(3):Plot pf corrsoion rate Vs inhibitor concntration, Figure(4): plot of corrosion rate Vs acid concentration and Figure(5):Plot of corrosion rate Vs immersed time respectively.

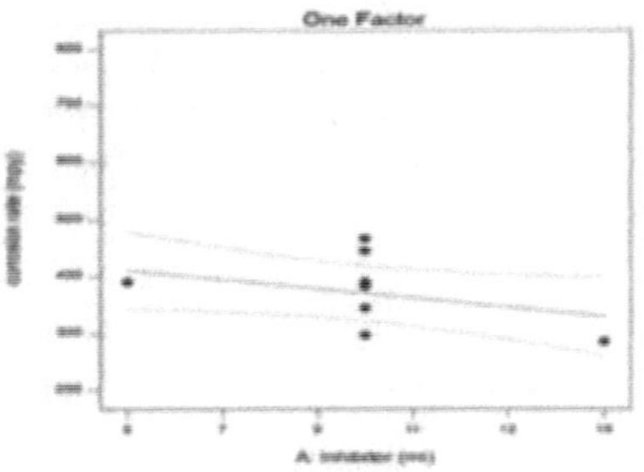

Figure(3)

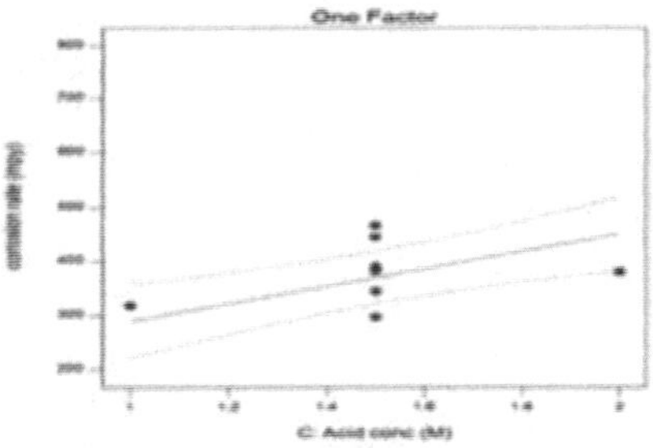

Figure(4)

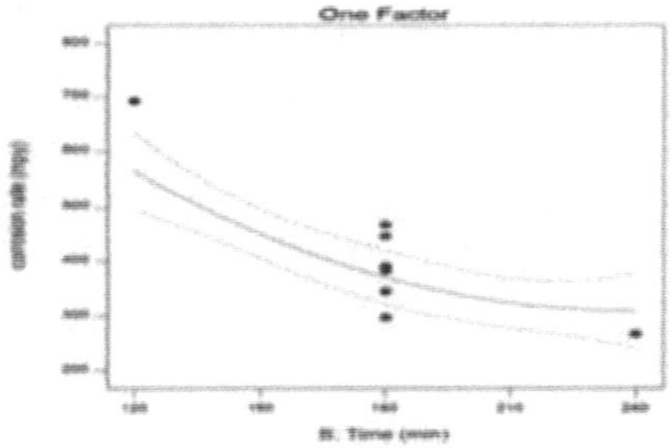

Figure(5)

The effects of two factors into the corrosion rate can be studied from the 3D plots as shown in Figure(6): 3D plot of corrosion rate Vs inhibitor concntration and acid concntration and Figure(7): 3D plot of corrosion rate Vs inhbitor concntration and immersed time.

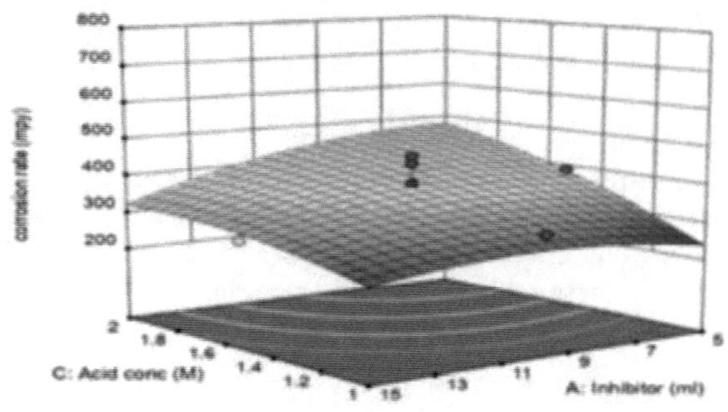

Figure(6)

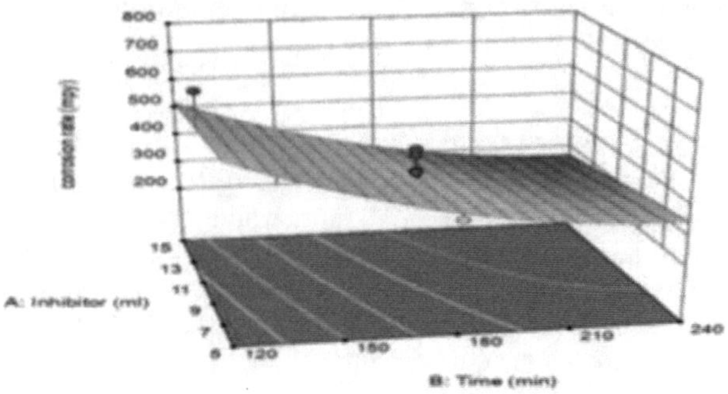

Figure(7)

The optimum conditions given by the RSM of 1M of H_2SO_4 with 5ml of berries juice extract inhibitor runs for 240 minutes gives 213.402mpy as a minimum corrosion rate but the experimental results at the same conditions gives 232.022mpy which means that they are too close as the suggested solution considered as an excellent optimum conditions.

SEM ANALYSIS

The SEM has been done for the blank mild steel sample for facilitating comparison as shown in Figure(8) which shows smooth mild steel surface compare to others. Rough surface in absence of berries juice extract is shown in Figure(9); where corrosion starting with the formation of H_2 bubbles as per the equation(3) which means that the surface of mild steel is exposed to the H_2SO_4 medium. In presence of 15 ml of berries juice extract, the surface of mild steel is protected by the adsorption of the organic compounds that are present in the berries juice extract where fewer cracks and pits can be noted and smooth surface is shown in Figure (10), which means that the mild steel surface is protected. (Raja, P. et al, 2008)

Figure(8)

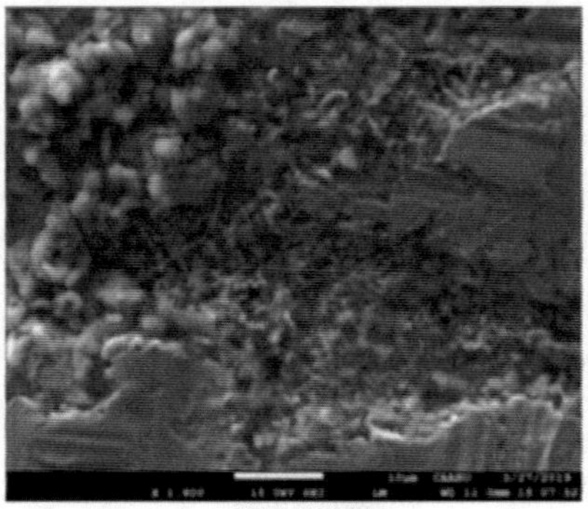

Figure(9)

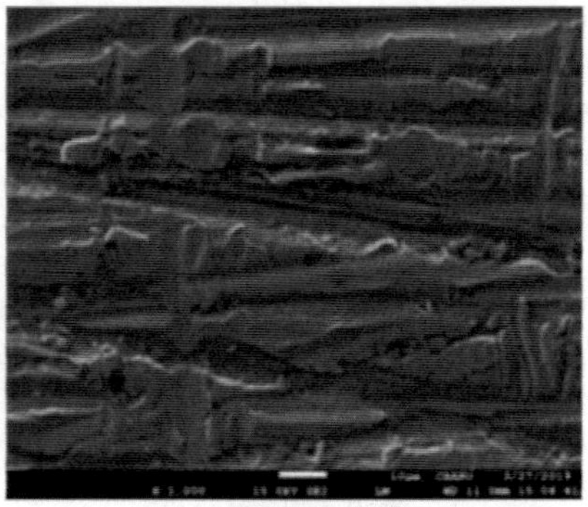

Figure(10)

EDS ANALYSIS

EDS has been used to study the chemical characterization or elemental analysis; Figure (11) shows the blank mild steel sample before being immersed in 1M H_2SO_4 in absence of berries juice extract where high Fe peak is shown, which indicates the presence of high amount of iron which reaches to 88 Wt% in the mild steel surface. Figure (12) shows the mild steel surface that has been immersed for 2 hours in 50 ml of 1M H_2SO_4 in absence of berries juice extract; the mild steel surface contains low amount of Fe (87.3 Wt%) compared to the blank mild steel sample. In Figure (13) the mild steel sample immersed for 2 hours in 50 ml of 1M H_2SO_4 in the presence of 5ml of berries juice extract, it shows high spark of Fe where the mild steel surface contains lower amount of Fe (84.8 Wt%) compared to Figure(11) and (12) which indicate lower iron dissolution rate in 1M H_2SO_4 medium containing berries juice extract (Alaneme, K. et al,2016). A high amount of oxygen and carbon are observed in presence of berries juice extract compared to the absence of berries juice extract. Si is shown in the absence and presence of berries juice extract which means that's it comes out to the surface during the reaction with sulfuric acid.

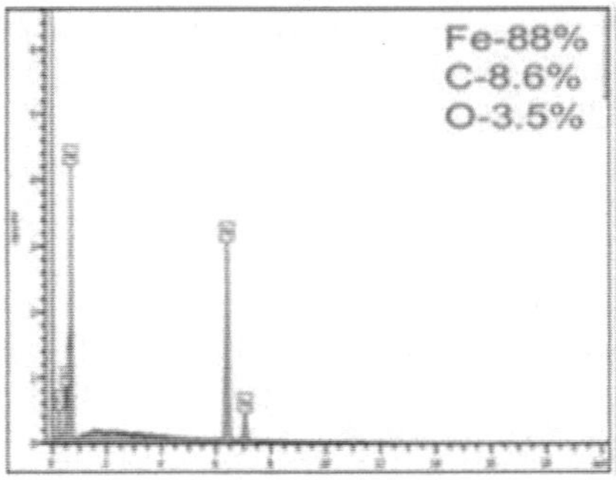

Figure(11)

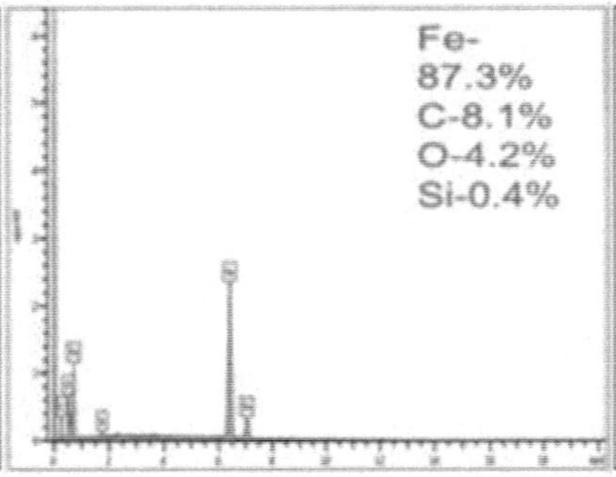

Figure(12)

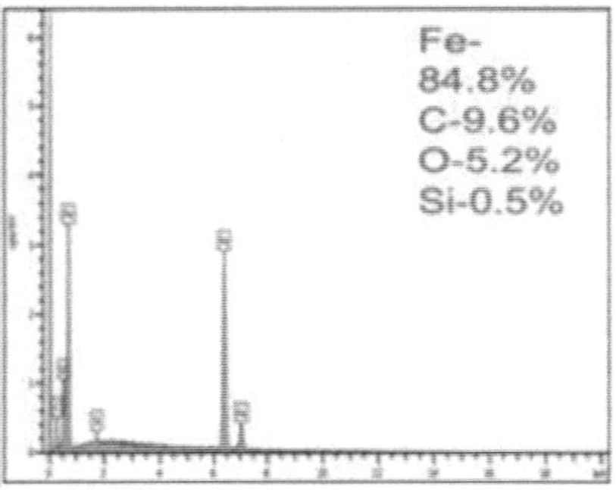

Figure(13)

GENETIC ALGORITHM OPTIMIZATION

The experimental result gives 288.367mpy as a corrosion rate. The software gives the minimum corrosion rate value of 188.866mpy from 50 runs as shown in Figure(14). A small difference can be noted as the genetic algorithm is not the exact fit with linear equations so the experimental and software results are too close to each other.

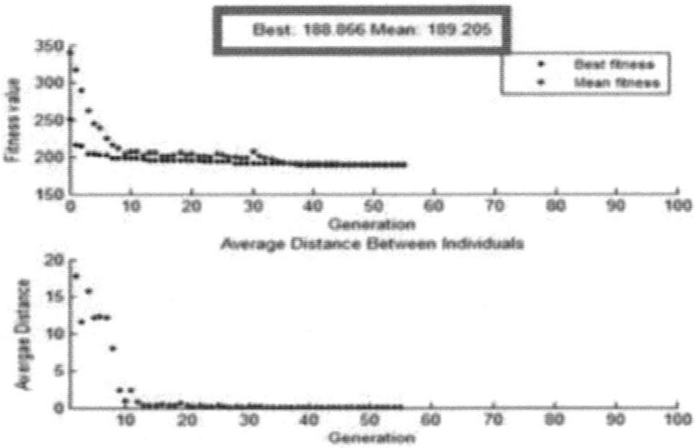

Figure(14): Genetic Algorithm optimization.

CONCLUSION

In this project, all the objectives have been achieved by designing and experimenting.

Firstly; the anticorrosive extract (berries) has been prepared, the prepared extract which contains 4 types of berries namely blueberries, blackberries, cranberries, and raspberries inhibit the mild steel corrosion in sulfuric acid medium with 41.427% efficiency in presence of 5 ml of berries extract.

Secondly, the mild steel corrosion in the presence and absence of berries juice extract were studied where the corrosion rate reaches to 1020.374mpy in absence of berries juice. By using 5 ml of berries juice extract the corrosion rate minimized to 232.022mpy.

Thirdly, the optimum conditions to inhibit the mild steel corrosion were studied by Design Expert V11. At the optimum conditions of 1M of H_2SO_4. 5ml of berries juice extract, 240 minutes and at room temperature the corrosion rate reaches to 232.022mpy as a minimum value.

The actual equation obtained from Design Expert V11 was later optimized using Genetic Algorithm. Genetic Algorithm shows that at 1.001M of H_2SO_4, 15 ml of berries juice extract and 237.208 minutes at room temperature, the corrosion rate is minimized to 188.866mpy.

Finally, mixing tank has been designed with 1.18m height and 2.5m diameter and a turbine agitator has been selected.

ACKNOWLEDGEMENT

I would like to thank god for giving me patience and power to work in my most important part of my life, my father, mother, my brother and my friends for their help and support. I would also like to express my sincere gratitude to Dr. Priy Brat Dwivedi for his support and help in developing my project idea, Mr. M.J.Varghese for his help in RSM analysis, Mr. Hatem Yazidi and Dr. Moammar Ali AlTubi for providing me the technical knowledge in Matlab optimization, Mr.Mohammed Salim Adeeb in providing me the technical help in the laboratory, Ganesh Nagraj Patil, Dr.Lakkim Setty

Rao, and Dr.Santosh Madhav Walke for their help in the calculations of design mixing tank. Last but not the least, I would like to thank National University of Science and Technology-College of Engineering for giving me this opportunity to complete the project in their laboratories and supporting me by providing me with the required chemicals.

REFERENCES

1.Alaneme,K. et al .,2016.Corrosion Inhibitory Properties of Elephant Grass(Pennisetum Purpureum) Extract Effect on Mild Steel Corrosion in 1M HCl Solution :*Alexandria Engineering Journal*.[e-journal]. 55 (8). p. 1069-1076. Available from: https://www.sciencedirect.com/science/article/pii/S1110016816300242.[Accessed: 8th May 2019].

2.Ambrish,S. et al .,2012. The Effect of Environmentally Bening Fruit Extract of Shahjan(Moringa Olifera) on the Corrosion of Mild Steel in Hydrochloric Acid Solution:*Journal of Chemical Engineering Communications*.[e-journal]. 199 (6). p. 63-77. Available from:https://www.tandfonline.com/doi/abs/10.1080/00986445.2011.570390.[Accessed: 6th May 2019].

3.Blumberg,J.,et al.,2013.Carnberries and Their Bioactive Constituents in Human Health:*Journal of Advances in Nutrition*.[e-journal]. 12 (7). p. 618-632. Available from: https://academic.oup.com/advances/article/4/6/618/4595566.[Accessed: 1th May 2019].

4.Bobinaite,R.,Viskelis,P.&Venskutonis,P.,2015.Chemical Composition of Raspberry Cultivars:*International Journal of Molecular Sciences*.[e-journal]. 16 (8). p. 18642-18663. Available from: https://www.researchgate.net/publication/306158828_Chemical_Composition_of_Raspberry_Rubus_spp_Cultivars.[Accessed: 1th May 2019].

5.Cafferty,E.,2010.*Introduction to Corrosion Science*.United States:Springer.
Roberge,P.,2012.*Handbook of Corrosion Engineering*.2nd edition.United States:McGraw-Hill company.
6.Dynamixianc,2015.*Optimal Mixer Design* .[Online]. Available from: https://www.dynamixinc.com/optimal-mixer-design. [Accessed: 14th December 2018]

7.Ibryamoval,Y. et al .,2009.Design of Mixers for Chemical and Biochemical Industry: *Trakia Journal of Sciences*.[e-journal]. 8 (6). p. 6-11. Available from: http://tru.uni-sz.bg/tsj/Vol8.Suppl.3.2010/Y.Ibryamova.pdf.[Accessed: 13th December 2018].

8.Kaume, L. et al.,2012. The Blackberry Fruit A Review on Its Composition and Chemistry, Metabolism and Bioavailability and Health Benefits:*Journal of Agricultural and Food Chemistry*.[e-journal]. 37 (7). p. 5716-5727. Available from: https://www.ncbi.nlm.nih.gov/pubmed/22082199.[Accessed: 6th May 2019].

9.Kumar.A,2012. Inhibition of Mild Steel Corrosion in HCl Solution Using Pipali(Piper Longum)Fruit Extract:*Arabian Journal for Science and Engineering*.[e-journal]. 38 (9). p. 85-97. Available from: https://link.springer.com/article/10.1007/s13369-012-0409-9.[Accessed: 9th May 2019].

10. Michalska,A.& Lysiak.G .,2015.Bioactive Compounds of Bluberries:*International Journal of Molecular Sciences*.[e-journal]. 16 (8). p. 18642-18663. Available from: https://www.ncbi.nlm.nih.gov/pmc/articles/PMC4581264/.[Accessed: 1th May 2019].

11.*Mixing Tank*,2018.[Online]. Available from: https://www.conceptdraw.com/examples/agitator-tank-symbol.[Accessed: 19th November 2018].

12. Mycourses,2016. *Response Surface Modelling-RSM*.[Online]. Available from:https://mycourses.aalto.fi/pluginfile.php/364340/mod_resource/content/1/BOS_Lecture3_2016.[Accessed: 4th May 2019].

13. Ostovari, L. et al.,2009.Corrosion Inhibition of Mild Steel in 1M HCl Soluition by henna extract:*Journal of Corrosion Science*.[e-journal]. 32 (9). p. 1935-1949. Available from: https://www.sciencedirect.com/science/article/pii/S0010938X09002157. [Accessed: 6th May 2019].

14. Petropedia,2018. *Mixing Tank*.[Online]. Available from: https://www.petropedia.com/definition/7742/mixing-tank.[Accessed: 11th December 2018].

15.Raja,P. et al.,2008.Inhibitive Effect of Black Pepper Extract on the Sulphuric Acid Corrosion of Mild Steel:*Journal of Materials Letters*.[e-journal]. 62 (9). p. 1977-2979. Available from: https://www.sciencedirect.com/science/article/abs/pii/S0167577X08001316.[Accessed: 8th May 2019].

16. Sinnot,R.&Towler,G.,2009.*Chemical Engineering Design*.5th edition.India:Elsevier.

17. Xianghong,L. et al.,2009.Inhibition Effect of 6-benzylaminopurine on the Corrosion of Cold Rolled Ssteel in H_2SO_4 Soluition:*Journal of Corrosion Science*.[e-journal]. 51 (9). p. 620-634. Available from: https://www.sciencedirect.com/science/article/pii/S0010938X09000080. [Accessed: 6th May 2019].

590434 Verification of Firefighters' Heuristics and Predictive Model Development through Big Data Analysis

Abstract:

The heuristics accumulated in the field activities of firefighters are reviewed through big data analysis of fire occurrences and emergency responses in Gyeonggi-do Province, Korea and expanded as predictive models to be utilized for proper fire prevention activities according to time, day, and target through quantitative modeling. Empirical rules with high sympathy are collected through direct interviews with firefighters. Among these, three types of high-priority rules are compared and analyzed for big data such as fire occurrence and damage history in Gyeonggi-do in 2018. Big data comparison analyses are conducted, including the number of fires and damages that occurred in Gyeonggi-do in 2018. Furthermore, fire occurrence patterns by region, day of the week, time of day, and building type are derived. In addition to using empirical rules that have been validated through research, relatively inexperienced firefighters also can make better decisions by relying on refined quantitative predictive models replacing empirical rules, including local government and time-based factors that reflect big fire occurrence data.

CONVERTING SHELTER-IN-PLACE LOCATIONS TO ZERO VULNERABILITY SAFE HAVENS

J. A. Rau
MineARC Systems America, LLC
4850 W. Ledbetter Drive
Dallas, Texas 75236
USA

A. Young
MineARC Systems, Pty Ltd
274 Welshpool Rd
Welshpool, WA 6106
Australia

M. McDermott
High Performance Building Solutions
7415 Folger Dr.
Charlotte, NC 28226
USA

CONVERTING SHELTER-IN-PLACE LOCATIONS TO ZERO VULNERABILITY SAFE HAVENS

ABSTRACT

While it is not uncommon to find shelter-in-place or safe haven locations designated across chemical facilities, what varies is the level of protection offered at these sites. Functionality can range from merely a muster point (shelter-in-place) to a positively pressurized room (safe haven). Either way, when determining the room's acceptable performance criteria, the industry generally applies ERPG-3[1] or a vulnerability factor. An acceptable vulnerability factor is often deemed the probability of a fatality being less than one percent.

In 2017, MineARC Systems and High Performance Building Solutions (HPBS) began working with a leading energy company to convert five shelter-in-place spaces to safe havens, which included additional life support systems that would ensure zero percent vulnerability. The objective of this paper is to examine the process to harden the space against contaminated air intrusion and what critical life support systems are essential for converting a shelter-in-place location to a safe haven.

INTRODUCTION

Emergency shelter-in-place (SIP) forms an integral part of any plant's broader emergency response plan. Explosions, fires, and the release of smoke or other toxic gases are some of the types of incidents that can occur despite high levels of planning and safety precautions[2].

Numerous government agencies and other organizations recommend shelter-in-place as a measure to reduce harm in the event of a chemical release. Any building may be used as a temporary measure to reduce health risk from exposure to the toxic materials by simply closing windows and doors and turning off ventilation fans and air conditioning/heating systems.

Three factors govern the effectiveness of a SIP location: 1) the tightness of the building or leakage rate, 2) the concentration of the toxic gas, and 3) the release duration. The tightness of the building is primarily based on building construction; however, temperature differential and wind speed also play a factor. Identification of the maximum toxic concentration occurs during a Facility Siting Study (FSS). FSSs are a hazard analysis that defines Maximum Credible Event (MCE) scenarios using a consequence-based approach to produce a quantitative risk assessment. Determining potential consequences is a necessary step in the process of developing a comprehensive safety plan. The FSS determines the expected release duration based on stored inventories of toxic materials. The higher the toxic concentration and longer the release duration, the less effective sheltering in place becomes.

While occupants may feel safe sheltering-in-place from the external toxic hazard, typical buildings have high leakage rates. A typical wood framed constructed building can have 3-5 air changes per hour (ACH50[3]), rendering it ineffective against most chemical releases. Additionally, sealing to an acceptable airtightness can be time consuming and expensive and even with professional air sealing it is very difficult to air seal existing structures to an acceptable tightness.

A case study for the conversion of five shelter-in-place spaces to zero percent vulnerability safe havens is presented. There are considerations for effectively sealing existing infrastructure and ensuring positive pressure to ensure contaminants do not infiltrate safe haven locations along with the critical element of life support. The critical elements of life support that are of primary importance to human survival in a safe haven are:

- Positive pressure
- Breathable air supply
- Supplemental O_2
- Removal of accumulated CO_2

---

[1] Emergency Response Planning Guidelines (ERPGs) are values developed by the AIHA Guideline Foundation's Emergency Response Planning Committee. ERPG-3 is the maximum airborne concentration below which it is believed nearly all individuals could be exposed for up to 1 hour without experiencing or developing life-threatening health effects.

2 The CSB is an independent federal agency charged with investigating industrial chemical accidents. Over the last 10 years they have completed investigations into 15 separate chemical release incidents.

[3] ACH50 is the number of times the air volume in a building changes per hour at 50 Pascal of pressure. Commercial air leakage rates are commonly reported at 75 Pascals, but this leakage study was conducted at 50 Pascals based on the targets included in the FSS report.

- Cooling and dehumidifying

CASE STUDY

In 2017 MineARC Systems and its project partner HPBS began working alongside a major energy company to convert five shelter-in-place locations into zero percent vulnerability safe havens.

The process started with the energy company completing its FSS, which included plume modeling to define the chemical release concentrations at each building location along with ideal shelter-in-place locations[4] and internal minimum positive pressure[5]. This information was used by MineARC to calculate indoor concentrations and occupant vulnerability to highlight that shelter-in-place alone would result in the vulnerability of greater than zero percent.

From there, HPBS conducted air infiltration testing following ASTM E779-10 to determine an air leakage baseline (see Table 6) for rooms that were designated to be sealed (pre-weatherization). A site assessment was then completed to review each location and determine its suitability for sealing and conversion to a safe haven. Firstly, the measured baseline infiltration rates were compared to the FEMA 453 estimates based on construction type[6] (see Figure 2). All surfaces of the room were examined for ease of conventional air sealing and mechanical equipment sealing. Rooms that included false floors, large amounts of electrical conduit, cabinetry, computer servers or windows etc. were deemed difficult to seal (i.e. control rooms). Secondly, the building or rooms open floor space was checked to ensure it could accommodate the specified number of people in an emergency. FEMA 453 section 1.6 recommends 20ft^2 per person for durations of less than 24 hours. With the FSS recommending 4 hours, it was deemed acceptable to have an unrestricted floor space of 6ft^2 per person.

Manual weatherization (air sealing) of the rooms included installing a cleanroom suspended ceiling system with gaskets and tiles that met the air permeance rate for an air barrier material, cleanroom airtight light fixtures with gaskets, sealing around electrical and plumbing, air sealing all visible air pathways to make sure the interior gypsum board was functioning as the air barrier, installing low-leakage HVAC dampers, installing an emergency shut off for the mechanical system and adding door weather-stripping and sweeps. Post-weatherization air leakage tests were conducted to determine the air infiltration rate of the SIP spaces after completion of the manual air sealing, cleanroom suspended ceiling system and HVAC work. The weatherization air sealing methods improved the air barrier integrity for all the proposed SIP spaces by between 57% and 65% (refer to Table 6).

Aerosol-based envelope sealing was used to reduce leakage rates even further. AeroBarrier® is a building envelope sealing technology that simultaneously measures and seals buildings. AeroBarrier® works by forcing an atomized sealant into orifices under pressure gradually sealing leaks. Rooms were sealed during night shift with personnel able to re-access the rooms by morning. The AeroBarrier® air sealing methods improved the air leakage for all the SIP spaces between 7% and 88% over the post manual weatherization leakage rates (refer to Table 7). AeroBarrier® is not effective for sealing large holes so manual air sealing is required for certain locations prior to using AeroBarrier®.

MineARC engineers once again calculated infiltration concentration rates along with the vulnerability rate. As the probability of a fatality was still higher than zero, MineARC then specified its AirBANK Pressure System and necessary carbon dioxide and oxygen control equipment.

SAFE HAVEN LIFE SUPPORT EQUIPMENT

Primary Life Supporting Atmosphere

A safe haven needs to provide a reliable breathable atmosphere independent of external toxic gas concentrations. Human breathing consumes oxygen (O_2) and expels carbon dioxide (CO_2), carbon monoxide (CO), water vapor, and heat. At complete rest, an average person exhales air at 17% O_2 and 3.2% CO_2. This breathing rate results in a 'resting' O_2 uptake of 0.3L/min of O_2 and a resting CO_2 discharge of 0.24L/min (Brake, Fellows, & Bates, 1999). Taking into consideration the above values, a person in a closed system such as a shelter-in-place succumbs to CO_2 poisoning well before O_2

4 Shelter-in-place location is beyond the scope of this paper, but should be determined based on the chemical hazard, facility layout, and location of personnel.

5 Positive pressure requirements are based on average wind speed at the facility. As wind speed increases so does the SIP leakage rate.

6 The FEMA 453 estimates are intended to estimate filter sizing when ASTM E779 air leakage testing is not conducted.

asphyxiation. Venter et al. (1998) found an average oxygen consumption for 12 entrapped individuals of 0.44 liters per minute, at STP, over 24 hours (including sleeping). A figure of 0.5 liters per minute over a non-sleeping entrapment period of 12 hours is a prudent, conservative design value. This corresponds to a metabolic rate of about 160W or 80W/m^2 for a typical person with a body skin area of about 2m^2, which is equivalent to a breathing rate of about 12.5 liters per minute of fresh air.

Consequently, life support systems are needed within a safe haven that can ensure positive pressure keeps contaminants outside, along with replacing the O_2 consumed, and removing exhaled CO_2, moisture, and heat to maintain the atmosphere within acceptable threshold limits.

Positive Pressure

Positive pressure is integral to prevent the ingress of airborne contaminants that may compromise the breathable atmosphere for occupants. Establishing a Class 1 Safe Room (FEMA 453) requires the sufficient release of filtered air from an external source, to form a positively pressurized zone.

Buildings are typically constructed to provide natural or forced ventilation to allow permanent occupancy and potentially combined with large exposed windows and surfaces which can present challenges for ensuring protection during a chemical release. When determining adequate positive pressure within a space, several structural factors need to be considered, including:

- Shelter volume
- HVAC (Heating, Ventilating, and Air Conditioning)
- Door and window tightness
- Existing opening such as vents and elevator shafts
- Indoor and outdoor temperature
- Wind speed and building orientation
- The external screening of the structure

Building design and multiple ingress points suggest the conversion of a smaller interior room within a building can provide a higher level of protection, than the whole building, as it is less subject to wind and buoyancy forces which induce the infiltration of external hazards.

Breathable air from compressed air cylinders venting directly into the SIP forms a positive internal pressure. Determining the amount of air required to reach the necessary level of positive pressure involves measuring the areas internal volume and the relationship between the internal and external atmospheric conditions.

After reaching required ACH (air changes per hour) or tightness, a positive-pressure management system must be sized for the expected shelter duration.

Breathable Air Supply

If the shelter-in-place is airtight, a supply of breathable air is necessary to support occupants over the required shelter duration. Unventilated SIPs that are tightly sealed cannot be occupied for long periods without the risk of high carbon dioxide levels. A secondary source of breathable air must be circulated within the shelter to meet occupancy over the required amount of time.

Clean, breathable air is supplied to a chemical, biological and radiological (CBR) safe room in a few ways: by trapping air inside the room and minimizing the air exchange, passing contaminated air through a filter (not practical for high concentration chemical releases or for chemicals that cannot be filtered) to purify it as it is supplied to the room, and the supplemental supply of oxygen through high pressure compressed air cylinders.

The air supply must contain enough oxygen in its mix, and not contain contaminants like carbon monoxide, carbon dioxide, water, or any other toxins that poison safe haven occupants during entrapment.

Table 1 lists international standards for flowrates for breathable compressed air for self-contained breathing apparatus (SCBA), supplied air respirators (SAR) and refuge chambers.

Table 1 – Standards for acceptable air supply requirements

Air supply requirements	Relevant Standard
Open-circuit positive-pressure SCBA, including all components and accessories except the air cylinder (shell) must resist the permeation and penetration of HD and GB chemical agents when tested for a period of 6 hours on an upper-torso manikin connected to a breathing machine operating at an airflow rate of 40 liters per minute (lpm), 36 respirations per minute, 1.1 liters tidal volume.	NIOSH 42 CFR Part 84 Respiratory Protective Devices
The flow rate should be at least 85L/min (3cfm) per person to remove expired CO2 and CO but at least 255L/min (9cfm) to provide adequate cooling and maintain comfortable conditions. The air inflow is assumed to be at a temperature of 26.5°C and 65% relative humidity	ITA Report No. 14 – Guideline for the provision of refuge chambers in tunnels under construction, May 2014
For manually operated blower type air-hose respirators, the blower shall be capable of delivering air to the space enclosed by the facepiece at a flow rate of at least 85 L/min.	AS/NZS 17:16:2003 Respiratory Protective Devices

The NIOSH standard of 40L/min is for SCBA apparatus (i.e. mouthpiece or face mask) where dilution of CO_2 is not a concern. Conversely, the ITA and Australian standards specify 85L/min and are concerned with both oxygen consumption and carbon dioxide dilution. It is important to note that three times the amount of air is required to maintain comfortable conditions inside a refuge chamber, when compared to simply breathing (Health & Safety Executive UK, 2006).

Unless the positive pressure system is supplying breathable air at a rate that maintains oxygen and compressed gas levels within acceptable ranges (dependent on safe haven occupancy and duration), a CO_2 scrubber and supplementary oxygen supply are required.

Supplemental O_2

O_2 is an essential requirement for cellular activity, and therefore, the absence of O_2 for relatively short periods results in death. Humans within an enclosed environment consume O_2 and if not replaced, would result in an irrespirable atmosphere. An O_2 concentration of less than 11% poses a high risk of death.

Retaining the oxygen levels between acceptable limits is critical. Various international standards outline acceptable O_2 limits for a breathable environment; examples are shown in Table 2.

Table 2 - Standards for allowable O_2 levels

Minimum O_2 Content (% by volume)	Maximum O_2 Content (% by volume)	Relevant Standard
18.5%	23.0%	Mines Safety and Health Administration 30 CFR Parts 7 and 75 Refuge Alternatives for Underground Coal Mines; Final Rule
19.5%	23.5%	Occupational Safety and Health Administration 29 CFR 1910.146 Permit-Required Confined Spaces for General Industry
19.5%	23.5%	Australian Standard AS 2865–2009 Safe working in a confined space

OSHA and Australian Standards for working in a confined space specify an acceptable lower limit of 19.5%, which is consistent with many other standards around the world. The slightly lower MSHA limit of 18.5% is based on a person at rest, not working, therefore, not likely to have trouble in breathing; conditions that are likely to be present in a refuge chamber.

O_2 itself is not flammable but supports and accelerates combustion. OSHA defines O_2 enriched mixtures or atmospheres as those containing more than 23.5% O_2 by volume (OSHA, 2013). In O_2-enriched atmospheres, the reactivity of O_2 significantly increases the risk of ignition and fire.

Within a safe haven, pure O_2 must be utilized in conjunction with a CO_2 removal system.

Removal of Accumulated CO_2

Protection within a safe haven is substantially increased by filtering CO_2 from the air. CO_2 is being produced continuously by the cells of our body and is a byproduct of respiration. The inability to expel this gas or an environment with high concentrations of CO_2 leads to CO_2 poisoning. Breathing air with more than 3% CO_2 can induce dizziness, headaches, and vision disturbance. Prolonged exposure at greater than 6% could result in unconsciousness and death. Listed in Table 3 are examples of international standards of satisfactory CO_2 removal requirements for a breathable environment.

Table 3 – Acceptable CO_2 removal requirements

CO_2 Removal Requirements	**Relevant Standard**
CO_2 removal components shall remove CO_2 at a rate of 1.08 ft^3 (30.5 L)/h/person.	Mines Safety and Health Administration 30 CFR Parts 7 and 75 Refuge Alternatives for Underground Coal Mines; Final Rule
OSHA has established a Permissible Exposure Limit (PEL) for CO2 of 5,000 parts per million (ppm) (0.5% CO2 in air) averaged over an 8-hour workday (time-weighted average or TWA.)	Occupational Safety and Health Administration
5.2.8 Atmospheric monitoring. Atmospheric monitoring There should be the capability to monitor temperature continuously, and for carbon monoxide, carbon dioxide, and oxygen in the chamber during use in both externally supported or standalone modes. The concentration of carbon dioxide should not exceed 10000 ppm.	ITA Report No. 14 – Guideline for the provision of refuge chambers in tunnels under construction, May 2014

The CO_2 that is exhaled by the occupants within the safe haven must be removed by, constant flushing by an external breathable air source (e.g. compressed air cylinders used for positive pressure) or by the regeneration of the internal atmosphere with a CO_2 absorbent (scrubber).

Different regulations around the world, state varied human production rates between 30–50L per hour per person. When considering an acceptable production rate; it must be noted that the rate of breathing is primarily triggered by the CO_2 content of inspired air with higher CO_2 levels triggering faster respiration rates (Brake, Fellows, & Bates, 1999) (i.e. if the maximum allowable CO_2 concentration inside the safe haven is 1%, a lower production rate is acceptable compared to a 3% CO_2 limit).

If CO_2 removal mechanisms are inadequate, the occupants can develop CO_2 toxicity by re-breathing their own exhaled CO_2.

Cooling and Dehumidifying

Thermal tolerability with a safe haven should be addressed during the design phase. Assumptions must be made in preparation for safe haven design in regard to the number of personnel, activity level, available hydration and heat production. The minimum airflow to maintain a comfortable environment is set by taking these assumptions into account as well as the input temperature and humidity of supplied air.

The healthy human body maintains its internal temperature around 37°C. An essential requirement for continued normal body function is that the deep body temperature is maintained within the narrow 1°C limit (Epstein, Moran, 2006). A change of body temperature exceeding 1°C occurs only during illness or when environmental conditions surpass the body's ability to cope with extreme temperatures.

When a person is subject to extreme heat and humidity over some time, the person may experience a very rapid progression of symptoms or exhibit a few of the less significant symptoms before falling victim to more severe forms of heat-related illness.

The acceptable cooling requirements for underground refuge chambers are defined in various international standards, examples of which are listed in the below table.

Table 4 – Standards for acceptable cooling requirements

Cooling Requirements	Relevant Standard
Simulated emergencies, in which a full complement of people has occupied a refuge chamber for a significant period, indicate that humidity and temperature can increase very rapidly to potentially heatstroke-inducing levels. Refrigerated air conditioning is strongly recommended for both externally supported and stand-alone refuge chambers to counter this potentially dangerous problem. Inevitably, this places a heavy demand on the stand-alone power supply, but there are systems available that can cope, and at an acceptable cost.	Department of Mines, Industry Regulations and Safety, 2005 Refuge Chambers in Underground metalliferous mines – guideline: Safety and Health Division, Department of Industry and Resources Western Australia, 32 pp
Apparent temperature in the fully occupied refuge chamber must not exceed 95 °F.	Mines Safety and Health Administration 30 CFR Parts 7 and 75 Refuge Alternatives for Underground Coal Mines; Final Rule
It should be possible to maintain an internal temperature not exceeding 30 °C wet bulb globe temperature for at least 24 hours	ITA Report No. 14 – Guideline for the provision of refuge chambers in tunnels under construction, May 2014

MSHA regulations specify a maximum internal apparent temperature[7] of 95°F (35°C). NIOSH specifies an equivalent maximum wet bulb temperature of 28.9°C (NIOSH, 2007).

The temperature maintained inside the safe haven is determined by the metabolic heat liberated by the occupants, dissipated to the external environment by the heat transfer characteristics of the room or building. The external temperature primarily influences the heat transfer while the humidity is the combination of moisture produced by the occupants (via exhaled air and sweat) and water liberated by CO_2 scrubbing chemicals.

It is important to remember that in most cases, before sheltering-in-place, the air conditioning system will need to be shut down (unless it is a split system). For short durations, artificial cooling is not a critical requirement unless external temperatures are extraordinarily high, and/or the safe haven's volume is small compared to its occupancy.

INFILTRATION TESTING

Background

High Performance Building Solutions Inc. (HPBS) was contracted to perform an air infiltration study for spaces identified as candidates to be used as shelter-in-place (SIP) in the event of an accidental chemical release. This SIP strategy uses the existing structure and its indoor atmosphere to separate individuals from a hazardous outdoor atmosphere temporarily.

A series of pre-weatherization air leakage tests were conducted to determine the baseline air infiltration rate of potential SIP spaces. Manual air sealing, a cleanroom suspended ceiling system and low-leakage HVAC dampers were installed to reduce the overall air infiltration of the SIP spaces and to close off air leakage pathways from non-protected spaces in the building through the HVAC system. Post-weatherization air leakage tests were conducted to determine the air infiltration rate of the SIP space after the completion of weatherization and HVAC work. The SIP spaces were additionally air sealed using the proprietary AeroBarrier® system to further reduce infiltration rates.

Scope

Air sealing work was completed, and low-leakage HVAC dampers were installed to reduce the overall air infiltration of the SIP spaces and to isolate the HVAC from non-protected spaces in the building. Post-weatherization air leakage tests were conducted to determine the air infiltration rate of the SIP space after the completion of air sealing, suspended ceiling and HVAC work. AeroBarrier® air sealing was conducted on SIP spaces as a pilot project on September 18 through September 21, 2017. Post AeroBarrier® air leakage tests were conducted on September 21 and September 22, 2017, to determine the air infiltration rate of the SIP space after the completion of the AeroBarrier® sealing.

7 Apparent temperature is the temperature equivalent perceived by humans, caused by the combined effects of air temperature and relative humidity.

Following ASTM E 779-10, pressurization air leakage tests were conducted on the following spaces recommended in the Facility Siting Study (FSS).

Table 5: SIP specifications

SIP Space	Occupancy	Floor Area (Square Feet)	Volume (Cubic Feet)
SIP Space 1	40	990	10,481
SIP Space 2	6	955	8,210
SIP Space 3	5	391	3,917
SIP Space 4	10	1,697	16,968
SIP Space 5	40	293	2,637

Air Leakage Test Results

Positive pressure air leakage test results are reported in four different ways:

CFM4 is the quantitative volume of air passing through an enclosure under a pressure difference of 4 Pascals (equivalent to a 5mph wind). This is a measurable physical quantity and the value recommended in the FSS to eliminate toxic gas ingress into the SIP spaces.

CFM25 is the quantitative volume of air passing through an enclosure under a pressure difference of 25 Pascals (0.1-in w.g.). FEMA 453 requires Class 1 Safe Havens be pressurized to 25 Pascals. This is a measurable physical property.

CFM50 is the quantitative volume of air passing through an enclosure under a pressure difference of 50 Pascals (0.2-in w.g.).

ACH50 is the quantitative volume of air passing through an enclosure under a pressure difference of 50 Pascals (0.2-in w.g.) normalized by the enclosure volume. ACH is the abbreviation for air changes per hour. It is the number of times the air volume in a building changes per hour at 50 Pascals of pressure.

Table 6 below shows the CFM50 and ACH50 for each of the SIP spaces pre-weatherization and post-weatherization along with the percentage reduction in leakage. SIP Space 1 and SIP Space 5 did not have pre-weatherization results as initially the whole building was tested. Smaller spaces were selected as the whole building was too large to feasibly convert to a SIP space (re-testing of the new SIP spaces was not completed). The control room, SIP space 4, was abandoned in favor of a smaller adjacent room due to high leakage post manual weatherization and the computers that could not be shut down, which prevented the use of AeroBarrier® in the space.

Table 6: Post-weatherization leakage results

Safe Haven Space	CFM50		ACH50		% Reduction
	Pre-Weatherization	Post-Weatherization	Pre-Weatherization	Post-Weatherization	
SIP Space 1	N/A	1,871	N/A	10.7	N/A
SIP Space 2	3,623	1,560	26	11.4	57%
SIP Space 3	385	135	6	2.1	65%
SIP Space 4	N/A	N/A	N/A	N/A	N/A
SIP Space 5	N/A	655	N/A	19.4	N/A

The AeroBarrier® air sealing was prescribed a target. According to the AeroBarrier® simultaneous air leakage calculations, all the spaces met and surpassed this airtightness target; AeroBarrier® sealing was halted at this point. Table 7 shows the CFM50 and ACH50 for each of the SIP spaces as tested by HPBS post AeroBarrier® along with the percentage reduction. The HPBS results were utilized to calculate the amount of positive pressure required to convert the rooms to positive pressure safe havens.

Table 7: Post AeroBarrier® leakage results

Safe Haven Space	CFM50		ACH50		% Reduction
	Post-Weatherization	Post AeroBarrier®	Post-Weatherization	Post AeroBarrier®	
SIP Space 1	1,871	455	10.7	2.6	76%
SIP Space 2	1,560	182	11.4	1.3	88%
SIP Space 3	135	125	2.1	1.9	7%
SIP Space 4	N/A	173	N/A	3.3	N/A
SIP Space 5	655	138	19.4	3.1	79%

The AeroBarrier® air sealing methods improved the air barrier integrity for all the proposed SIP spaces between 7% and 88% over the post manual weatherization leakage rates. Air leakage was still observed behind many electrical penetrations including HVAC thermostats, and fire and smoke alarms. These were covered with plastic to protect them from the AeroBarrier® Sealant but had hidden holes behind them.

Figure 1 displays the post AeroBarrier® leakage rate in comparison to the FEMA 453 estimates for air leakage rates based on construction types. FEMA 453 estimates are reported in square feet of floor area at 25 Pascals and do not take into consideration the ceiling height and volume of the space.

Figure 1: Post AeroBarrier® results compared to FEMA 453 estimates

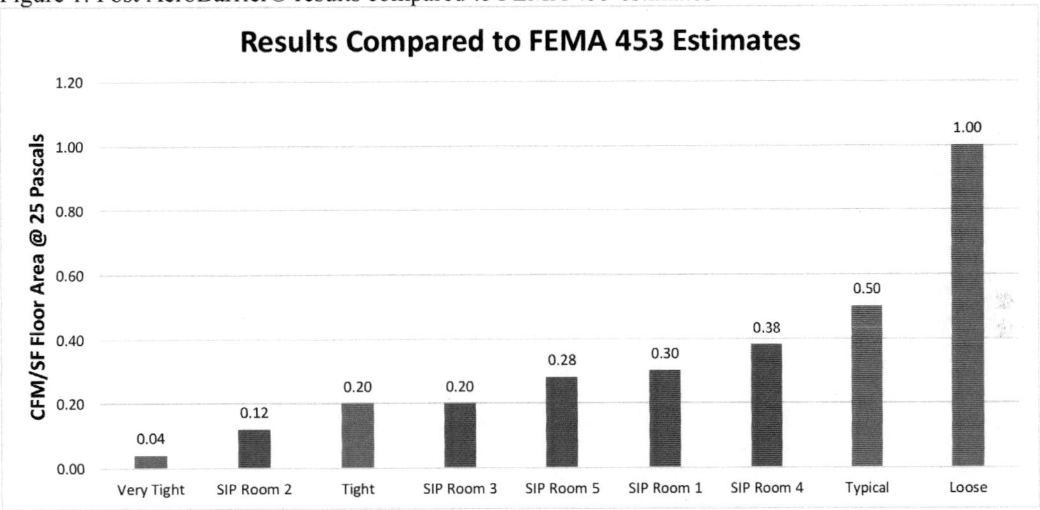

Figure 2 is an estimated leakage rate based on construction type to estimate the filtration and positive pressure system size under the FEMA 453 standard. All the SIP locations were between tight and typical except for SIP Room 2, which was a small prefabricated metal constructed portable building.

Figure 2: Estimated leakage rate per square foot of floor space to achieve 0.1 iwg (25Pa) overpressure

Construction Type	cfm per square foot of floor area
Very tight: 26-inch thick concrete walls and roof with no windows	0.04
Tight: 12-inch thick concrete or block walls and roof with tight windows and multiple, sealed penetrations	0.20
Typical: 12-inch thick concrete or block walls with gypsum wall board ceilings or composition roof and multiple, sealed penetrations	0.50
Loose: Wood-frame construction without special sealing measures	1.00

Table 8 shows the combined AeroBarrier®, and manual air sealing methods improved the airtightness for all the proposed SIP spaces between 68% and 95% over the original leakage rates.

Table 8: Air infiltration improvement for SIP spaces

Safe Haven Space	**ACH50**			**% Reduction**
	Pre-Weatherization	**Post-Weatherization**	**Post AeroBarrier®**	
SIP Space 1	N/A	10.7	2.6	76%
SIP Space 2	26.5	11.4	1.3	95%
SIP Space 3	5.9	2.1	1.9	68%
SIP Space 4	N/A	N/A	3.3	N/A
SIP Space 5	N/A	19.4	3.1	84%

Figure 3 compares the ACH50 pre-weatherization, post-weatherization and AeroBarrier® air infiltration results.

Figure 3: Pre-weatherization, post-weatherization and AeroBarrier® comparison

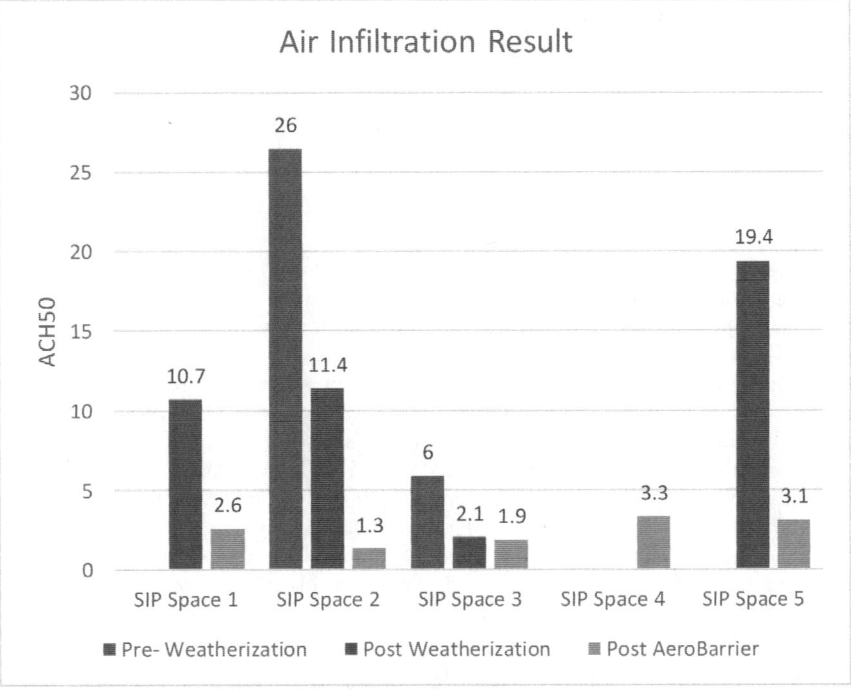

Life Support Systems

MineARC re-calculated the vulnerability for each SIP space based on the post AeroBarrier® leakage rates. All SIP spaces still had a vulnerability of 100% and therefore required the installation of independent life support systems to be deemed zero percent vulnerability safe havens.

Figure 4 shows the toxic gas concentration, duration of the release, and vulnerability for each SIP space. As the largest MCE was a full tank rupture the toxic concentrations were extremely high, but the duration was short at one minute.

Figure 4: SIP space vulnerability post AeroBarrier®

	TOXIC GAS				
Safe Haven Location	Worst Case Threat Scenario	Toxic External [ppm]	Duration of Reslease [min]	ACM	Vulnerability
SIP Space 1	Tank Rupture	585,000	1	0.032	100%
SIP Space 2	Tank Rupture	652,000	1	0.042	100%
SIP Space 3	Tank Rupture	660,000	1	0.007	100%
SIP Space 4	Tank Rupture	412,000	1	0.042	100%
SIP Space 5	Tank Rupture	600,000	1	0.028	100%

The leakage rate at 4 Pascals of pressure difference is slightly more than would occur under a 5 mile per hour wind used in the dispersion model in the FSS report based on climatic data for the facility (to maintain positive pressure and neutralize air infiltration). Post-weatherization infiltration rates in cubic feet per minute at 4 Pascals of pressure for each of the SIP spaces is shown in Table 9.

Table 9: Post AeroBarrier® air leakage results at CFM4

Safe Haven Space	Volume (Cubic Feet)	CFM4
SIP Space 1	10,481	99
SIP Space 2	8,210	32
SIP Space 3	3,917	24
SIP Space 4	3,110	45
SIP Space 5	2,637	21

The air leakage test results provide a mathematical formula that can be used to calculate the amount of breathable compressed air cylinders (437ft^3) that must be added to each candidate SIP space to maintain a 4 Pascal positive pressure to overcome a 5mph design wind coincident with a toxic release.

MineARC Systems was contracted to install life support systems to convert the SIP spaces to positive pressure safe havens. The MineARC AirBANK Positive Pressure System provides rapid pressurization, which is activated and maintained using the AirBANK Control via a simple human-machine interface (HMI) touch screen. MineARC's integrated Aura-FX Gas Monitor ensures breathable air automatically remains within acceptable limits. Additionally, the installation of supplementary oxygen and carbon dioxide scrubbing systems in safe havens, are as needed.

Figure 5 shows the number of 437 cubic feet cylinders required for each safe haven location. The minimum breathable air per person is 2.5cfm (70L/min) per person based on MineARC's experience and testing within the refuge chamber industry.

Figure 5: Quantity of cylinders required for 4Pa positive pressure

	SAFE HAVEN SPECIFICATIONS		POSITIVE PRESSURE (@ 4Pa)						
Safe Haven Location	Occupancy	Duration	CFM	CFM per Floor Space	ACH	No. Cylinders	No. Cylinder Racks	Breathable Air per Person [CFM]	Breathable Air per Person [LPM]
Safe Haven Space 1	40	4	99	0.10	0.6	55	5	2.5	70.8
Safe Haven Space 2	6	4	32	0.03	0.2	18	2	5.4	152.9
Safe Haven Space 3	5	4	24	0.06	0.4	14	2	4.8	135.9
Safe Haven Space 4	10	4	45	0.14	0.9	25	3	4.5	127.4
Safe Haven Space 5	40	4	21	0.07	0.5	12	2	0.6	17.0

The maximum CO_2 concentration is at 1% and O_2 between 19.5% and 23.5% based on OSHA limits (refer to Table 2). Due to the volume and number of people in Safe Haven Space 5, it required supplemental oxygen and carbon dioxide scrubbing to maintain the internal gas concentrations within specified limits. Three sodium chlorate oxygen candles and two AirGEN CO_2 scrubbers were included in the design.

Figure 6: Internal gas concentrations in safe havens

	SAFE HAVEN SPECIFICATIONS		POSITIVE PRESSURE (@ 4Pa)			02 & CO2 CONCENTRATIONS (@ 4Pa)			
Safe Haven Location	Occupancy	Duration	CFM	CFM per Floor Space	ACH	Indoor O_2 %	No. O_2 Candles	Indoor CO_2 %	No. AirGEN Scrubbers
Safe Haven Space 1	40	4	99	0.10	0.6	20.45	0	0.50	0
Safe Haven Space 2	6	4	32	0.03	0.2	20.84	0	0.17	0
Safe Haven Space 3	5	4	24	0.06	0.4	20.77	0	0.23	0
Safe Haven Space 4	10	4	45	0.14	0.9	20.66	0	0.31	0
Safe Haven Space 5	40	4	21	0.07	0.5	18.67	3	2.04	2

In 2018 MineARC completed installation and training on the five safe haven spaces. Each safe haven was tested during the commissioning process to ensure it could maintain acceptable positive pressure.

CONCLUSIONS

The conversion of shelter-in-place (SIP) locations to zero percent vulnerability safe havens is a four-step process consisting of the following:

1) Identifying the correct SIP location based on ease of sealing and small volume based on the required occupancy.
2) Manual air sealing of SIP spaces.
3) AeroBarrier® sealing of SIP spaces.
4) Installation of life support equipment.

Identifying SIP spaces that can seal from the outset is crucial to a successful result. In the case study, three of the five original locations were abandoned in favor of rooms in adjacent locations due to high leakage rates or reducing the size to make the project more cost effective. Control rooms pose the most difficulty in trying to seal due to the false floor and a myriad of cable tray and pass-throughs installed for the computer systems. The smaller the room and the more robust the construction (i.e. prefabricated metal or concrete buildings), the better it seals.

Manual air sealing needs to be overseen by someone with familiarity in building tightness. While most of the tasks are general construction, standard contractors are not trained to work at the level of precision needed to ensure proper sealing. Even a tiny orifice or ceiling tile not sealed correctly can ruin a leakage test result. HPBS and MineARC Systems have developed a system for ensuring airtightness in construction methodology and by avoiding un-sealable spaces or spaces that are not cost effective to seal.

For future construction projects, incorporation of SIP spaces should occur during the design phase. Air barrier commissioning is standard in new construction and should include the following tasks at a minimum: design review of the plans with specifications for the air barrier and mechanical system - specifically as these relate to spaces designated for use as safe havens; quality assurance inspections during construction to ensure a continuous air barrier is in place; and final air leakage testing of the continuous air barrier to ensure the leakage rate of the safe haven spaces are within acceptable limits.

Complementary life support equipment is essential to sustain life within an airtight environment. For personnel to remain within the sealed safe haven, the environment must be habitable. Air is to be free of toxins such as carbon dioxide and acceptable levels of oxygen maintained. Positive pressure systems reduce the ingress of toxins and provide some or all of the required source of breathable air. Additional equipment such as compressed air, oxygen cylinders and air scrubbers are used to meet the shortfall. Furthermore, internal conditions must consider the human condition, supporting workable body temperature and humidity.

Reducing the risk of harm to personnel is a priority for emergency response planning. While the likelihood of an MCE occurring may be low, the risk to on-site personnel in the event of an incident is high. Converting shelter-in-place (SIP) locations to zero vulnerability safe havens improves emergency response.

REFERENCES

ASTM E779-10, *Standard Test Method for Determining Air Leakage Rate by Fan Pressurization*, ASTM International, West Conshohocken, PA, 2019, www.astm.org

Brake, Fellow, & Bates (1999). *Criteria for the design of emergency refuge stations for an underground metal mine*. Aus. IMM Proceedings, No.2, 1999, pp 1–7.

Department of Mines and Petroleum, 2013, *Refuge chambers in underground mines — guideline: Resources Safety*, Department of Mines and Petroleum, Western Australia, 41 pp.

Epstein, Y., & Moran D. (2006). *Thermal Comfort and the Heat Stress Indices,* Industrial Health, 44, 388–398.

FEMA 453. (2006). *Design Guidance for Shelters and Safe Rooms.* Federal Emergency Management Agency

Health & Safety Executive UK (2008). *Guidance and Information on the role and design of safe havens in arrangements for escape from mines.* Retrieved from http://www.hse.gov.uk/pubns/mines08.pdf

ITA Working Group 5 (2016). *Provisions for refuge chambers in tunnel projects under construction.* ITA Report N°014 - V2 / March 2018

Mines Safety and Health Administration (2008). *30 CFR Parts 7 and 75 Refuge Alternatives for Underground Coal Mines; Final Rule.*

National Institute for Occupational Safety and Health (2007). *Research Report on Refuge Alternatives for Underground Coal Mines.* Retrieved from http://www.cdc.gov/niosh/docket/archive/pdfs/NIOSH-125/125 ResearchReportonRefugeAlternatives.pdf

Occupational Safety & Health Administration (2013). *Permit-required confined spaces.* Retrieved from http://www.osha.gov/pls/oshaweb/owadisp.show_document?p_table=STANDARDS&p_id=9797

Suming Du. (2014). *On the effectiveness of shelter-in-place as a measure to reduce harm from atmospheric releases. Journal of Emergency Management, Vol. 12, No, 3, May/June 2014.*

Venter. J. (1998). *Portable Refuge Chambers: Air or Tomb in Underground Escape Strategies.* School for Mechanical and Materials Engineering, Potchefstoom University, Republic of South Africa

Worksafe Australia National Standard (2005). *Australian Standard AS: NZS 1716:2003 Respiratory protective devices.*

Yantosik. G. (2006). *Shelter-in-Place Protective Action Guide*. The Chemical Stockpile Emergency Preparedness Program (CSEPP). U.S. Department of Energy

589008 How the Field of Human Factors Helps Solve the Riddle of Serious Injuries and Fatalities

Abstract:

Process and occupational safety are often treated as distinct endeavors, with the former tending to focus on facility system function and the latter on individual worker wellbeing. Though other differences certainly exist between the two, both endeavors are generally purposed to prevent incidents that can cost lives, health, time, and money, in addition to employee morale and public trust. The overarching thesis of this paper is that the field of Human Factors is uniquely positioned to effect positive change in both of these major safety categories.

The backdrop for our argument is the recent finding that while injury and illness incidence rates published by the Bureau of Labor statistics have decreased in recent years, the occurrences of serious injuries and fatalities have declined at a significantly slower rate. We discuss possible explanations for this finding but note that regardless of the preferred explanation, the trend should motivate scientists, engineers, and other safety professionals to revisit long-held accident causation theories holding that near misses – these are incidents that could have resulted in serious injuries and fatalities but did not – are precursors of future serious incidents. While this idea still has merit, the just-mentioned statistics warrant reconsideration of the relations between serious and less-serious incidents to explain why the rates of the two are not decreasing in lockstep. We suggest these data indicate additional (human) factors are involved.

To address this riddle and to support our general thesis, we (a) review core tenets of process safety, occupational safety, and the field of Human Factors, (b) point out that the human is always an element common to both process and occupational safety contexts, and (c) discuss how preventing or mitigating human error can benefit both. We offer specific examples from both safety contexts to show how Human Factors perspectives and techniques can help. Examples touch on standards, policies, and other administrative controls (e.g., procedures), human-machine interfaces of both workplace hardware (e.g., valves) and software (e.g., SCADA systems), environments (e.g., physical, regulatory, and cultural), and communications between personnel (e.g., elements of team resource management). We conclude that fundamental Human Factors considerations can help prevent and mitigate human performance gaps and therefore increase the rate of reduction in serious injuries and fatalities in both process and occupational safety.

589233 Initial Conditions Used in Atmospheric Dispersion Models:Issues with Commonly Used Assumptions for Gas Jets and Plumes

Abstract:

When simulating jets and plumes, all atmospheric dispersion models require initial conditions that appropriately parameterize the behavior of a gas release to the atmosphere. Often, process conditions are such that loss of primary containment of a gas will result in choked flow conditions. A great deal of experimental and theoretical consideration has been given to the flow from stagnation conditions to choked flow conditions, but less experimental work has centered on describing the flow through the depressurization process to atmospheric pressure where atmospheric dispersion models would be applicable. Traditional modeling of the depressurization process has taken the simplified approach of assuming a (depressurized) "top hat" profile (laterally constant) with application of the principles of conservation of mass, momentum, and energy. Such assumptions are known to result in predicted depressurized velocities that exceed the (sonic) velocity at the point of loss of containment, and because of the energy balance, the associated temperatures are significantly less than temperatures at the choked conditions. Physically, these effects could possibly self-cancel because as jet velocity slows, the decrease in kinetic energy of the jet will increase the jet temperature. However, no atmospheric dispersion model used for routine hazard assessment purposes accounts for kinetic energy in the energy balance. The traditional modeling approaches of the depressurization process are reconsidered based on analysis of available data, and an alternative approach to modeling the depressurization process is proposed. Example cases showing the important differences between the two approaches are discussed.

Fault Tree Uncertainty Analysis

Raymond "Randy" Freeman
S&PP Consulting
12303 Lake Shore Ridge
Houston, TX 77041
Cell: 713 408 0357
Rafree@yahoo.com

For Presentation at the 2020 Spring National AIChE Meeting
Houston, Texas

March 30, 2020

ABSTRACT

Fault tree analysis (FTA) is a widely used methodology in the process industries. FTA is used for the development of failure mechanisms, computation of failure frequencies and the determination of the probability of failure on demand of safety systems. Much of the data used in a FTA study are uncertain. For example, the failure rate of a pump is often not known with great precision. Likewise the failure rates of instrumentation are often known only within some defined limits. The common practice, used by analysts in the quantification of a fault tree, is to use the most likely or best guess as to the needed failure rate data. The use of best guess values as data inputs to the quantification of a fault tree creates uncertainty in the computed results.

This paper presents a general methodology for the determination of the impact of uncertainty on the results of a fault tree study. The general methodology is based on the mathematics of propagation of error and variance contribution analysis. For a fault tree that contains 10 event minimal cut sets, the methodology will compute the variance in the probability (or frequency) of the top event with a maximum error of 5%. For 5 event minimal cut sets the error in variance of the top event probability (or frequency) will be no more than 2%. An example is presented to illustrate the application of the fault tree uncertainty analysis methodology to a real world problem.

Keywords

Quantitative Risk Assessment, Fault Tree Analysis, Uncertainty Analysis

1. Background

Fault tree analysis is a widely used method for representation of failure mechanisms in chemical plants. The methodology is described in the CCPS CPQRA guideline book [Ref. 1] and is extensively discussed in Chapter 9 of Lees [Ref. 2]. Lacking in these discussions is a systematic method to quantify the uncertainty of the results of the fault tree calculations.

Fault tree analysis has also been used in the aircraft and aerospace industries, nuclear power plant review and is recognized by the OSHA Process Safety Management (PSM) standard [Ref. 3] as a method for process hazards analysis.

2. Review of Error Propagation and Variance Contribution Analysis (VCA) Methodology

The mean and variance of a function of random variables can be approximated using the method described by Haugen [Ref. 4] and applied by Freeman [Ref. 5, 6, 7]. Define an arbitrary function of a set of random variables, x_i, as:

Let

$$Y = F(x_i) \qquad \text{(Eq 1)}$$

The mean of Y can be estimated using the following approximation:

$$E(Y) = F[\,E(x_i)] \qquad \text{(Eq 2)}$$

Where:

E(Y) = expected value of random variable Y = mean of Y

$E(x_i)$ = expected value of random variable x_i = mean of x_i

The variance of Y can likewise be estimated as:

$$V(Y) = \sum_{i=1}^{n}\left[\frac{\partial Y}{\partial x_i}\right]^2 V(x_i) \qquad \text{(Eq 3)}$$

Where:

$V(Y)$ = variance of random variable Y as defined above in Equation 1

$V(x_i)$ = variance of random variable x_i as defined above in Equation 1

Note that the variance is simply the square of the standard deviation. Using the variance will simplify the mathematics that is described below. The contribution of each independent random variable to the overall variance in the function is:

$$V(Y \text{ from } x_i) = \left[\frac{\partial Y}{\partial x_i}\right]^2 V(x_i) \qquad \text{(Eq 4)}$$

The relative contribution of each term to the overall variance V(Y) is a measure of the importance in the uncertainty in the particular random variable, x_i. In effect, this is a sensitivity analysis combined with a uncertainty evaluation. The variance contribution combines the sensitivity in the answer to changes in the uncertain random variable, x_i, with a measure of the uncertainty in the random variable, x_i. The overall variance in Y is found by summing the sensitivity weighted variances from each random variable.

Numerical Estimate of the Sensitivity

Let us return to the fundamental definition of the derivative.

As before, set y = F(xi)

$$\frac{\partial y}{\partial x} = \lim_{\Delta xi \to 0} \left[\frac{F(xi + \Delta xi) - F(xi)}{\Delta xi}\right] \qquad \text{(Eq 5)}$$

We are interested in a numerical estimate of the derivative. Therefore, we will make the following approximation.

$$\frac{\partial y}{\partial x} \sim \left[\frac{F(xi + \Delta xi) - F(xi)}{\Delta xi}\right] \qquad \text{(Eq 6)}$$

Linear Functions

Let us examine the approximation for a simple linear function.

Let Y = AX + B (Eq 7)

What is the sensitivity of Y to a change in variable X? By simple calculus:

$$\frac{\partial y}{\partial x} = A \qquad \text{(Eq 8)}$$

We can also find the sensitivity using the numerical approximation as:

$$\frac{\partial y}{\partial x} \sim \left[\frac{F(xi+ \Delta xi)- F(xi)}{\Delta xi}\right] \qquad \text{(Eq 9)}$$

$$\frac{\partial y}{\partial x} \sim \left[\frac{(A(xi+ \Delta xi)+ B)-(A(xi)+ B)}{\Delta xi}\right] \qquad \text{(Eq 10)}$$

$$\frac{\partial y}{\partial x} \sim \left[\frac{(A(xi+ \Delta xi))-(A(xi))}{\Delta xi}\right] \qquad \text{(Eq 11)}$$

$$\frac{\partial y}{\partial x} \sim \left[\frac{((A\, xi+A\, \Delta xi))-(A(xi))}{\Delta xi}\right] \qquad \text{(Eq 12)}$$

$$\frac{\partial y}{\partial x} \sim \left[\frac{A\, \Delta xi}{\Delta xi}\right] \qquad \text{(Eq 13)}$$

$$\frac{\partial y}{\partial x} = A \qquad \text{(Eq 14)}$$

Note that for a simple linear equation the numerical result for the sensitivity is exactly the same as the analytical expression. There is no restriction on the size of the perturbation (Δxi) used in the calculations. Previously, Freeman [Ref 7] has suggested using a 10% perturbation in xi to do the sensitivity calculations. However any size of perturbation will work and generate the correct sensitivity.

This is a general result. As long as the function can be expressed as a linear function of a number of variables, xi, the sensitivity can be directly calculated by a simple perturbation of the function. The resulting sensitivity will be exactly the same as would be obtained using analytical methods.

Non-Linear Expressions

In the quantification of a fault tree, the top event frequency is computed using failure rates of the component devices and time that measures the interval between system validations. In a minimal cut set of two or more basic events, the time variable may be the same. For example the test interval for a valve may be the same for a pressure transmitter. A minimal cut set involving both the valve and the pressure transmitter would create a 2rd order term in the test interval. If the test interval is uncertain, how do we compute the sensitivity of the top event frequency with respect to the test interval? We will now explore how to deal with this type of problem when completing the uncertainty analysis.

As before, let

$$y = F(xi) = Axi^n + B \qquad \text{(Eq 15)}$$

What is the sensitivity of Y to a change in variable X? By simple calculus:

$$\frac{\partial y}{\partial x} = nAxi^{n-1} \qquad \text{(Eq 16)}$$

Let us evaluate the numerical approximation for this case. If n=2, the function of interest is:

$$y = F(xi) = Axi^2 + B \qquad \text{(Eq 17)}$$

From basic calculus, the exact analytical expression for the sensitivity is:

$$\frac{\partial y}{\partial x} = 2\ A(xi) \qquad \text{(Eq 18)}$$

If xi = 1

$$\frac{\partial y}{\partial x} = 2\ A \qquad \text{(Eq 19)}$$

Again the approximation of the derivative is:

$$\frac{\partial y}{\partial x} \sim \left[\frac{(A(xi+\Delta xi)^2+B)-(Axi+B)}{\Delta xi}\right] \qquad \text{(Eq 20)}$$

$$\frac{\partial y}{\partial x} \sim \left[\frac{F(xi+\Delta xi)-F(xi)}{\Delta xi}\right] \qquad \text{(Eq 21)}$$

Inserting the equation for F(xi) we obtain:

$$\frac{\partial y}{\partial x} \sim \left[\frac{(A(xi+\Delta xi)^2+B)-(Axi+B)}{\Delta xi}\right] \qquad \text{(Eq 22)}$$

$$\frac{\partial y}{\partial x} \sim \left[\frac{(A\ (xi^2+2\Delta xi+\Delta xi^2)\ +B)-(Axi+B)}{\Delta xi}\right] \qquad \text{(Eq 23)}$$

$$\frac{\partial y}{\partial x} \sim \left[\frac{A\ (xi^2+2\Delta xi+\Delta xi^2)\ -Axi}{\Delta xi}\right] \qquad \text{(Eq 24)}$$

$$\frac{\partial y}{\partial x} \sim A\left[\frac{(xi^2+2\Delta xi+\Delta xi^2)\ -xi}{\Delta xi}\right] \qquad \text{(Eq 25)}$$

Setting xi =1, and using a 10% perturbation as $\Delta xi = 0.1\ xi = 0.1$

$$\frac{\partial y}{\partial x} \sim A\left[\frac{(1^2+(2)(0.1)1+0.1^2\ )\ -1}{0.1\ (1)}\right] \qquad \text{(Eq 26)}$$

$$\frac{\partial y}{\partial x} \sim A\left[\frac{(2)(0.1)+0.01}{0.1}\right] \qquad \text{(Eq 27)}$$

$$\frac{\partial y}{\partial x} \sim A\left[\frac{0.21}{0.1}\right] \qquad \text{(Eq 28)}$$

$$\frac{\partial y}{\partial x} = 2.1\ A \qquad \text{(Eq 29)}$$

Note that the numerical estimate (Eq 29) is approximately 5% greater than the analytical value (Eq 19). As shown in Table 1, for higher order equations the error will increase as the size of the perturbation becomes more important. Table 2 shows the impact of reducing the perturbation size to 1%. For 10th order variables (say a minimal cut set with 10 basic events) the error is 5%. For almost all engineering evaluations, a maximum error of 5% should be adequate.

For cases where the test frequency, repair time, or mission time appears in a minimal cut set 3 or more times, I suggest that the following perturbation in time be used to compute the sensitivity of the top event:

$$\Delta time = 0.01\ time \qquad \text{(Eq 30)}$$

The sensitivity of the top event frequency to all uncertain variables can be computed using a perturbation of 1%.

$$\Delta xi = 0.01\ xi \qquad \text{(Eq 31)}$$

3. Fault Tree Analysis

Fault trees are often used to analyze potential failures of an engineered system. A fault tree is a failure logic diagram that shows the relationship of device or system failures leading to an accident. Fault tree analysis is covered in detail in the CCPS CPQRA book [Ref. 1]. A fault tree is constructed by starting with a top event or condition and asking the question:

What conditions lead to this outcome?

Each condition in the next layer down is then analyzed by asking the same question. What conditions lead to this outcome? The logic of how these intermediate conditions (termed intermediate events) are related is expressed using AND and OR logic gates. An AND gate output is true if and only if (iff) all of the inputs to the AND gate are all true simultaneously. An OR gate output is true iff one of the inputs to the OR gate is true.

Once the fault tree logic structure is created, numerical data can be used to calculate the frequency or probability of the top event of concern occurring. For fault trees where a basic event is found in multiple branches of the tree, such as power failure, the logic structure of the tree must be simplified to remove the impact of the repeated events on the top event of concern. This simplification is done using Boolean Algebra. Each resulting grouping of basic events can cause the top event to occur. These groupings of basic events are called minimal cut sets. As a simple example consider the Boolean Algebra representation of a fault tree :

$$T = BE11 + BE1 \bullet BE2 + BE1 \bullet BE3 + BE2 \bullet BE9 + BE3 \bullet BE7 \qquad \text{(Eq 32)}$$

Where:

T = Fault Tree Top Event

BE11 = Basic Event 11

BE1 = Basic Event 1

BE2 = Basic Event 2

BE3 = Basic Event 3

BE7 = Basic Event 7

BE9 = Basic Event 9

+ = Boolean Algebra Addition Symbol

• = Boolean Algebra Multiplication Symbol

For this example the top event T, the minimal cut sets are:

BE11

BE1 BE2

BE1 BE3

BE2 BE9

BE3 BE7

For example the existence of basic events BE1 and BE3 guarantees that the top event T will occur. Publically available standard computer software such as SAPHIRE [Ref. 8] may be used to find the minimal cut sets of a large fault tree.

The representation of the top event in equation 32 can now be used to compute the probability or frequency of occurrence of the top event. If the devices describe by the basic events are non-repairable and are tested after a period of time TI, the probability of top event, T, may be calculated as:

$$\text{Prob (T)} = \lambda_{11}\,TI/2 + (\lambda_1\,TI/2)(\lambda_2\,TI/2) + (\lambda_1\,TI/2)(\lambda_3\,TI/2) + (\lambda_2\,TI/2)(\lambda_9\,TI/2) + (\lambda_3\,TI/2)(\lambda_7\,TI/2) \qquad \text{(Eq 33)}$$

Where:

λ_{11} = failure rate of device 11 in basic event BE11, failures per unit time

λ_1 = failure rate of device 1 in basic event BE1, failures per unit time

λ_2 = failure rate of device 2 in basic event BE5, failures per unit time

λ_3 = failure rate of device 3 in basic event BE3, failures per unit time

λ_7 = failure rate of device 7 in basic event BE7, failures per unit time

λ_9 = failure rate of device 9 in basic event BE9, failures per unit time

TI = test interval, time

Equation 33 assumes that the device failure rates are small and the product λ TI < 0.1. The other major assumption is that the values of all of the failure rates and test intervals are known. In reality the failure rate data are not known exactly with only a range or probability distribution representing the state of knowledge.

4. Example Problem

The following example is taken from the ISA technical report on fault tree analysis [Ref. 9]. This example should not be considered as a recommendation by either ISA or AIChE. The example is presented to allow for the demonstration of the methods that can be used in a fault tree study to quantify the uncertainty in the resulting calculated failure frequency or failure probability.

Consider the interlock on the intermediate storage tank T101 shown on Piping and Instrument Diagram in Figure 1. The interlock block diagram is shown in Figure 2. The interlock is intended to prevent an abnormal condition in the tank which could lead to a uncontrolled release of the tank contents. The process hazards analysis team has recommended that the interlock be designed to safety integrity level 2 (SIL 2) with a target probability of failure on demand (PFD) of no greater than 1E-2 or a risk reduction factor (RRF) of 100. Does the proposed interlock shown on Figure 1 meet the SIL 2 target? What is the uncertainty in the predicted PFD of the interlock?

A fault tree for the failure of this interlock is shown in Figure 3. A minimal cut set analysis has been completed and the resulting minimal cut sets are presented in Table 3. We can now write the Boolean Algebra equation that represents this fault tree as:

$$T = PE + TS1 \bullet TS2 + LS1 \bullet LS2 + FT1 \bullet FT2 + FT2 \bullet FT3 + FT1 \bullet FT3 + BV1 \bullet BV2 + BV1 \bullet SOL1 + BV2 \bullet SOL2 + SOL1 \bullet SOL2 + PT1 \bullet PT2 \qquad \text{(Eq 34)}$$

Assuming that the interlock is non-repairable until tested at time TI and using lamda-time (λT) approximation of the failure rate, the Boolean representation of the fault tree may now be converted to a failure probability model as:

$$\begin{aligned} Prob(T) = {} & Prob(PE) + (\lambda_{TS1}\, TI/2)(\lambda_{TS2}\, TI/2) + (\lambda_{LS1}\, TI/2)(\lambda_{LS2}\, TI/2) + (\lambda_{FT1}\, TI/2)(\lambda_{FT2}\, TI/2) + \\ & (\lambda_{FT2}\, TI/2)(\lambda_{FT3} TI/2) + (\lambda_{FT1}\, TI/2)(\lambda_{FT3}\, TI/2) + (\lambda_{BV1}\, TI/2)(\lambda_{BV2}\, TI/2) + \\ & (\lambda_{BV1}\, TI/2)(\lambda_{SOL1} TI/2) + (\lambda_{BV2}\, TI/2)(\lambda_{SOL2}\, TI/2) + \\ & (\lambda_{SOL1}\, TI/2)(\lambda_{SOL2}\, TI/2) + (\lambda_{PT1}\, TI/2)(\lambda_{PT2}\, TI/2) \qquad \text{(Eq 35)} \end{aligned}$$

Where:

Prob(PE) = failure probability of electronic logic solver, assumed to be constant at 5E-3.

λ_{TS1} = failure rate of temperature switch 1, failures per unit time

λ_{TS2} = failure rate of temperature switch 2, failures per unit time

λ_{LS1} = failure rate of level switch 1, failures per unit time

λ_{LS2} = failure rate of level switch 2, failures per unit time

λ_{FT1} = failure rate of flow transmitter 1, failures per unit time

λ_{FT2} = failure rate of flow transmitter 2, failures per unit time

λ_{FT3} = failure rate of flow transmitter 3, failures per unit time

λ_{BV1} = failure rate of block valve 1, failures per unit time

λ_{BV2} = failure rate of block valve 2, failures per unit time

λ_{SOL1} = failure rate of solenoid valve 1, failures per unit time

λ_{SOL2} = failure rate of solenoid valve 2, failures per unit time

λ_{PT1} = failure rate of pressure transmitter 1, failures per unit time

λ_{PT2} = failure rate of pressure transmitter 2, failures per unit time

TI = test interval, time

The electronic logic solver PE is assumed to have a constant failure probability of 5E-3.

Failure rate data for the other devices are present in Table 4. The mode (most likely value) of the failure rate is set equal to the point value taken from the ISA example [Ref. 9]. The triangular probability distribution (Appendix A) was assumed to represent the failure rate data for the equipment items. The upper and lower limits are based on the failure rate data of Appendix 4 of Smith [Ref. 10].

Using the device failure rates represented by the Table 4 column labeled as the mode and the minimal cut sets presented in Table 3, the ISA technical report [Ref.9, page 35] computes an interlock PFD as 7.5E-3 (RRF of 133). This would satisfy the requirement for a SIL 2 interlock {PFD of 1E-2 or a RRF of 100}. Now what is the uncertainty in the interlock PFD?

The first thing to remember is that the PFD should be calculated based on the mean value of the failure rate lambda (λ), not the mode of the failure rate. For a data set, the mean is the best single point representation of the data set. Table 5 presents the PFD of each minimal cut set using the mean of the device failure rate lambda (λ). The total of the PFD becomes 1.47E-2 (RRF of 68). This would not satisfy a SIL 2 interlock requirement.

Now we will compute the variance of the predicted probability of top event, T. To compute the variance of the probability of the top event, T, we will need to compute the sensitivity of the top event to each of the variables.

Sensitivity Using Numerical Perturbation Calculation

We will use the ISA interlock example previously analyzed by Freeman (Ref. 7). Using public data sources Freeman found that the probability of failure on demand (PFD) was 1.49E-2. To use the perturbation method for the evaluation of the sensitivities requires evaluating the impact of small changes in the uncertain input parameters. To enable these calculations to be completed numbers using 8 significant digits were used. The calculations were completed using Microsoft Excel which uses double-precision floating-point arithmetic compliant with IEEE 754 specification. Excel nominally carries 15 significant figures in calculations. We will complete the calculations with Excel displaying 5 significant digits and round the result at the end of the calculations to 4 significant digits. We will illustrate the calculations by looking at the Flow Transmitter, FT1.

a. First we recalculate the PFD of the example and find it to be 1.4594E-02. This is the base number that we will use in the perturbation calculations.
b. We now make a small change in mean failure rate of FT1. The mean failure rate is 8E-6 failures/hour. A 1% perturbation is used or an increase of 0.08E-6 failures/hour.

c. The perturbed value of the failure rate of FT1 becomes
   (8 + 0.08) * 1E-6 failures/hour or 8.08E-6 failures/hour.
d. The perturbed value of the failure rate of FT1 is now used to re-calculate the PFD of the interlock. The resulting re-calculation finds that the PFD is 1.4618E-02
e. The change in the PFD from the base number due to the perturbation of the failure rate of FT1 is now computed as:

   delta PFD = 1.4618E-02 - 1.4594E-02 = 2.4556E-05

f. The sensitivity of the PFD to a change in the failure rate of FT1 is now computed by dividing the change in the PFD by the change FT1 failure rate as:

   sensitivity = (2.4556E-05)/(0.08E-6 failures/hour) = 307.0 hours
   Result is rounded to 4 significant digits.

We now note that in the paper by Freeman (Ref. 7) that equation 32 presents the sensitivity for FT1 as:

$$\frac{\partial Prob(T)}{\partial \lambda \mathrm{FT1}} = 307.0 \text{ hr} \qquad \text{(Eq 32 of Ref 7)}$$

The answers are exactly the same. The sensitivity calculations for all of the devices are presented in Table 6. Note that in every case the sensitivity found using the analytical expression and the numerical estimate are the same. This is a general result which will be true for all basic event equipment failure rates found in a fault tree minimal cut set.

Once found, the sensitivities can now be used to compute the variance contribution of each particular device to the overall variance of the top event PFD in the same manner as Freeman previously presented.

The sensitivities of the top event probability are summarized in Table 6. We may now compute the variance contribution of each uncertain variable using Equations 1 – 4 in the methodology section of this paper. The relative variance contribution is expressed as a percent of the total variance of the top event probability. The top event variance is found to be 8.24E-6 (failures per hour)^2 or equivalently the top event standard deviation is 2.78E-3 failures per hour.

In this example the variance (measure of uncertainty) is dominated by the flow transmitters. The flow transmitters account for almost 2/3 of the uncertainty in the top event probability. Failures of the flow transmitters appear in 3 of the minimal cut sets. Since the uncertainty in the transmitter failure rates is large, the flow transmitters should contribute a significant amount to the variance of the fault tree top event.

We can use the variance in the top event to define the likelihood of achieving SIL 2 performance of the interlock. The calculations are done using the normal probability distribution. The normal distribution is tabulated as the standard normal distribution using a normalization factor Z.

The standard normal factor, Z, [Ref. 11] is defined as:

$$\mathrm{Z} = \left[\frac{xi - E(x)}{\sigma}\right] \qquad \text{(Eq 36)}$$

Where:

σ = standard deviation. Note that the variance of a random variable is the square of the standard deviation of the random variable.

E(x) = Expected value of the random variable x_i

For the example, the expected value of the probability of the top event of concern was calculated as

$$E[Prob(T)] = 1.47E\text{-}2 \qquad \text{(Eq 37)}$$

From Table 4, the standard deviation in the probability of the top event is

$$\sigma_T = 2.78E\text{-}3 \qquad \text{(Eq 38)}$$

The target PFD for a SIL 2 interlock must not be worse than 1E-2 (RRF of 100).

We compute the normal distribution Z factor at the SIL 2 target value as

$$Z = (0.01 - 0.0147)/2.78E\text{-}3 \qquad \text{(Eq 39)}$$

$$Z = -1.69065 \qquad \text{(Eq 40)}$$

Tabulations of standard normal distribution are presented in most statistics books such as Meyer [Ref. 11]. For this value of Z, the corresponding probability that the interlock will achieve SIL 2 performance is 4.55 percent. We can also compute the risk reduction that the proposed interlock design will achieve.

What is the interlock probability of failure on demand (PFD) that we can be 95% certain that the interlock will provide? At the 95% level, the corresponding Z factor is 1.65 [Ref 11]. We now compute the corresponding PFD of the interlock that corresponds to this Z factor.

$$Z = \left[\frac{x95\% - E[Prob(T)]}{\sigma}\right] \qquad \text{(Eq 41)}$$

In this case:

$x_{95\%}$ = the PFD that we can be confident that the interlock will achieve

$$\sigma_T = 2.78E\text{-}3 \qquad \text{(Eq 42)}$$

$$E[Prob(T)] = 1.47E\text{-}2 \qquad \text{(Eq 43)}$$

$$Z = 1.65 \qquad \text{(Eq 44)}$$

Rearranging equation 40 to find $x_{95\%}$ we obtain:

$$x_{95\%} = E[Prob(T)] + Z\,\sigma_T \qquad \text{(Eq 45)}$$

$$x_{95\%} = 1.47E\text{-}2 + (1.65)\,(2.78E\text{-}3) = 0.0193 \qquad \text{(Eq 46)}$$

or a 95% certain risk reduction factor of

$$RRF = 1/PFD = 1/0.0193 = 52 \qquad \text{(Eq 47)}$$

By the same logic there is 5% chance that the risk reduction is a PFD of 0.01 or an RRF of 99. In summary, this interlock is likely to perform as a mid-range SIL 1 interlock, not the SIL 2 interlock desired by process hazards analysis team.

5. Methodology for Fault Tree Uncertainty Analysis

We can now generalize the methods used in the above example to a proposed method for the evaluation of the uncertainty in a fault tree. The following step-by-step procedure is modified from that that previously published by Freeman (Ref 7) to indicate the use of numerical methods to compute the sensitivity. Step 8 shown in *italics* indicates the modification..

1. Create the fault tree using standard methods outline in the CPQRA book [Ref 1].
2. Determine minimal cut sets by hand or using standard computer software such as SAPHIRE [Ref 12].
3. Define the needed failure rate data for each basic event.
4. Define those basic events that are considered to be uncertain.
5. For the basic events that are uncertain, define the probability distribution and associated parameters needed to numerically define the probability distribution. Appendix B of this paper presents the needed description of 4 commonly used probability distributions. In many cases the triangular distribution will be selected. The minimum, maximum and mode (most likely) parameters will be needed for the triangular distribution.
6. Compute the mean of each of the uncertain variable
7. Compute the probability or frequency of the fault tree top event of interest using the mean value for each uncertain variable.
8. *Compute the sensitivity of the top event probability or frequency using the numerical methods outlined in this paper.*
9. Compute the variance of each of the uncertain variables
10. Compute the variance contribution of for each of the uncertain variables to the top event probability or frequency using equation 6.
11. Compute the total variance of the top event of interest probability or frequency by summing all of the contributions determined in Step 10.
12. Compute the variance contribution percent of each uncertain variable by dividing the variable contribution (Step 10) by the total variance of the top event of interest (Step 11).
13. Define the level of risk the project is willing to take. What chance will the project management accept for potential failure of the interlock to achieve the desired risk reduction? In the above example, I have used a 5% risk of failure or a 95% certainty that the interlock will achieve a risk reduction factor of 52. In the example, there is a very low probability that the interlock will achieve a SIL 2 PFD of 0.01 or an RRF of 100.

Alternately you can report the 90% range for the top event of interest. For the example the 90% range starts at the 5% RRF of 99 and there is a 95% certainty that that an RRF of at least 52 will be achieved. The expected risk reduction for this interlock is an RRF of 68.

6. Conclusions

A general procedure has been presented to quantify the uncertainty in calculations of failure frequencies using the fault tree methodology. The uncertainty analysis procedure is based on the application of propagation of error and variance contribution analysis techniques to the minimal cut sets created during a fault tree study. The procedure is simple and can be incorporated into standard fault tree analysis programs such as SAPHIRE [Ref 8]. This procedure is based on perturbation calculation of the frequency of the top event from changes in the uncertain variables. Incorporation of a numerical perturbation method into standard fault tree software such as SAPHIRE would be easy to implement and would provide the means for evaluation of the uncertainty in the quantified fault tree results.

For the example interlock presented in this paper, the uncertainty in the device failure rates degrades the predicted interlock SIL from meeting SIL 2 requirements to becoming a mid range SIL 1 interlock. This is a general result. Uncertainty in design parameters will reduce the likelihood that the equipment will achieve desired results. The design engineer has two choices:

- Accept the conclusion that the interlock will not perform as SIL 2
- Re-design the interlock using more reliable components to achieve the SIL 2 target

The variance contribution analysis shown in Table 7 indicates that the major contributor to the uncertainty in the interlock performance is the Flow Transmitters. Flow transmitters certified to SIL-2 or SIL-3 performance with lower failure rates could improve the calculated PFD of the interlock.

7. References

1. Center for Chemical Process Safety , *Guidelines for Chemical Process Quantitative Risk Analysis – 2rd edition*, American Institute of Chemical Engineers, New York, 2000.
2. Frank P Lees, Loss Prevention in the Process Industries – 2rd Edition, Butterworth, ISBN 0-7506-1547-8, Oxford, UK, 1996
3. US Department of Labor, Occupational Safety and Health Administration, CFR 1910.119 - Process Safety Management of Highly Hazardous Chemicals, 1910.119(e)(2)(vi).
4. Edward B. Haugen, *Probabilistic Approaches to Design*, John Wiley, New York, 1968
5. Raymond. Freeman and Angela Summers, "Evaluation of Uncertainty in Safety Integrity Level Calculations," *Process Safety Progress,* Published online in Wiley Online Library (wileyonlinelibrary.com) DOI 10.1002/prs. 11805, 2016
6. Raymond. Freeman, "General Method for Uncertainty Evaluation of Safety Integrity Level Calculations – Part 2 Analytical Methods," *Process Safety Process*, Published online in Wiley Online Library (wileyonlinelibrary.com) DOI 10.1002/prs 111915, October 2017
7. Raymond. Freeman ,"Error Propagation and Uncertainty Analysis: Application to Fault Tree Analysis", Process Safety Progress, Published Wiley Online Library (wileyonlinelibrary.com/journal/prs) DOI: https://doi.org/10.1002/prs.12080, June 29, 2019

8. US Nuclear Regulatory Commission, "Systems Analysis Programs for Hands-on Integrated Reliability Evaluations (SAPHIRE) – Version 8," NUREG/CR-7039, Vol 3, June 2011.
9. ISA, "Safety Instrumented Functions (SIF) – Safety Integrity Level (SIL) Evaluation Techniques Part 3: Determining the SIL of a SIF via Fault Tree Analysis, ISA Technical Report ISA-TR-84.00.02-2002 – Part 3, Approved 17 June 2002.
10. David J. Smith, *Reliability Maintainability and Risk – 8th Edition*, ISBN 978-0-08-096902-2, Elsevier, London, 2011
11. Paul L. Meyer, Introductory Probability and Statistical Applications, John Wiley, New York, 1972

Table 1. Comparison of Analytical Sensitivity with Numerical Approximation with Perturbation of 10% in xi for the equation: $y = F(xi) = Axi^n + B$

Order of Equation n	Equation for Analytical Sensitivity $\frac{\partial y}{\partial x} = nAxi^{n-1}$	Sensitivity with xi = 1 and A=1	Numerical Sensitivity with $\Delta xi = 0.1\ xi$	Error %
1	$\frac{\partial y}{\partial x} = 1A(xi)^0$	1	1	0
2	$\frac{\partial y}{\partial x} = 2A(xi)^1$	2	2.1	5
3	$\frac{\partial y}{\partial x} = 3A(xi)^2$	3	3.31	-10%
4	$\frac{\partial y}{\partial x} = 4A(xi)^3$	4	4.64	-16%
5	$\frac{\partial y}{\partial x} = 5A(xi)^4$	5	6.11	-22%
6	$\frac{\partial y}{\partial x} = 6A(xi)^5$	6	7.72	-29%
7	$\frac{\partial y}{\partial x} = 7A(xi)^6$	7	9.49	-36%
8	$\frac{\partial y}{\partial x} = 8A(xi)^7$	8	11.44	-43%
9	$\frac{\partial y}{\partial x} = 9A(xi)^8$	9	13.58	-51%
10	$\frac{\partial y}{\partial x} = 10A(xi)^9$	10	15.94	-59%

Error = (Analytical- Numerical)/Analytical

Table 2. Comparison of Analytical Sensitivity with Numerical Approximation with Perturbation of 1% in xi for the equation: $y = F(xi) = Axi^n + B$

Order of Equation n	Equation for Analytical Sensitivity $\frac{\partial y}{\partial x} = nAxi^{n-1}$	Sensitivity with xi = 1 and A=1	Numerical Sensitivity with $\Delta xi = 0.1\, xi$	Error %
1	$\frac{\partial y}{\partial x} = 1A(xi)^0$	1	1	0
2	$\frac{\partial y}{\partial x} = 2A(xi)^1$	2	2.01	-1%
3	$\frac{\partial y}{\partial x} = 3A(xi)^2$	3	3.03	-1%
4	$\frac{\partial y}{\partial x} = 4A(xi)^3$	4	4.06	-2%
5	$\frac{\partial y}{\partial x} = 5A(xi)^4$	5	5.10	-2%
6	$\frac{\partial y}{\partial x} = 6A(xi)^5$	6	6.15	-3%
7	$\frac{\partial y}{\partial x} = 7A(xi)^6$	7	7.21	-3%
8	$\frac{\partial y}{\partial x} = 8A(xi)^7$	8	8.29	-4%
9	$\frac{\partial y}{\partial x} = 9A(xi)^8$	9	9.37	-4%
10	$\frac{\partial y}{\partial x} = 10A(xi)^9$	10	10.46	-5%

Error = (Analytical- Numerical)/Analytical

Table 3. Basic Events in Minimal Cut Sets For Example

Cut Set No	BE 1	BE 2
1	PE	-
2	TS1	TS2
3	LS1	LS2
4*	FT1	FT2
5*	FT2	FT3
6*	FT1	FT3
7*	BV1	BV2
8*	BV1	SOL1
9*	BV2	SOL2
10*	SOL1	SOL2
11	PT1	PT2

* Minimal cut Set with same basic event as another cut set

Table 4. Lamda (λ) Failure Rate Data With Uncertainty Limits

Device	Minimum Failures per 1E6 Hours	Mode Failures per 1E6 Hours	Maximum Failures per 1E6 Hours	Mean Failures per 1E6 Hours	Lamda Variance (Failures per 1E6 Hours)^2
Flow Transmitters FT1, FT2 and FT3	1.0	2.9	20.0	8.0	18.3
Pressure Transmitters PT1 and PT2	1.0	2.3	20.0	7.8	18.8
Temperature Switch TS1 and TS2	3.0	7.6	20.0	10.2	12.9
Level Switch LS1 and LS2	2.0	4.6	20.0	8.9	15.8
Block Valves BV1 and BV2	0.2	2.3	10.0	4.2	4.4
Solenoid Valves SOL1 and SOL2	1.0	2.3	8.0	3.8	2.3

Table 5. Probability of Minimal Cut Sets Based on Mean Failure Rates

Cut Set No	BE 1	BE 2	PFD based on mean λ
1	PE	-	5.00E-03
2	TS1	TS2	2.00E-03
3	LS1	LS2	1.52E-03
4	FT1	FT2	1.23E-03
5	FT2	FT3	1.23E-03
6	FT1	FT3	1.23E-03
7	BV1	BV2	3.38E-04
8	BV1	SOL1	3.38E-04
9	BV2	SOL2	3.38E-04
10	SOL1	SOL2	2.77E-04
11	PT1	PT2	1.17E-03

Table 6. Sensitivity of Example System PFD to Basic Event Failure Rates Based on a 1% Perturbation

Basic Event Label	Device Type	Mean Failures per 1E6 Hours	Failure Perturbation of 1% of Mean Failures per 1E6 Hours	Perturbed Failure Rate Failures per 1E6 Hours	Total System PFD Using Perturbed Failure Rate of Basic Event	Change in Total System PFD	Sensitivity of PFD to Change in Failure Rate Hours
FT1	Flow Transmitter	8	0.08	8.08	1.461848E-02	2.455603E-05	307.0
FT2	Flow Transmitter	8	0.08	8.08	1.461848E-02	2.455603E-05	307.0
FT3	Flow Transmitter	8	0.08	8.08	1.461848E-02	2.455603E-05	307.0
PT1	Pressure Transmitter	7.8	0.078	7.878	1.460560E-02	1.167179E-05	149.6
PT2	Pressure Transmitter	7.8	0.078	7.878	1.460560E-02	1.167179E-05	149.6
TS1	Temperature Switch	10.2	0.102	10.302	1.461389E-02	1.995945E-05	195.7
TS2	Temperature Switch	10.2	0.102	10.302	1.461389E-02	1.995945E-05	195.7
LS1	Level Switch	8.9	0.089	8.989	1.460912E-02	1.519596E-05	170.7
LS2	Level Switch	8.9	0.089	8.989	1.460912E-02	1.519596E-05	170.7
BV1	Block Valves	4.2	0.042	4.242	1.460037E-02	6.445958E-06	153.5
BV2	Block Valves	4.2	0.042	4.242	1.460037E-02	6.445958E-06	153.5
SOL1	Solenoid Valves	3.8	0.038	3.838	1.459976E-02	5.832058E-06	153.5
SOL2	Solenoid Valves	3.8	0.038	3.838	1.459976E-02	5.832058E-06	153.5

Table 7. Variance of Top Event Probability Due to Uncertain Device Failure Rates

Device No	Device Type	Label	Sensitivity of Top Event Probability to Device Failure Rate $Hour^{-1}$	Variance of Device Failure Rate (Failures per 1E6 $Hour)^2$	Variance Contribution to Top Event Variance	Contribution to Top Event Variance %
1	Logic Solver	PE	-	0	0.00E+00	0%
2	Temperature Switch	TS1	195.7	12.9	4.94E-07	6%
3	Temperature Switch	TS2	195.7	12.9	4.94E-07	6%
4	Level Switch	LS1	170.7	15.8	4.60E-07	6%
5	Level Switch	LS2	170.7	15.8	4.60E-07	6%
6	Flow Transmitter	FT1	307	18.3	1.72E-06	21%
7	Flow Transmitter	FT2	307	18.3	1.72E-06	21%
8	Flow Transmitter	FT3	307	18.3	1.72E-06	21%
9	Block Valve	BV1	153.5	4.4	1.04E-07	1%
10	Block Valve	BV2	153.5	4.4	1.04E-07	1%
11	Pressure Transmitter	PT1	149.6	18.8	4.21E-07	5%
12	Pressure Transmitter	PT2	149.6	18.8	4.21E-07	5%
13	Solenoid valve	SOL1	153.5	2.3	5.42E-08	1%
14	Solenoid valve	SOL2	153.5	2.3	5.42E-08	1%

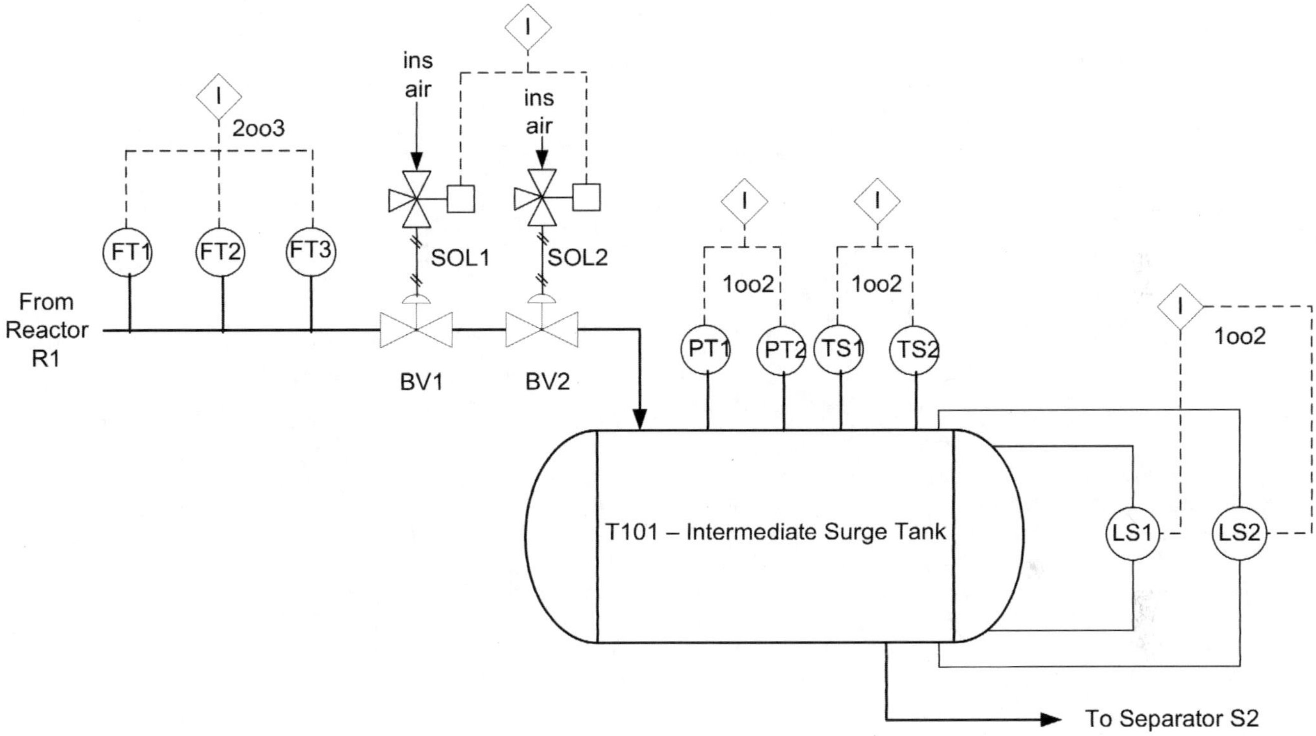

Figure 1. P&ID Diagram for Example

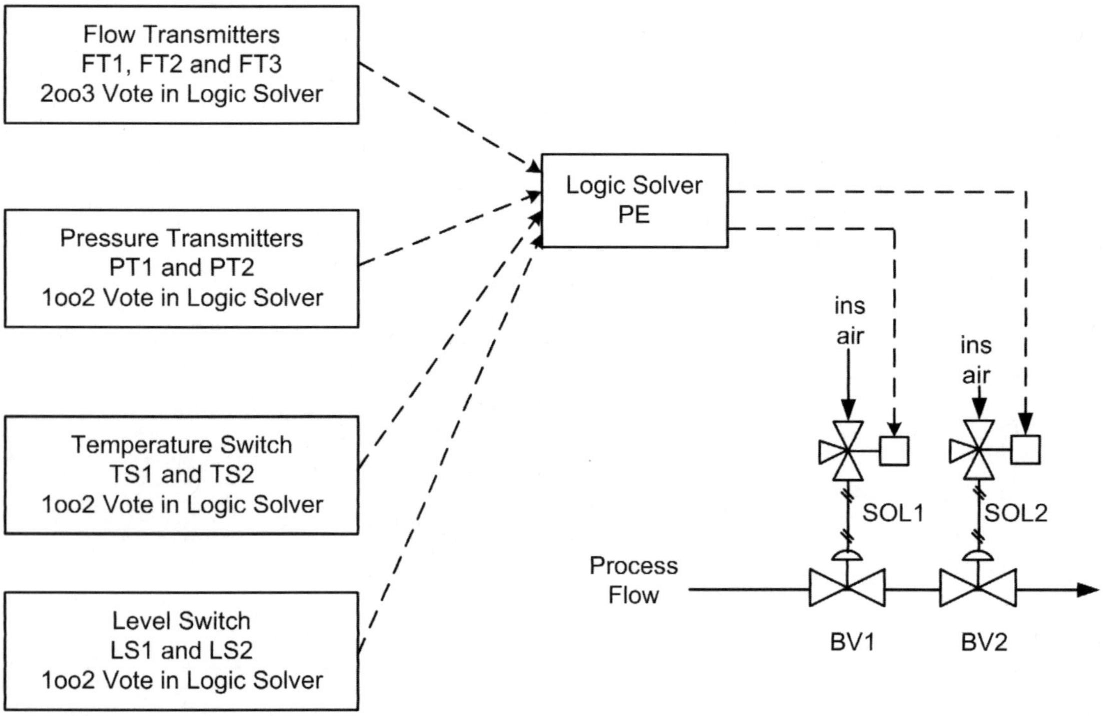

Figure 2. Interlock Block Diagram

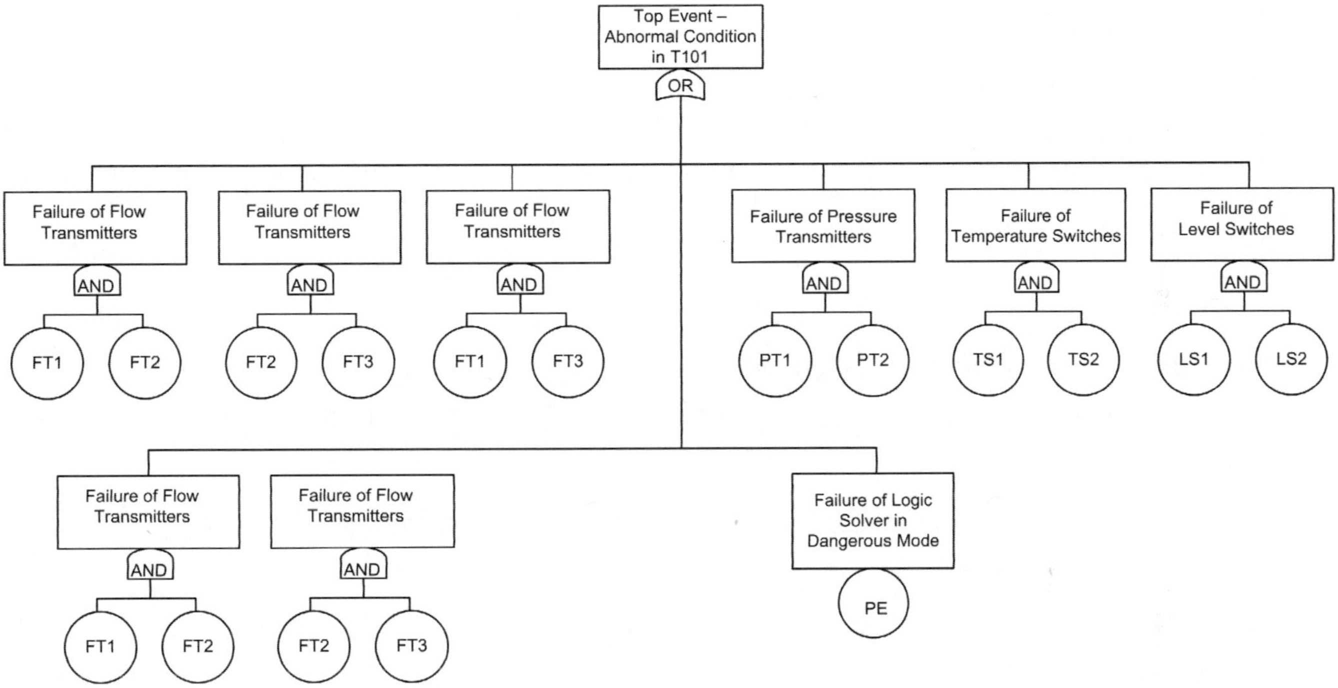

Figure 3. Fault Tree for Example Interlock

Appendix A – Commonly Encounter Probability Distributions

There are four probability distributions used to represent the failure rate of equipment items. These are:

1. Uniform Distribution
2. Triangular Distribution
3. Normal Distribution
4. Log-Normal Distribution

This appendix reviews the properties of these four probability distributions.

1. Uniform Distribution

The uniform distribution [Ref. A-1, pg 74; A-2, pg 687-688] is used when a variable is only known within a defined range. Figure A-1 presents a plot of the probability density function (pdf) for the uniform distribution. The distribution is defined using two parameters, A and B.

Where:

A = minimum value of the random variable

B = maximum value of the random variable

The mean or expected value of a random variable, x, that follows the uniform distribution is:

Mean = Expected Value – $E(x) = (A+B)/2$ (Eq A-1)

The variance of a random variable, x, that follows the uniform distribution is:

Variance = $V(x) = (B-A)^2/12$ (Eq A-2)

The standard deviation of a random variable, x, that follows the uniform distribution is:

Standard Deviation = $(\text{Variance})^{1/2} = (B-A)/(12)^{1/2}$ (Eq A-3)

2. Triangular Distribution

The triangular distribution [Ref. A-2, pg 686-687] is used when a variable is known to lie within a defined range and a "best guess" or estimate can be made as to the most likely value to the variable. Figure A-2 presents a plot of the probability density function (pdf) for the triangular distribution. The distribution is defined using three parameters, A. B and C.

Where:

A = minimum value of the random variable

B = maximum value of the random variable

C = most likely (mode) of the random variable

The mean or expected value of a random variable, x, that follows the triangular distribution is:

Mean = Expected Value – $E(x) = (A+B+C)/3$ (Eq A-4)

The variance of a random variable, x, that follows the triangular distribution is:

Variance = $V(x) = [A^2 + B^2 + C^2 - AB - AC - BC]/18$ (Eq A-5)

The standard deviation of a random variable, x, that follows the triangular distribution is:

Standard Deviation = $(\text{Variance})^{1/2} = \{[A^2 + B^2 + C^2 - AB - AC - BC]/18\}^{1/2}$ (Eq A-6)

3. Normal distribution

The normal distribution [Ref A-2, pg 665-666] is used when a variable when data analysis finds the normal distribution is the best model to describe the spread in the measured variable. Figure A-3 presents a plot of the probability density function (pdf) for the normal distribution. The distribution is defined using two parameters, μ and σ.

Where:

μ = mean of the measured data for variable

σ = standard deviation of the measured data for variable

The expected value of a random variable, x, that follows the normal distribution is:

Expected value = mean = $E(x) = \mu$ (Eq A-7)

The variance of a random variable, x, that follows the normal distribution is:

Variance = square of standard deviation = $V(x) = \sigma^2$ (Eq A-8)

The standard deviation of a random variable, x, that follows the normal distribution is:

Standard deviation = σ (Eq A-9)

4. Lognormal Distribution

The lognormal distribution [Ref. A- 2, pg 658-659] is used when a variable when data analysis finds the lognormal distribution is the best model to describe the spread in the measured variable. Figure A-4 presents a plot of the probability density function (pdf) for the lognormal distribution. Many variables in nature are lognormally distributed (for example the height of people). The log normal distribution is used when the variable of interest has a known physical lower limit of zero. A variable, x, is lognormally distributed if ln(x) is normally distributed. The lognormal distribution is defined using two parameters, μ_y and σ_y.

Let the original data be defined by the variable x with mean and standard deviation as:

μ_x = mean of the measured data for variable x

σ_x = standard deviation of the measured data for variable x

Define a new variable y as:

$y = \text{Ln}(x)$ (Eq A-10)

Then variable x is said to be lognormally distributed if y is normally distributed. We can now write the mean and standard deviation of y (μ_y and σ_y.) [Ref. A-3, A-4 and A-5] as

$$\mu_y = Ln\left[\mu_x^2 /((\sigma_x^2 + \mu_x^2)^{1/2})\right] \quad \text{(Eq A-11)}$$

$$\sigma_y = \left[Ln(\sigma_x^2 / \mu_x^2 + 1)\right]^{1/2} \quad \text{(Eq A-12)}$$

Where:

Ln is the natural logarithm (base e) of the argument

Appendix A References:

A-1. P. L. Meyer, *Introductory Probability and Statistical Applications*, 2nd Edition, Addison-Wesley, Reading Mass, Library of Congress Catalog no. 75-104971, 1972, pp. 74

A-2. D. Vose, *Risk Analysis–A Quantitative Guide*, 3rd edition, Wiley, Chichester, West Sussex, England, ISBN 978-0-470-51284-5, 2008, pp. 686-687

A-3. "Lognormal mean and variance - MATLAB lognstat". www.mathworks.com. Retrieved 27 December 2018

A-4. M. Mood, F. A. Graybill, and D. C. Boes. *Introduction to the Theory of Statistics.* 3rd ed., New York: McGraw-Hill, ISBN 0-07-042864-6, 1974. pp. 117.

A-5. E.L. Crow and K. Shimizu, Lognormal Distributions – Theory and Applications, Marcel Dekker, New York, ISBN 0-8247-7803-0, 1988, pp. 9

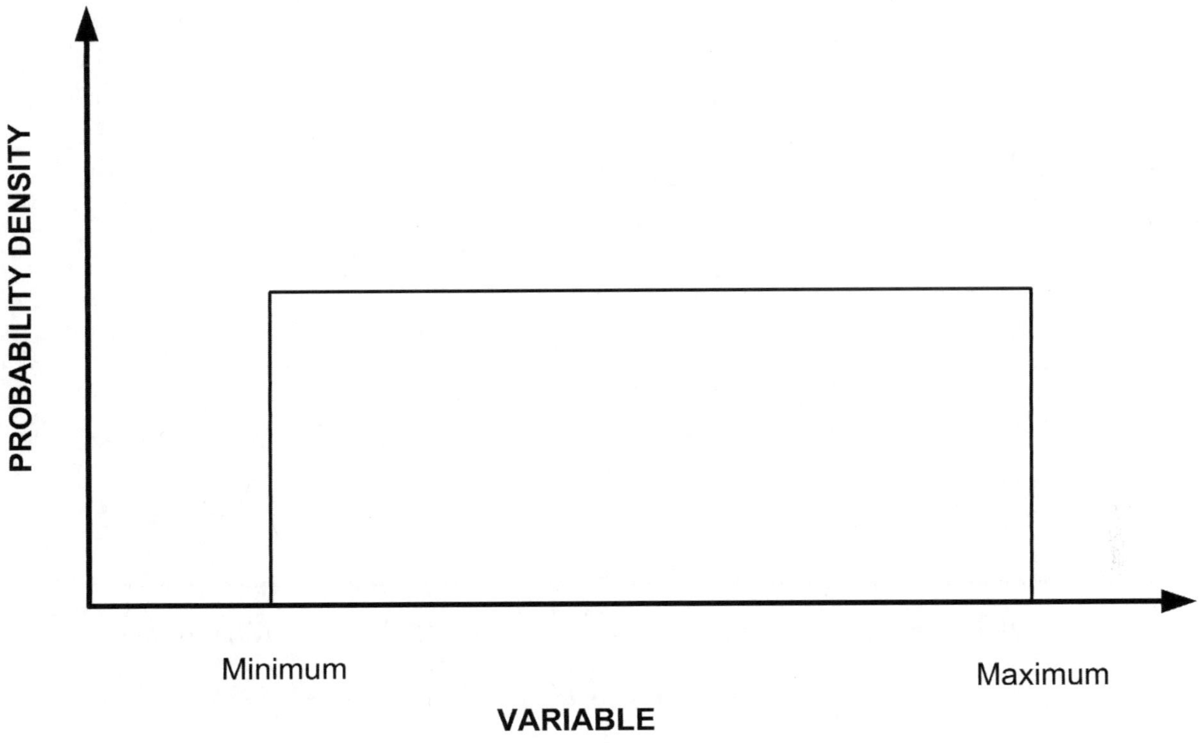

Figure A-1. Uniform Probability Distribution

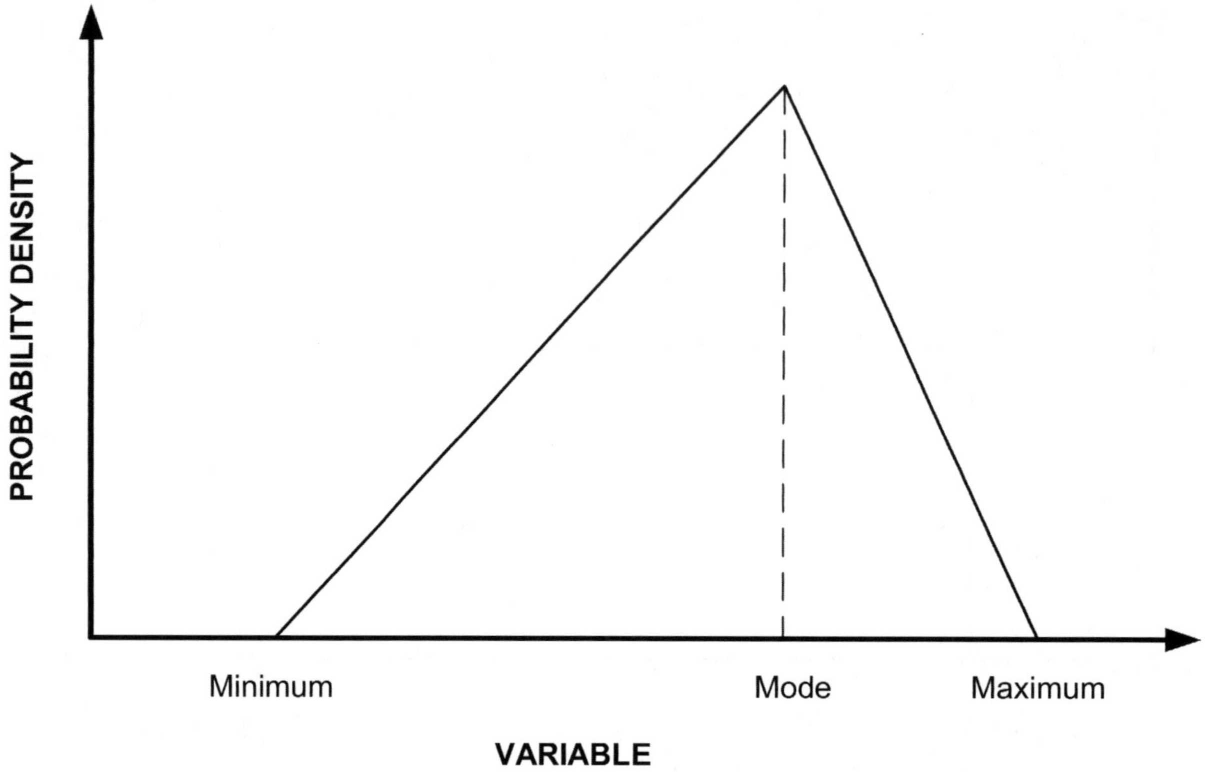

Figure A-2. Triangular Probability Distribution

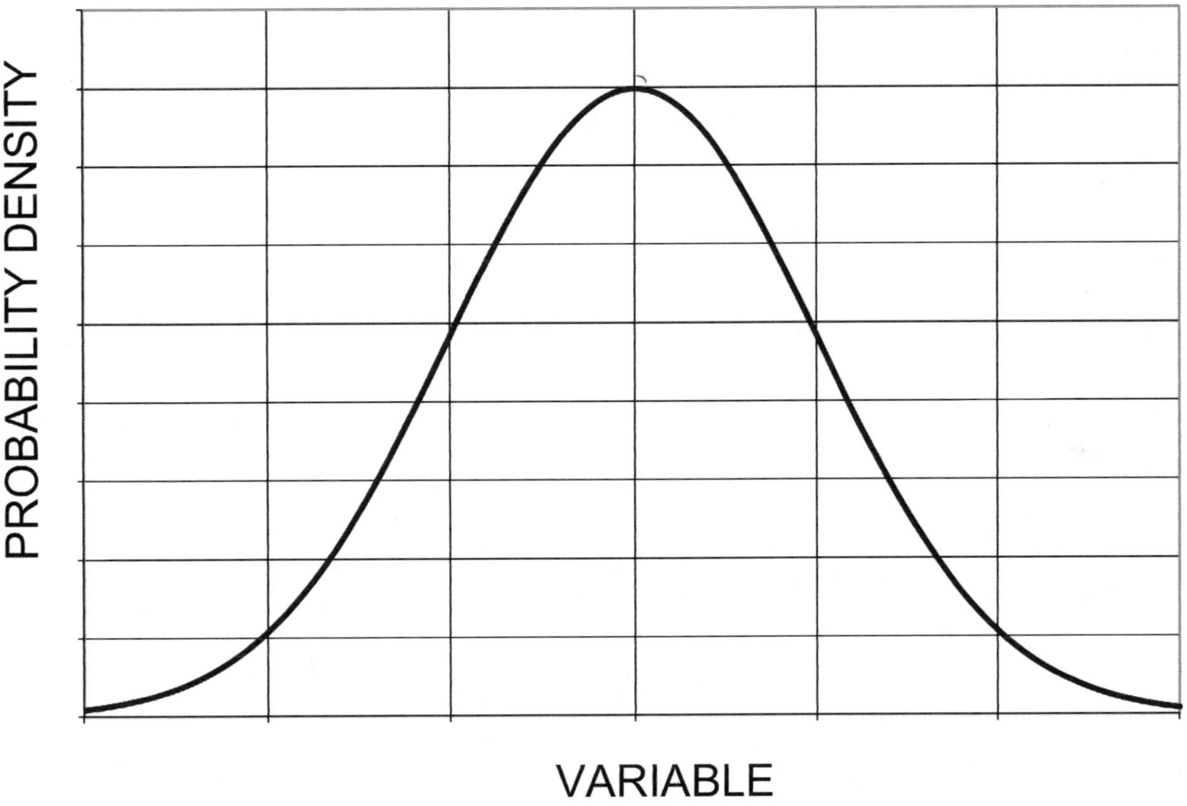

Figure A-3. Normal Probability Distribution

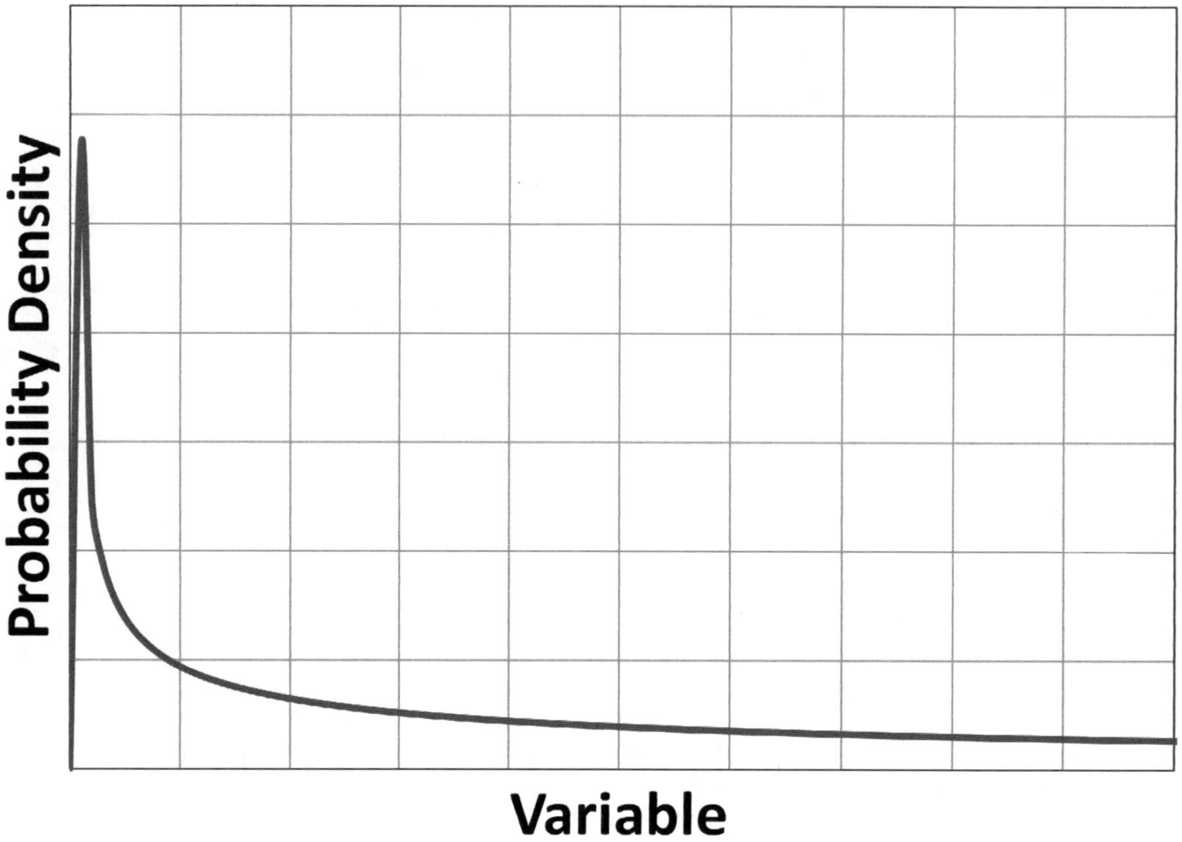

Figure A-4. Lognormal Probability Distribution

590424 Making the Most of Your PHA

Abstract:

A Process Hazard Analysis (PHA) is required to address OSHA Process Safety Management (PSM) and EPA Risk Management Program (RMP) regulations. Understanding the potential process hazard is widely recognized as a foundational element of PSM and RMP, and in addition to simply meeting regulatory requirements, can be a critical mechanism for flushing out potential vulnerabilities early in the design process for capital projects. Although an important activity, PHA is also recognized as a sizeable investment in time and energy for an Owner/Operator. So why not make the most of it? Other important activities at plants such as Mechanical Integrity, Alarm Rationalization, and Management of Change are able to stay focused and possibly be streamlined (i.e., reduced cost) based on the ability to use the results of a quality PHA having the necessary information and usability of documentation. So why not structure the PHA and documentation to be able to capitalize on potentially significant cost-savings and possible program integration?
The thrust of this paper and conference presentation will be to identify tips for conducting and documenting the PHA to maximize its ability to be used for supporting other important programs such as MOC, MI, and Alarm Rationalization. The objective will be to allow PHA Coordinators to be able to readily apply these tips to bring about improvements and cost-savings on these interrelated programs.

589814 Theoretical Analysis of the Viability of Isopropanol-Cyclohexane Solvent Systems for Biomatter Lipid Extraction

Abstract:

A theoretical alternative solvent combination for the extraction of lipids from biomatter is the Water, Isopropanol, and Cyclohexane instead of the more common combination used in the Bligh and Dyer Method (Water, Methanol, and Chloroform). While the Bligh and Dyer Method is considered effective, toxicological concerns have been raised over its use of both Chloroform and Methanol. The alternative method utilizes solvents that do not pose as high as a toxicological concern while theoretically maintaining a similar level of practical viability. The alternative method's extraction efficacy was judged against the performance of the Bligh and Dyer method using the simulation software Aspen Plus V10®. The equilibrium data generated from Aspen Plus V10® was then graphed into tertiary diagrams for data analysis using the graphing software SigmaPlot.

592166 Techno-Economic Analysis of the Conversion of CH4 and CO2 Intoacetic Acid Using Microwave Plasma

Abstract:

Methane and carbon dioxide together represent 90% of greenhouse gases [1], and one way of diminishing their emissions is through the development of pathways that convert them into added value chemicals. Acetic acid is one of such chemicals, and efforts at the laboratory scale have been carried out to obtain alternative ways of producing it. However, there are very few studies that evaluate the possibility of implementing these processes at a larger scale. In this paper, techno-economic and sustainability analyses were used to assess the feasibility of upscaling two pathways to produce acetic acid: (1) by direct conversion of methane and carbon dioxide and (2) by indirect conversion, using syngas (obtained from dry methane reforming) as an intermediate. For the direct method a microwave non-catalytic plasma reactor was considered as the main driver in the process simulation [2]. For the indirect method, a zeolite catalytical single step reaction unit was considered [3]. APEN Plus was used to simulate the reaction and purification systems of both methods, and alongside literature values and plant cost estimation techniques, the capital and operational expenses were calculated. Finally, the feasibility of both projects is assessed on carbon sequestration mass balances and economic metrics. The main sensitivity variables that need to be improved in order to obtain a minimum ROI threshold value were also evaluated.

Key words: carbon management, process design, microwave plasma reactor

[1] EPA. *Overview of Greenhouse Gases.* United States Environmental Protection Agency. 2017.

[2] Wang, L., Yi, Y., Wu, C., Guo, H., Tu, X. *One-Step Reforming of CO2 and CH4 into High-Value Liquid Chemicals and Fuels at Room Temperature by Plasma-Driven Catalysis.* Angewandte Chemie. 2017.

[3] Lopez, L., Velasco, J., Montes, V., Marinas, A., Cabrera, S., Boutonnet, M., Jaras, S. *Synthesis of Ethanol from Syngas over Rh/MCM-41 Catalyst: Effect of Water on Product Selectivity.* Catalysts. 2015.

589668 Model Predictive Control Design for the Allam Power Plant Integrated with an Air Separation Unit

Abstract:

Since electricity and heat generation account for around 25% of this greenhouse gas emission, high efficiency power generation technology such as the Allam cycle which emits essentially no CO_2 and NOx is much needed. Being an oxy-fuel power cycle, the Allam power plant is inherently integrated with an air separation plant (ASU). In this study Aspen Plus Dynamics was used to model the integrated Allam power plant/Air Separation Unit (ASU) with a high degree of heat and work integration based on the steady state work of Net Power. In our previous work, regulatory and advanced PID controls were implemented and tested for various set point tracking and disturbance rejection scenarios. However, PID controllers have limitations for various reasons. Consequently, a model predictive control (MPC) was designed in this study to optimize and improve the performance of the integrated unit and to test its performance in terms of dynamic behavior and robustness to model uncertainties.

589672 Explore the Function of Ion-Exchange Membrane (IEM) in Membrane Capacitive Deionization (mCDI) Cell, an Alternative Technique in Brackish Water Desalination Field

Abstract:

In the arid west, the freshwater supply of many communities is limited, leading to increased interest in tapping brackish water resources. Although reverse osmosis is the most common technology to upgrade saline waters, there is also interest in developing and improving alternative technologies. Here we focus on Capacitive Deionization (CDI), which has attracted broad attention as an energy efficient desalination technology, especially for ion extraction from low salinity brackish water. The objective of this work is to explore the function of ion-exchange membrane (IEM) in enhancing the performance of CDI cell. In this study, a fully coupled two-dimensional model describing ion transport and adsorption behavior in Membrane Capacitive Deionization (mCDI) cell was developed and solved numerically, taking advantage of Darcy's Law, Nernst-Planck equation, and Donnan equilibrium considering hydraulic dispersion in channel, non-ideal membrane, and modification of ion diffusivity in porous media. The model behavior was supported by the experimentally observed variations in transient effluent concentration under different operating conditions. The results showed that ion flux was significantly improved because of the IEM, leading to a faster desalination rate. The equilibrium adsorption amount is increased due to the permselectivity of the IEM, resulting in extra ion accumulation in electrode's macropores during desalination process. The cut-off time (optimal time to switch polarity to generate highest energy efficiency) of mCDI was shorter with a higher salt removal percentage compared to CDI under the same operating conditions. A sensitivity study of the thickness and the fixed charge density (starting from 500mM) of the IEM was done separately, and neither of them were found to be the dominant influencing factors under a low concentration range of brackish water (below 100mM) in this study.

590103 An Integrated Process for the Production of Propylene Glycol from Shale Gas

Abstract:

Propylene glycol (PG) is produced on a large scale primarily for the production of environmentally safest and least toxic industrial and domestic products such as polyester resins, thermoset plastics, polyurethanes, food additives and antifreeze. In this research, a novel integrated plant is developed for the production of PG from shale gas. This integrated approach has the benefit of safer operating conditions since the intermediate propylene oxide, which is explosive, does not need to be stored and transported. Furthermore, there are potential economic benefits from integration. The overall plant is simulated in the Aspenplus environment, and a variety of process conditions are tested at steady-state to optimize the production of PG. Heat-integration tools are utilized for energy-saving and capital cost reduction opportunities. A comparative economic assessment based on existing plant information indicates that the use of process integration techniques has the potential to reduce costs significantly.

Keywords: Process Integration, Heat Integration, Propylene Glycol

592456 Monitoring Methods for Fault Detection and Diagnosis: Algorithms and Applications

Abstract:

Many industrial processes collect an abundance of data from different sensors. These sensors often measure a wide variety of physical properties in order to ensure that these parameters are adhering to expected values. This is essential to ensure plant and operator safety, increase economic benefits, and maintain product quality.

A number of algorithms have been developed in order to improve existing fault detection and diagnosis performance. These algorithms integrate a number of different data-driven driven tools and methods. Multiscale wavelet-based representation of data can be used in order to handle data that is autocorrelated, non-Gaussian, and noisy. Hypothesis testing methods such as the Generalized Likelihood Ratio (GLR) technique can be used in order to provide the best possible detection for a fixed false alarm rate. Moreover, certain model-based methods have been developed in order to monitor process drifts and degradations in the process model, even when a process is operating under control.

This work will discuss the different tools utilized in development of the algorithms, their features, and practical applications, including the benchmark Tennessee Eastman Process (TEP).

589902 Selective Separation of Industrial Relevant Metal Ions from High Tds Water Using Wafer-Enhanced Electrodeionization (WE-EDI)

Abstract:

There is a need for an efficient and economical process of ion-selective separation. In hydraulic fracturing, for instance, several million gallons of water are used in a site, most of which are disposed of by deep well injection taking the water from the ecosystem and causing earthquakes. However, due to the high amount of ions in wells at fracking sites, pressure driven processes for water recovery are impossible to use. This project will demonstrate an emerging environmentally friendly Wafer Enhanced Electrodeionization (WE-EDI) technology which attempts to eliminate the shortcomings that come with normal EDI and separates effluent streams into concentrated ions and clean water. In WE-EDI, ion-exchange resins are replaced by wafers that consist of cation and anion exchange beads and a polymer binding agent. In this study, selective separation of these ion exchange wafers and the effect of electroosmotic drag on the membrane structure have been performed for various anions and cations (Na^+, K^+, Ca^{2+}, Cl^-) and improved selectivity and higher water recoveries have been observed while significant modeling has taken place to predict the selective separation as a strong function of the equilibrium constants between the ions and the beads. This project is supported by the Membrane Science and Technology (MAST) Center and is conducted with the collaboration and guidance of industry mentors.

593303 Influence of Particles on Amine Losses during CO_2 Capture: A Process Simulation Coupled Aerosol Dynamics Model _

Abstract:

Influence of Particles on Amine Losses during CO_2 Capture: A Process Simulation Coupled Aerosol Dynamics Model

David Dhanraj and Pratim Biswas

Aerosol and Air Quality Research Laboratory

Department of Energy, Environmental and Chemical Engineering

Washington University in St. Louis

St. Louis, MO 63130, USA

Efforts to reduce the concentration of atmospheric CO_2 have been initiated due to escalating concerns of global warming and climate change. Amine-based post-combustion CO_2 capture offers an economically viable retrofit for power-plants in combating these emissions. However, solvent losses have been a major challenge to commercializing this technology. Recent experiments[1] show that vapor-based and aerosol-based solvent losses are the key mechanisms of amine solvent loss during CO_2 capture. While vapor-based amine losses can be abated by employing techniques like water-washing, there has been little work to curtail aerosol-based solvent losses. This is primarily because the underlying mechanisms governing these losses are still not well understood. Although there are a few numerical models proposed for this system, there are no models that consider aerosol phenomena such as coagulation or particle loss. Moreover, these models have not been fully validated through experiments.

Furthermore, there are a few techniques that are being developed to tackle solvent losses. It is known that reducing particle number concentration in the flue gas will reduce amine emissions. Conventional Electrostatic Precipitators (ESP) are however inefficient in capturing particles in the size range 0.1 – 1.0 μm. Amongst different approaches demonstrated, introducing a pre-charging stage and a photo-ionization stage has shown potential in increasing the charging efficiency.

Therefore, the broad objectives of this work are primarily focused on two facets of the amine loss problem. The first is the development of a model that integrates aerosol phenomena with a process engineering model that accounts for the reactive mass transfer process in a packed-bed absorber. The second is the design, fabrication and testing of a bench-scale Photoionization Enhanced Two-staged Electrostatic Precipitator (PI-ESP) at Abbott Power Plant. In essence, with the PI-ESP aimed at reducing the particulate concentration in the flue gas, and the model relating the influence of particulate concentration in the flue gas to amine emissions, this work aims to illuminate the underlying mechanisms governing amine emissions, and demonstrate a solution to tackle the problem.

References

1. International Journal of Greenhouse Gas Control, vol. 85, pp. 86-99, 2019.
2. International Journal of Greenhouse Gas Control, vol. 19, pp. 138-144, 2013.

614718 Spatially-Oriented Control in Batch Crystallization Processes Using Population Balance Modeling

Abstract:

Crystallization is an important separation and purification process with relevant applications in the food, pharmaceutical, fine chemical and fertilizing industries. In these processes, modeling and process control for crystallization are focused on crystal size distribution (CSD) and shape of the crystals in order to make it possible to achieve the desired quality and purity of the product of interest. The control of crystal size and shape is of considerable importance for industrial crystallization processes, as crystal morphology has great influence on downstream steps and also from the point of view of product specification criteria. The consideration of crystal shape in modeling of crystallization processes is also important due to its impact on product effectiveness, such as the bioavailability and tablet stability in pharmaceuticals. Control of the crystallization process is can be facilitated with online measurements, which provide information to be used in the modeling of the process. However, if online measurements are not available, the strongly nonlinear dynamics of these processes makes difficult to establish relations between some inputs (e.g. temperature) and the final desirable characteristics of the crystal product. In this context, image analysis has been shown to be a suitable technique to monitoring the process [1].

In our present work, a population balance model-based scheme for controlling the size and shape of crystals is presented. The modeling was performed using a two-dimensional population balance (2D-PBM) in terms of the two characteristic dimensions of the crystals, describing the evolution of crystal size and shape distributions along the batch time. To identify the nucleation, growth and dissolution kinetics parameters using the 2D-PBM approach along the length and width directions, QICPIC-LIXELL's high speed dynamic image analysis equipment was used for the online measurement of the CSD, shape parameters and aspect ratio distribution of crystals during the experiments. In this way, the developed model allowed for the prediction of the crystal size and shape distributions during growth-dissolution cycles.

The developed framework was employed to obtain optimal temperature policies for controlling the crystal size and shape distributions that meet target criteria for size, morphology and productivity using the dynamic image analysis along the runs. Based on the concept of using measurements along batch crystallization processes as a trajectory in a phase space [2], a trajectory-endpoint control problem is presented with an oriented control scheme that is applied to the shape of the crystals. The scheme was used to control the batch cooling crystallization of potassium dihydrogen phosphate. The results demonstrate the application of the control policies for temperature to produce crystals of the desired average shape and size in different batch times. Thus, using the first principle modeling coupled to the developed trajectory tracking control, the proposed framework has the potential to be applied to industrial crystallization processes for shape control and to be extended to polymorph control investigations.

References:

[1] Eisenschmidt, H.; Bajcinca, N.; Sundmacher, K. (2016) Optimal Control of Crystal Shapes in Batch Crystallization Experiments by Growth-Dissolution Cycles. Crystal Growth & Design, v. 16, p. 3297-2803.

[2] Griffin, D. J., Grover, M. A., Kawajiri, Y., Rousseau, R. W. (2015) Mass-count plots for crystal size control. Chemical Engineering Science, v. 137, p. 338-351.

614717 Predicting The Flow of Various Sizes and Moisture Contents of Glass Beads

Abstract:

Whether it is pneumatic conveying, die filling in a tablet press, or storage in hoppers, powders will be subjected to some level of stress consolidation during their handling, affecting the flow of these materials. Flow issues are likely to occur over a wide range of particle-scale parameters like particle size, particle size distribution (PSD), particle shape and moisture level. For effective product development and process design, it is important to study the flowability of granular materials and develop scale-up equations which can be used on an industrial scale. The present work investigated the effects of particle sizes between 5 – 600 microns and moisture levels from 0 – 20% on the flowability and shear properties of silica beads using the Freeman FT4 Rheometer. A mathematical model in terms of a flowability descriptor called the flow function (FF) was developed, based on the material's yield point. The yield point for various particle size distributions and moisture levels was evaluated using a shear cell test. The FF for dry glass beads increased with particle size, following a linear relationship, as expected. On the other hand, the flow function for wet glass beads followed a power law relationship, where FF decreased with increasing moisture content. However, as the moisture level reached 20%, the flow function value escalated again due to supersaturation. The linear and power functions were combined to develop an empirical relation, which can be used to predict the flowability of glass beads, based on the particle size and moisture levels up to 20%.

AUTHOR INDEX

AIChE
120 Wall Street, FL 23
New York, NY 10005-4020

ISBN 978-1-7138-2350-6

9 781713 823506

CURRAN ASSOCIATES INC.
proceedings.com